The Radon Transform and Some of Its Applications

The Radon Transform and Some of Its Applications

STANLEY R. DEANS
Department of Physics
University of South Florida

A Wiley-Interscience Publication
JOHN WILEY & SONS
New York • Chichester • Brisbane • Toronto • Singapore

Library of Congress Cataloging in Publication Data:

Deans, Stanley R. (Stanley Roderick), 1937–
The Radon transform and some of its applications.

"A Wiley-Interscience publication."
Includes bibliographical references and index.
1. Radon transforms. I. Title.

QA649.D4 1983 516.3′6 83-1125
ISBN 0-471-89804-X

Printed in the United States of America

10 9 8 7 6 5 4 3 2 1

JOHANN RADON
laid its foundations

RON BRACEWELL and ALLAN CORMACK
showed how it could be used

Preface

Numerous situations arise when it is both possible and desirable to determine (reconstruct) certain structural properties of an object or substance utilizing data obtained by methods that for all practical purposes leave the entity under investigation in an undamaged and undisturbed state. The basic mathematical framework common to a large class of such reconstruction problems was developed by Johann Radon in 1917. Radon's original paper, which went almost unnoticed for half a century, is now famous; however, it is not readily available in most libraries and it is in German. A translation of this original work, which forms the basis for what has come to be known as the Radon transform, appears in Appendix A.

The purpose of this book is twofold:

(i) To provide basic information about properties of the transform itself, complete with examples that illustrate some of the more subtle points.

(ii) To document a wide variety of applications, along with extensive guidance to literature related to the transform.

These objectives are accomplished in a way that can be understood by individuals with a basic undergraduate background in mathematics such as that normally acquired by medical students, engineers, or physicists. In keeping with these goals, this work helps fill the gap between highly rigorous treatments and the applications. More specifically, this book is for the theorist who is interested in applications and for the experimentalist who wishes to gain a better understanding of the theoretical foundations, but is not trained as a professional mathematician.

The entire discussion is for the Radon transform on Euclidean space, mainly $\mathbb{R}^2$ and $\mathbb{R}^3$, with ample reference to more general treatments. Chapter 1 is primarily a survey of some of the more important areas of application where the transform emerges in a natural fashion. It is interesting that major developments in these areas did not come as a result of Radon's work. In fact, all of these areas were highly developed before connections with the Radon transform were recognized. In retrospect, its relevance is apparent in several different fields. The formal definition of the transform is given in Chapter 2. The reader with absolutely no previous knowledge of the transform may wish to read this rather brief chapter first. Chapter 3 is on basic properties. Many examples are given to illustrate these properties and enhance general understanding. The connection with certain other transforms is discussed in Chapter 4. Chapters 5, 6, and 7 are on inversion, but with a different emphasis in each case. Some additional applications and properties of a slightly more technical nature are discussed in Chapter 8. Appendix A contains the translation mentioned earlier. Appendix B is a collection of some basic facts about generalized functions, and Appendix C contains standard information about some of the special functions used throughout the book.

It is a pleasure to thank A. B. Brill, T. F. Budinger, and G. T. Gullberg for stimulating my interest in this area. I am grateful to P. A. Carruthers, T. F. Budinger, D. T. Jacobs, and W. Miller, Jr., for providing convenient working conditions during various phases of the writing. I appreciate the photographs and figures generously provided by several of my colleagues and the friendly guidance of librarians at Los Alamos, Berkeley, Stanford, and the University of Minnesota. Special thanks are reserved for R. Lohner, who translated Radon's classic paper; W. F. Ames, who provided valuable suggestions and continued interest; Mary Vavrik, who helped with the typing; and Nancy Pope, who helped with the graphics, reviewed the manuscript, suggested significant improvements, and provided general encouragement.

STANLEY R. DEANS

Tampa, Florida
April 1983

Contents

Chapter 4. Relation to Other Transforms, 96

Chapter 5. Inversion, 108

Chapter 6. Recent Development of Inversion Methods, 125

Chapter 7. Series Methods, 151

Chapter 8. More Properties, Applications, and Generalizations, 184

Chapter 1

Major Fields of Application

[1.1] INTRODUCTION

The Reconstruction Problem

During the past decade there has been renewed interest in what we shall call the *reconstruction problem*. This is the problem of determining the internal structure or, more precisely, some property of the internal structure of an object without having to cut, crack, or otherwise macroscopically damage the object. Various probes, including X rays, gamma rays, visible light, microwaves, electrons, protons, neutrons, heavy ions, sound waves, and nuclear magnetic resonance signals have been used to study a large variety of objects whose sizes vary over an enormous range, from complex molecules studied by the electron microscopist to distant radio sources studied by the radio astronomer. Among the vast number of objects that fall between these extremes the human body or some particular organ in the body is an especially important case. The success of the new technology associated with medical imaging by use of computerized tomography (CT)* is largely responsible for much of the current interest in a variety of reconstruction methods. Not only do these methods and this technology hold much promise in medicine, but recent success there suggests interesting possibilities for expanded use of similar techniques in other fields.

Mathematical Framework

The appropriate unifying mathematical framework for a large class of reconstruction problems is the Radon transform on Euclidean space (see Fig. 1.1). The reason for this can be illustrated by the correspondence:

$$\left\{ \begin{matrix} \text{INTERNAL} \\ \text{DISTRIBUTION} \end{matrix} \right\} \rightarrow \text{acted on by probe} \rightarrow \{\text{PROFILE}\}$$

$$\{f\} \rightarrow \text{acted on by Radon transform} \rightarrow \{\check{f}\}.$$

*The term CAT is also used, for computer assisted tomography or computerized axial tomography. The reader not already familiar with CT may wish to consult one or more of the introductory articles: Gordon, Herman, and Johnson (1975), Ledley (1976), Swindell and Barrett (1977), and Scudder (1978).

Some *internal distribution* of the object, perhaps density or something closely related to density, is acted on by (or interacts with or emits) some probe. The probe can be detected to produce what we call a *projected distribution* or *profile*. Often, the internal property of the object may be identified with or approximated by some function f and the profile may be identified with or approximated by the Radon transform of f, designated by

$$\check{f} = \mathscr{R}f. \tag{1.1}$$

More precisely, each profile is identified with a *sample* of the Radon transform. Knowledge of all possible profiles constitutes full knowledge of the Radon transform (see Fig. 1.2).

Remark 1. The reader who feels a need at this point for a more detailed definition of the Radon transform may consult Chapter 2. Briefly, if f is a function defined on n-dimensional Euclidean space $\mathbb{R}^n$, the Radon transform of f is determined by integrating f over all hyperplanes of dimension $n - 1$. Thus, if f is defined on the plane $\mathbb{R}^2$, then $\check{f}$ is determined by the line integrals of f. And if f is defined on $\mathbb{R}^3$ then $\check{f}$ is determined by the surface integrals of f over two-dimensional planes. □

There is a fundamental connection between the Radon transform and the Fourier transform. Although much more will be said about this in later chapters, it is worth brief consideration here for purposes of context when discussing applications. Let $f = f(x, y)$ be a function of two real variables. The Radon transform of f is designated by $\check{f} = \mathscr{R}f$, and the two-dimensional

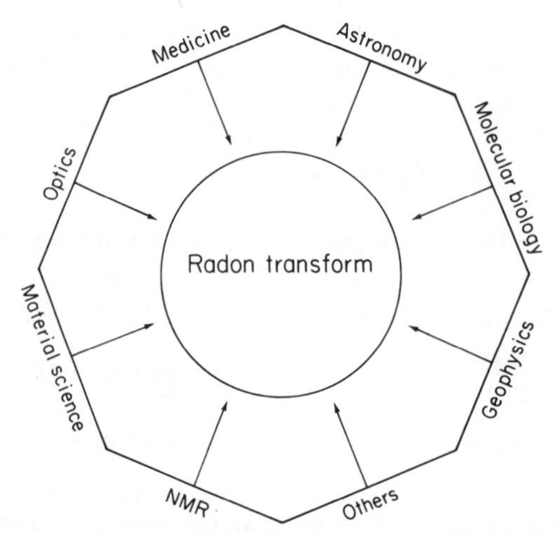

Figure 1.1 Reconstruction problems from diverse fields may be united within the framework of the theory of the Radon transform on Euclidean space.

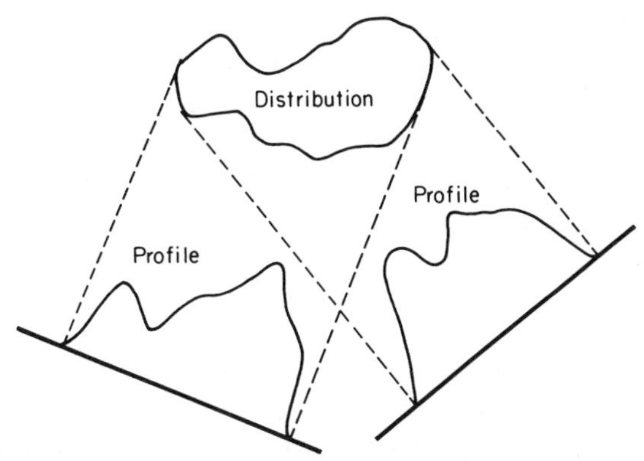

Figure 1.2 Distribution and two profiles. The distribution may be identified with f and each profile with a sample of the Radon transform $\check{f}$.

Fourier transform of f is designated by $\tilde{f} = \mathcal{F}_2 f$. Some important interrelationships developed by Bracewell (1956a) are shown in Fig. 1.3.* Here the one-dimensional Fourier transforms are always on the radial variable, and the convention is that used by Lighthill (1958) and Bracewell (1978). For example, if $\check{f}$ is expressed as a function of the radial coordinate p and angular coordinate ϕ, then

$$\tilde{f}(u, v) = \int_{-\infty}^{\infty} \check{f}(p, \phi) e^{-i2\pi q p}\, dp$$

with $u = q\cos\phi$ and $v = q\sin\phi$. For a justification of the interrelations illustrated in Fig. 1.3, see Examples 3 and 4 of [4.2].

In many situations (especially when f is defined on $\mathbb{R}^2$), it is possible to measure profiles directly and the problem is to use these measurements to approximate the desired distribution. Correspondingly, in terms of (1.1) the problem is to solve for f when given $\check{f}$. Building on some ideas of Minkowski (1904–06) and Funk (1913, 1916) and private communication with G. Herglotz, this problem was solved by Johann Radon (1917). A translation of this important but relatively inaccessible work appears in Appendix A.

Remark 2. Radon was primarily concerned with $\mathbb{R}^2$ and $\mathbb{R}^3$. The major work on extending Radon's work to $\mathbb{R}^n$ was done by John (1934) and Mader (1927). Other important early papers include those by Bockwinkel (1906) and Uhlenbeck (1925). It is clear from the observations of Bockwinkel and Uhlenbeck that the Dutch physicist H. Lorentz knew the inversion formula for the three-dimensional case around the turn of the century.† Additional historical

*In this connection, see Remark 2.

†It is a pleasure to thank Professors Cormack and Grünbaum for conversations and material regarding this observation.

remarks appear in the Introduction of the book by John (1955). Most of the early applications were of a mathematical nature, especially in the area of partial differential equations. These applications and the literature on such applications are discussed in detail by John (1955) and Bureau (1955). One notable paper that does not fall into this category is the article by Birkhoff (1940) on drawings composed of uniform straight lines. This work is especially interesting since it contains the seeds of ideas relevant to using the Radon transform in connection with many current problems of image reconstruction from multiple projections.

One further point: Although we indicated earlier that there is a fundamental connection between the Radon transform and the Fourier transform, it is not clear to the author just who first recognized this connection. However, it was apparently known by John (1934) in differential equations and by Cramér and Wold (1936) in statistics. It was Rényi (1952) who established the connection between Radon's work and that of Cramér and Wold. Later, Gilbert (1955) sharpened some of Rényi's results. No doubt there were others who knew, but it was Bracewell (1956a) who truly illuminated the connection—and at that time he was not even aware of Radon's work! □

The importance of the general unifying framework, whereby the unknown internal distribution is identified with f and the observable profiles identified with $\check{f}$ can not be overemphasized. The problems one may wish to solve may be quite diverse and may involve a wide variety of objects and probes, but once cast into the unifying framework they are all subject to the same type of analysis. In some cases, the mathematical framework itself serves to help define

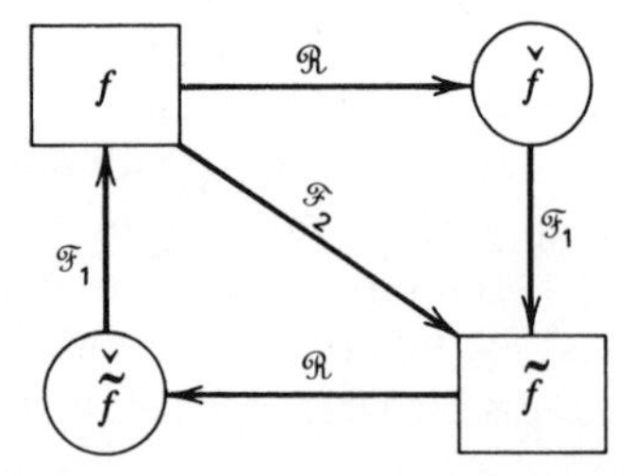

Figure 1.3 Some relations between Radon and Fourier transforms of a function of two real variables. Explicitly, the definitions are

$$\mathscr{F}_2 f = \tilde{f}(u, v) = \int_{-\infty}^{\infty} \int_{-\infty}^{\infty} f(x, y) e^{-i2\pi(ux+vy)}\, dx\, dy,$$

and

$$\mathscr{R} f = \check{f}(p, \phi) = \int_{-\infty}^{\infty} \int_{-\infty}^{\infty} f(x, y)\, \delta(p - x\cos\phi - y\sin\phi)\, dx\, dy,$$

where δ is the Dirac delta function [Lighthill (1958); Bracewell (1978)].

the problem and a mechanism for solution. As a result of this unification, an improvement in one area, such as recent breakthroughs in CT, can have an immediate effect in seemingly unrelated fields.

Active and Passive Approaches

The type of reconstruction problem that arises when external probes are used to collect information about projections has been called the *active* approach to reconstruction from projections by Professor T. F. Budinger of the University of California at Berkeley. Another approach, which Budinger refers to as the *passive* approach, does not require external probes initially. There is a probe involved, but it comes from the object itself, for example, a radioactive object where the desired property f is some internal isotope distribution or nuclear magnetic resonance signals that yield information about proton density. The active and passive approaches will be described in greater detail in a later discussion of various probes utilized and objects that have been studied more or less successfully.

Rediscoveries

Before discussing various fields where the Radon transform emerges in a natural way, it must be stated that in nearly all areas of application the early investigators were not aware of Radon's original paper [Radon (1917)].* Consequently, there are many "rediscoveries" of Radon's results throughout the applied literature. These rediscoveries ended around 1972 when Soviet and American authors [Shtein (1972); Vainshtein and Orlov (1972); Vest (1973); Cormack (1973)] pointed out that Radon's work was fundamental to the problem of reconstruction from projections. Cormack (1980b) further alluded to this in his 1979 Nobel Prize address. In a historical observation on this point, Marr (1982) notes that Pincus (1964) apparently was the first person to develop a reconstruction algorithm with knowledge of the available material in the mathematics literature, including Radon's 1917 paper.

Although Radon's 1917 work was virtually unknown in applied areas prior to the early 1970s, it was certainly appreciated by mathematicians, and already had been put to major use in books by John (1955) and Gel'fand, Graev, and Vilenkin (1966). For other mathematical uses, see the references under "Mathematical and Generalizations" in [1.10].

Approximations

Finally, a few remarks must be made about the identification of f and $\check{f}$ with physical quantities. Since $\check{f}$ is identified with a measured quantity, it only represents an approximation. Even worse, since the probe must be applied a

*In most applied areas, the relevance of the Radon transform has been recognized in retrospect.

finite number of times, $\check{f}$ is not even approximated in a continuous fashion. Consequently, any determination of f is at best only an approximation to the desired distribution and at worst bears no resemblance to the desired distribution. In the following summaries of application areas, not much will be said about this since the purpose is to emphasize concepts rather than details of implementation and treatment of data. However, keep in mind that many difficulties are associated with reconstruction problems simply because the function $\check{f}$ is not known exactly. See [6.1] for further discussion.

Overview

In [1.2], X-ray transmission computed tomography (TCT) is discussed. This has been referred to as CT or CAT earlier. Sometimes it is useful to use TCT to emphasize the *transmission* aspect when there may be confusion with emission computed tomography (ECT) discussed in [1.3]. Often CT is used to represent either TCT or ECT. Also, the terms CT and TCT may be used when X rays are not used as the probe; for example, protons [Hanson (1979b)], neutrons [Koeppe, Brugger, Schlapper, Larsen, and Jost (1981)], weak electrical currents [Price (1979a, b)], microwaves [Forgues, Goldberg, Smith, and Stuchly (1980)], or heavy ions [Holley, Henke, Gauger, Jones, Benton, Fabrikant, and Tobias (1979)] may be used. Fortunately, the context usually eliminates any confusion. Another way out of this problem is to use RT (for reconstructive tomography) to replace all these abbreviations. There is good reason for this since, as emphasized by Wade, Mueller, and Kaveh (1979), the reconstruction or inversion problem is a mathematical problem first. It is the complexity which leads to the use of a digital computer rather than something basic in principle. Although RT is in some sense more technically correct, it does not appear in most of the current literature and for that reason it has not been used here either.

The X-ray CT problem is *the* prototype application for active reconstruction problems, and it is the area that has received the most publicity over the past few years. The importance and diversity of this kind of problem are expressed by Nalcioglu, Cho, and Knoll (1979) in their "Foreword" to the *Workshop on Physics and Engineering in Computerized Tomography* held at Newport Beach, California, January 17–19, 1979.

> Rapid advances made during the last five years in the field of computerized x-ray tomography, also known as CT, have added a new dimension to the field of medical imaging. Development of CT is the most significant advancement in the medical imaging field since the discovery of x-rays by Roentgen in 1895.
>
> As we are aware at this time, the introduction of computerized tomography has brought medical imaging into the realm of advanced physics and engineering utilizing the most sophisticated computer and radiation detector technology available.

During the past five years this new technology has witnessed further developments in the field of medical imaging and other physical sciences. The organizing committee felt that due to the diversity of the applications and recent advances made in the field, a meeting in which the physics and engineering aspects of computerized tomography would be discussed was overdue.

The conference under these premises was held January 17–19, 1979 in Newport Beach, California, and was attended by 150 scientists and engineers from around the world. Sessions were organized into broad categories of mathematical methods, engineering developments, new methods and systems, emission tomography, nuclear magnetic resonance, ultrasound and x-ray computerized tomography. Comments, questions, and a panel discussions indicated the high level of interest in all of these imaging modalities. It is apparent that the potentially useful methods in medical imaging are rapidly becoming more diverse and will encompass an even greater spectrum of technologies in the future.

This meeting followed three previous conferences on a similar theme. Programs at UCLA in 1973, Brookhaven in 1974, and Stanford in 1975 all stressed the technical aspects of medical imaging. It is the feeling of this Organizing Committee that future conferences should be held at least every two years to keep pace with the rapid advances taking place in the field.

The Organizing Committee would like to thank the following companies for their generous contributions: Biomedia, Inc., Cardiac Medical Systems Corp., E. I. Du Pont De Nemours and Company, EMI Medical Inc., Floating Point Systems, Inc., Harshaw Chemical Co., Omnimedical, EG & G ORTEC, Philips Medical Systems, Inc., Searle Diagnostics, Inc., S & S X-Ray Products, Inc., and Varian Associates.

A further indication of the importance is reflected in the explosion of numbers of papers on various aspects of CT, and the important recent book by Herman (1980b). New journals* dedicated to both basic and clinical aspects of CT have appeared. And in 1979 the Nobel Prize in physiology or medicine was shared by Godfrey N. Hounsfield of the Central Research Laboratories of EMI and Allan M. Cormack, a physicist from Tufts University, for pioneer work in the field. Their Nobel Prize addresses are available [Hounsfield (1980); Cormack (1980b)] and informative background discussions have been given by Di Chiro and Brooks (1979, 1980). Their discussion rightly emphasizes that other individuals also had pioneer roles in the solution of the reconstruction problem in applied areas and thus in the ultimate development of CT.† Notable among these are Ronald N. Bracewell, Lewis M. Terman Professor of Electrical Engineering at Stanford University, who developed the inversion equations in connection with strip integration problems in radio astronomy [Bracewell (1956a)]; William H. Oldendorf, Professor of Neurology and Psychiatry at UCLA School of Medicine and Senior Medical Investigator at Brentwood

*Radon's inversion formula in its original form (Theorem III in Appendix A) appears on the cover of one of these: *Journal of Computer Assisted Tomography*.

†Also, remarks by Gordon (1979) are directed to this important point.

Medical Center, who devised and patented an apparatus for reconstruction from projections using X rays [Oldendorf (1961)]; and David E. Kuhl, Professor of Radiology and Director of the Division of Nuclear Medicine at the Hospital of the University of Pennsylvania, the undisputed pioneer in the field of radioisotope imaging and emission computed tomography [Kuhl and Edwards (1963)]. Finally, in Japan, Shinji Takahashi studied rotation techniques as applied to radiology as early as 1946 [see Takahashi (1969) and references contained therein on pp. 307–314].

In [1.3], the prototype passive reconstruction problem is discussed: ECT (for emission computed tomography). The major objective in ECT is to describe the location and concentration (or intensity) of some radioactive isotope by studying the emitted photons from an object. (In nuclear medicine, the "object" is some organ or region of the human body after injection of an appropriate radiopharmaceutical or radionuclide.) This is in contrast to TCT in [1.2], where the goal is to determine the distribution function of the attenuating medium. The basic difference is illustrated in Fig. 1.4. As with TCT, the development of ECT has been very rapid over the past decade. Interestingly though, pioneers such as Kuhl and Edwards (1963) were active in this area of reconstruction several years prior to the development of X-ray computed tomography. For an excellent recent review of ECT, see Budinger, Gullberg, and Huesman (1979).

In [1.4], ultrasonic computed tomography (UCT) is discussed. In UCT, the ultrasound probe is used in the *transmission* mode, as X rays in X-ray CT, to obtain projection data needed for transverse section reconstruction. This use of ultrasound is not to be confused with the common "pulse-echo" imaging

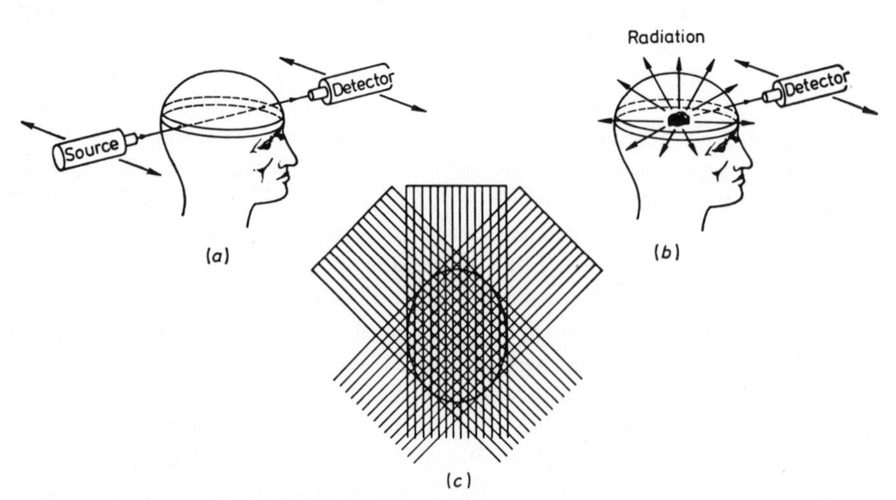

Figure 1.4 Reconstructive tomography as applied to (*a*) transmission imaging, and (*b*) emission imaging. (*c*) A typical scanning pattern consists of linear translations at successive angular increments. Reprinted with permission [Brooks and Di Chiro (1976a)]. Courtesy of R. A. Brooks.

technique based on *reflection* of ultrasonic signals, which is currently dominant in clinical practice [Maginness (1979); Havlice and Taenzer (1979)]. Use of ultrasound in medicine is steadily increasing, and medical UCT is no exception. Improved ultrasonic instrumentation along with the absence of radiation hazard are making ultrasound an increasingly attractive medical diagnostic, therapeutic, and surgical tool [Wells (1977a, b)].

In [1.5], some of the applications in the field of astronomy are discussed. Most of these are in the area of radio and radar astronomy with a few examples from optical and X-ray astronomy. Due largely to the development of image reconstruction methods in strip scanning, aperture synthesis, radar astronomy, and occultation studies, radio maps of the sky and solar system bodies exist with a resolution comparable to those from the best optical telescopes [Ryle (1972); Pasachoff and Kutner (1978)]. Recent surveys of aperture synthesis methods, image formation and improvements, and developments in radar astronomy may be found in van Schooneveld (1979) and Pettengill (1978), and a discussion of several telescope facilities may be found in the September 1973 issue of *Proceedings of the IEEE* indicating the worldwide nature of the effort in radio astronomy [Bracewell, Colvin, D'Addario, Grebenkemper, Price, and Thompson (1973); Baars, van der Brugge, Casse, Hamaker, Sondaar, Visser, and Wellington (1973), Christiansen (1973); Roger, Costain, Lacey, Landecker, and Bowers (1973); Erickson (1973); Hills, Janssen, Thornton, and Welch (1973); Delannoy, Lacroix, and Blum (1973); Swarup and Bagri (1973); Hackenberg, Grahl, and Wielebinski (1973); Reid, Clauss, Bathker, and Stelzried (1973)].

In sharp contrast to the astronomical application of [1.5], the reconstruction problem in electron microscopy is discussed in [1.6].* Here, both the need and possibility for reconstruction methods arise because the transmission electron microscope has a large depth of focus compared to the structural details of interest. The image (electron micrograph) may be regarded as an enlarged two-dimensional parallel projection of the three-dimensional specimen, much as with ordinary X-ray shadow images. Thus, by using reconstruction methods, an appropriate set of projections (micrographs) may be used to approximate the three-dimensional structure of the object being studied with the electron microscope. Recent reviews of electron microscope analysis of the three-dimensional structure of various molecules have been given by Crowther and Klug (1975), Vainshtein (1978), Hoppe and Typke (1978), and Hoppe and Hegerl (1980).

In [1.7], nuclear magnetic resonance (NMR) imaging is discussed. Although some of the fundamental ideas behind using NMR in the medical field were provided by Damadian (1971, 1972) and a practical NMR image was displayed

*The 1982 Nobel Prize in chemistry was awarded to Aaron Klug of the Medical Research Council Laboratory, Cambridge, England, for pioneer work in this field.

by Lauterbur (1973) several years ago, the feasibility of using this imaging modality in the important medical imaging field is just beginning to be clear [Hawkes, Holland, Moore, and Worthington (1980); Hoult (1980c)]. During the intervening period cross-sectional images that represent spatial variations of the NMR signal have been produced by a variety of techniques, many of which are compared by Bottomley (1979) and Brunner and Ernst (1979). Some of these methods do not make use of reconstruction from projections; however, reconstruction techniques have been employed by the Nottingham group in their recent success in NMR medical imaging [Moore, Holland, and Kreel (1980); Holland, Moore, and Hawkes (1980); Holland, Hawkes, and Moore (1980)]. Unlike other methods of medical imaging, NMR imaging uses no ionizing radiation.* For this reason and in view of the recent success of the Nottingham group and others [see Crooks, Hoenninger, Arakawa, Kaufman, McRee, Watts, and Singer (1980); Hansen, Crooks, Davis, DeGroot, Herfkens, Margulus, Gooding, Kaufman, Hoenninger, Arakawa, McRee, and Watts (1980); Partain, James, Watson, Price, Coulam, Rollo (1980); Wolfe, Crooks, Brown, Howard, and Painter (1980)], there is a cautious optimism regarding the future of NMR imaging in the medical field [Evens (1980); Hoult (1980c); Oldendorf (1982)] and several companies are developing machines for clinical usage.†

Use of NMR in biology and medicine may be traced back to the first biological NMR experiment, which was performed by Felix Bloch soon after the discovery of NMR [Andrew (1980b)]. Professor Bloch placed his finger in the probe coil of the first NMR device and obtained a strong proton NMR signal [Bloch, Hansen and Packard (1946); Bloch (1978)].

In [1.8], optical applications are discussed, including the areas of aerodynamics, fluid flow, and plasma physics. In most cases considered, the probe is a coherent beam of light from a laser that passes through some transparent object, and the reconstructed distribution is the refractive index. The transform data may be collected by interferometric methods; recently, holographic interferometry is often used. The relatively new field of holographic interferometry was discovered around 1965 [Horman (1965); Powell and Stetson (1965); Heflinger, Wuerker, and Brooks (1966)] soon after the development of off-axis holography [Leith and Upatnieks (1962, 1964)], which followed the pioneer work in holography by Gabor (1948–51). An excellent treatment of holographic interferometry is given by Vest (1979), complete with a full discussion of the application of the Radon transform in optics.

Finally, in [1.9], geophysics and several other areas are discussed, and in [1.10], many of the references are categorized according to major area of application.

*For an analysis of the potential health hazards of NMR imaging, see Budinger (1979b).

†See the "NMR industry progress report," *RNM Images* [formerly *Radiology/Nuclear Medicine Magazine*], **12** (3), June 1982, 44–45.

[1.2] MEDICINE; X-RAY COMPUTED TOMOGRAPHY

If a narrow (small cross-sectional area) beam of monoenergetic X-ray photons passes through some homogeneous material, the beam intensity is observed to decrease* in accord with the equation

$$I = I_0 e^{-\mu x}, \tag{2.1}$$

where I_0 is the input intensity (number of photons per second per unit cross-sectional area) and I is the observed intensity after the beam passes a distance x through the material. The *linear attenuation coefficient* μ depends on both the density of the material ρ and the nuclear composition characterized by the atomic number Z,

$$\mu = \mu(\rho, Z).$$

In the following discussion, this dependence on ρ and Z will not be shown explicitly.

If the X-ray beam passes through two different materials, distance x_1 through medium 1 characterized by μ_1 and distance x_2 through medium 2 characterized by μ_2, the fractional decrease in intensity is given by

$$\frac{I}{I_0} = e^{-\mu_1 x_1 - \mu_2 x_2}.$$

For several media the relation is

$$\frac{I}{I_0} = \exp\left[-\sum_j \mu_j x_j\right].$$

If $\mu = \mu(x)$ is a continuous (or piecewise continuous) function of x, the sum goes over to an integral along the beam path L,

$$\frac{I}{I_0} = \exp\left[-\int_L \mu(x)\,dx\right]. \tag{2.2}$$

Now, consider a transverse section through some three-dimensional object. If the transverse section plane is perpendicular to the z axis (of an orthogonal coordinate system), then the (variable) linear attenuation coefficient in the

*X-ray photons interact with matter in several ways [Segrè (1977)]. At photon energies in the neighborhood of 70 keV, which are typical for CT scanners, the combined effects of scattering (Compton effect and Rayleigh scattering) and absorption (photoelectric effect) result in exponential attenuation of the beam as it passes through matter.

plane may be written as a function of two variables x and y, which coordinatize the xy plane,

$$\mu = \mu(x, y).$$

If the narrow beam traverses the xy plane as indicated in Fig. 1.5, the fractional decrease in intensity is given by

$$\frac{I}{I_0} = \exp\left[-\int_L \mu(x, y)\, ds\right], \tag{2.3}$$

where the line integral is along the beam path L. The natural logarithm of (2.3) yields a single *projection*,

$$P = -\log\left(\frac{I}{I_0}\right) = \int_L \mu(x, y)\, ds. \tag{2.4}$$

By moving the source and detector as indicated in Fig. 1.6, it is possible to obtain a set of projections or *profile* for the angle ϕ,

$$P(p, \phi) = \int_L \mu(x, y)\, ds. \tag{2.5}$$

By comparison with the definition of the Radon transform given in (2.3) of [2.2], it is clear that for continuous p and ϕ the profiles $P(p, \phi)$ may be identified with $\check{f}(p, \phi)$ provided the attenuation coefficient distribution $\mu(x, y)$ is identified with $f(x, y)$. In X-ray CT, the projection profiles are measured for

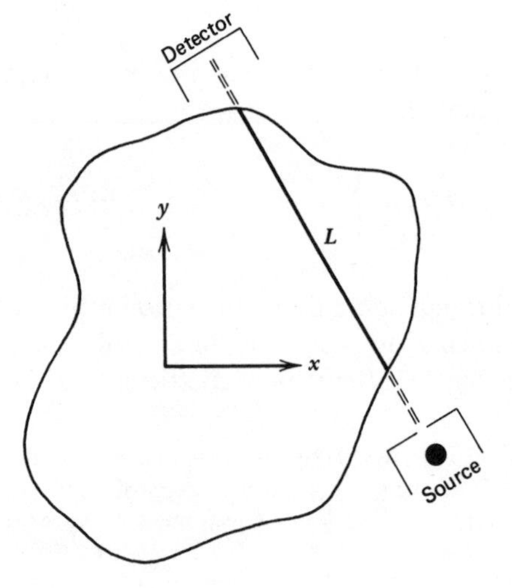

Figure 1.5 The beam passes through the region characterized by $\mu(x, y)$ along the line L.

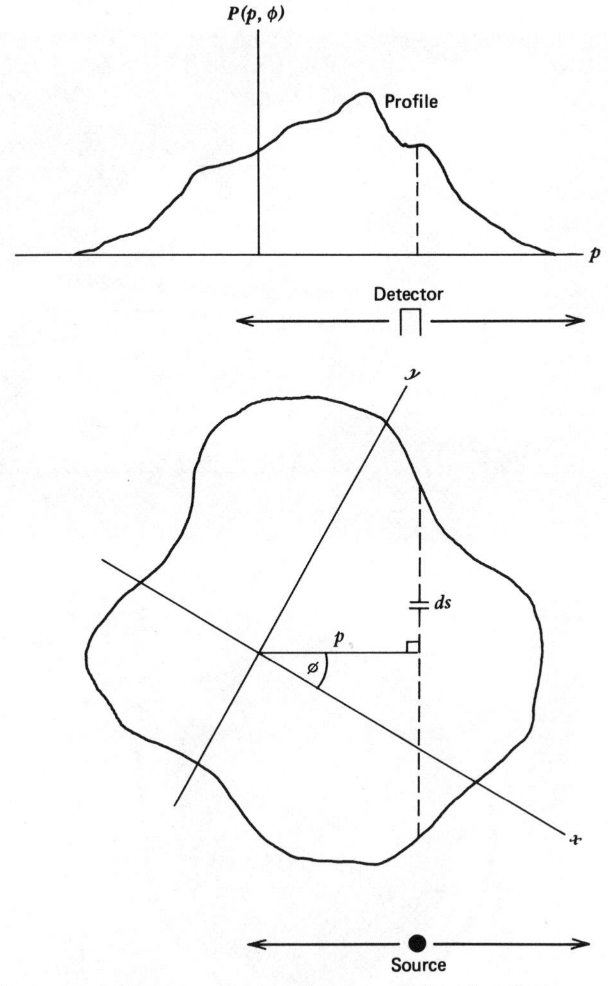

Figure 1.6 The source and detector move together (p varies) so that the beam covers the entire region leading to the profile $P(p, \phi)$ for fixed angle ϕ.

various incremental values of the angle ϕ. This constitutes a sampling of the Radon transform. Then an appropriate inversion or reconstruction algorithm (see Chapter 6) is applied to recover an approximation to the attenuation coefficient distribution over a transverse section of some portion of the human body. By stacking several transverse sections, the two-dimensional information may be converted to three-dimensional information. Some of the instrumentation and samples of various transverse section reconstructions are shown in Fig. 1.7.

Advances in CT scanning have been tied closely to the development of faster and more efficient reconstruction algorithms in conjunction with techno-

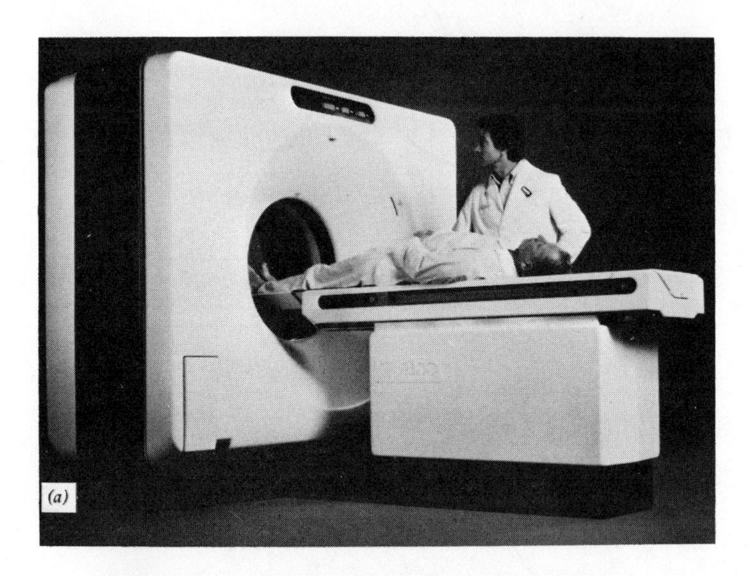

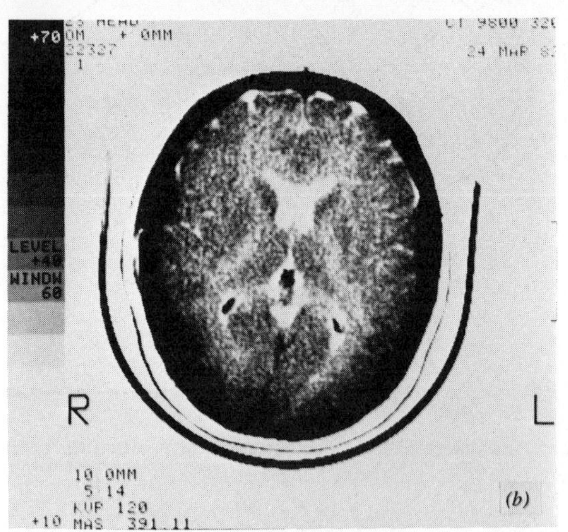

Figure 1.7 (a) The General Electric CT 9800 Computed Tomography System permits high-quality head and body imaging with scans as fast as 1.3 seconds. (b), (c), (d) Typical transverse sections. Photo courtesy General Electric Company.

logical improvements in the scanning hardware. Some of these developments and various algorithms are discussed further in Chapter 6. Improvements have been rapid since the first pictures obtained by Hounsfield in 1970.* The

*As a point of historical interest, see the Report of March 1970, Kreel (1977), which contains the earliest pictures on body scanning and an original handwritten note from G. N. Hounsfield regarding some of the pictures. Also, see Hounsfield (1980).

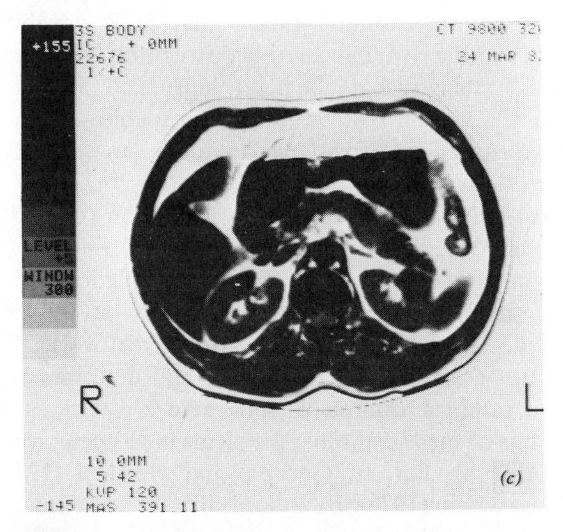

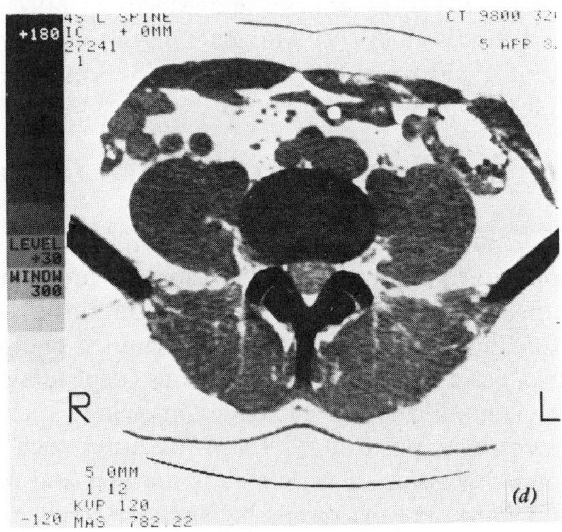

Figure 1.7 (*Continued*).

quality of the reconstructed image has improved greatly while computation times, once measured in terms of several minutes, have been reduced to a few seconds per transverse section. Even with the current impressive results, improvements and evaluations of CT performance are still important areas of investigation; see McCullough (1980), and references in Chapter 6 listed under "Improvements" etc. Major advances are under way and the development of a new generation scanner called the DSR (for dynamic spatial reconstructor) will provide two very powerful new dimensions to computerized tomography: high temporal resolution and synchronous truly three-dimensional scanning. Wood, Kinsey, Robb, Gilbert, Harris, and Ritman (1979) of the Biodynamics Research Unit of the Mayo Clinic expect that DSR capabilities will allow noninvasive study of the dynamic anatomic structural/functional relationships of moving organ systems such as the heart, lungs, and circulatory system. This new generation CT system will be uniquely capable of simultaneous three-dimensional reconstructions of vascular anatomy and circulatory dynamics in any region of the body.

In addition to long-range work on new generation CT systems such as the DSR, many recent research efforts in the medical imaging field have been related to the improvement of existing systems and algorithms. This is certainly understandable since the ideal situation presented at the beginning of this section is corrupted by many complexities of the real world. For example, the X-ray beam is not monoenergetic, the beam and detectors have finite width, the profiles are sampled discretely, and there are various noise problems. Recent discussions of these and other problems have been given by Brooks and Di Chiro (1976a, b, c), Barrett, Gordon, and Hershell (1976), Tanaka and Iinuma (1976), Kowalski (1977a, b), Bracewell (1977), Huesman (1977), Chesler, Riederer, and Pelc (1977), Duerinckx and Macovski (1978, 1979a, b), Kak (1979), Verly and Bracewell (1979), Morgenthaler, Brooks, and Talbot (1980), Joseph, Hial, Schulz, and Kelcz (1980). Other references appear in [1.10].

[1.3] MEDICINE; EMISSION COMPUTED TOMOGRAPHY

Here emission computed tomography (ECT) will refer to the usage in nuclear medicine, as opposed to other types of emission CT problems. The goal is to determine the distribution (location and concentration) of some radioactive isotope inside the human body by studying the emitted photons.

There are two basic types of ECT problems, depending on whether the isotope (radionuclide) utilized is a single photon emitter, such as iodine-131 or technetium – 99m, or a positron e^+ (or β^+) emitter such as carbon-11 or oxygen-15.* For a discussion of various radionuclides and radiopharmaceuticals used in ECT work, see the review by Budinger, Gullberg, and Huesman (1979).

When a β^+ emitter is used, the ejected positron loses most of its energy over a distance of a few millimeters, almost comes to rest, and promptly annihilates with a nearby electron resulting in the formation of two gamma rays ($e^+ + e^- \rightarrow \gamma + \gamma$). Since the annihilation occurs essentially at rest in the laboratory coordinate system, each γ ray photon has an energy of 0.511 MeV and the photons are always 180° apart; that is, they travel in opposite directions along the same straight line.† If a ring of detectors is placed around the patient and

*Single photon ECT was introduced by Kuhl and Edwards (1963). For a brief historical background on the concept of imaging positron emitting radionuclides, see Budinger, Derenzo, Gullberg, and Huesman (1979); see also the review by Ter-Pogossian, Raichle, and Sobel (1980).
†Since the reaction occurs with the e^+ and e^- not quite at rest, the angle is $180 \pm \frac{1}{4}°$ [Budinger, Gullberg, and Huesman (1979)].

two of the detectors simultaneously record 0.511 MeV gamma rays, we know the radionuclide was somewhere on a line between the detectors, as indicated in Fig. 1.8. For single photon emitters, the ring-of-detectors concept, or other geometrical designs, can still be used but, of course, there will be no coincidence counting, and collimators must be used to determine direction.

In either case, single γ or positron, *if* the attenuating properties of the medium can be neglected, the reconstruction problem can be related directly to the Radon transform. Since the direction of the photons is known, the data (number of counts) recorded by a detector aimed along the line defined by and ϕ as indicated in Fig. 1.9 is just a measure of the total radionuclide concentration along that line. Thus $f(x, y)$ is identified with concentration at point (x, y) and the Radon transform of f is identified with number of counts $\check{f}$,

$$\check{f}(p, \phi) = \int_L f(x, y)\, ds. \tag{3.1}$$

In most situations, attenuation cannot be neglected in either single photon or positron ECT [Budinger and Gullberg (1977); Budinger, Derenzo, Gullberg, Greenberg, and Huesman (1977); Phelps (1977); Brownell, Correia, Zamenhof (1978)]. However, in positron ECT, a correction can be made in the data to compensate for attenuation and thus cast the problem back into the realm of Radon transform theory. Thus one cannot measure $\check{f}$ directly, but the data that still depend on the variables p and ϕ can be modified by multiplicative corrections to produce $\check{f}$ so that (3.1) still holds [Budinger, Gullberg, and Huesman (1979)].

In single photon ECT, the correction is not so simple and the appropriate transform is not the Radon transform, but what is known as the *attenuated* or

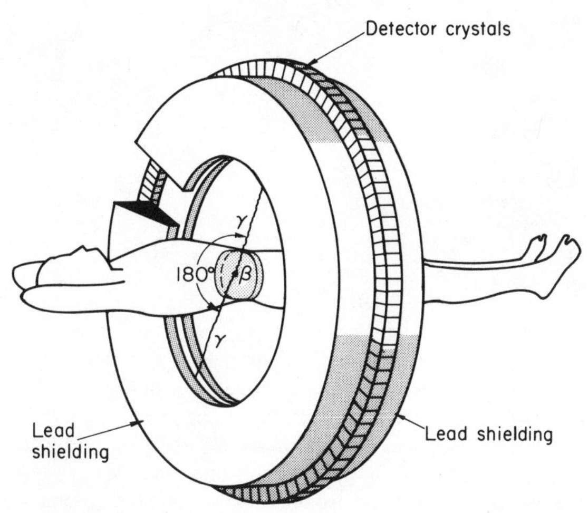

Figure 1.8 Arrangement of detector ring and shielding for imaging positron labeled compounds within the body. Courtesy of T. F. Budinger.

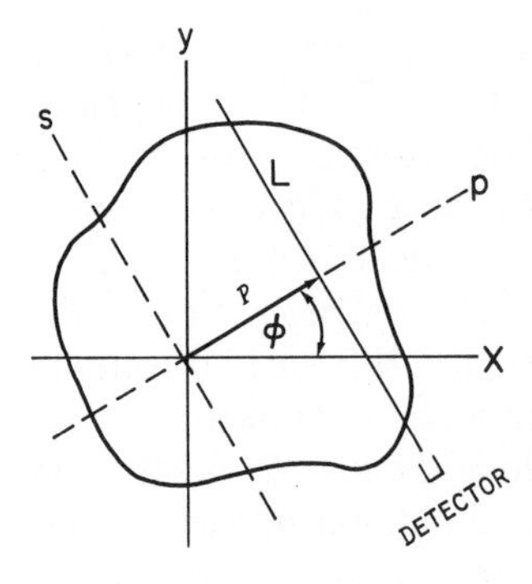

Figure 1.9 Typical coordinates for ECT imaging.

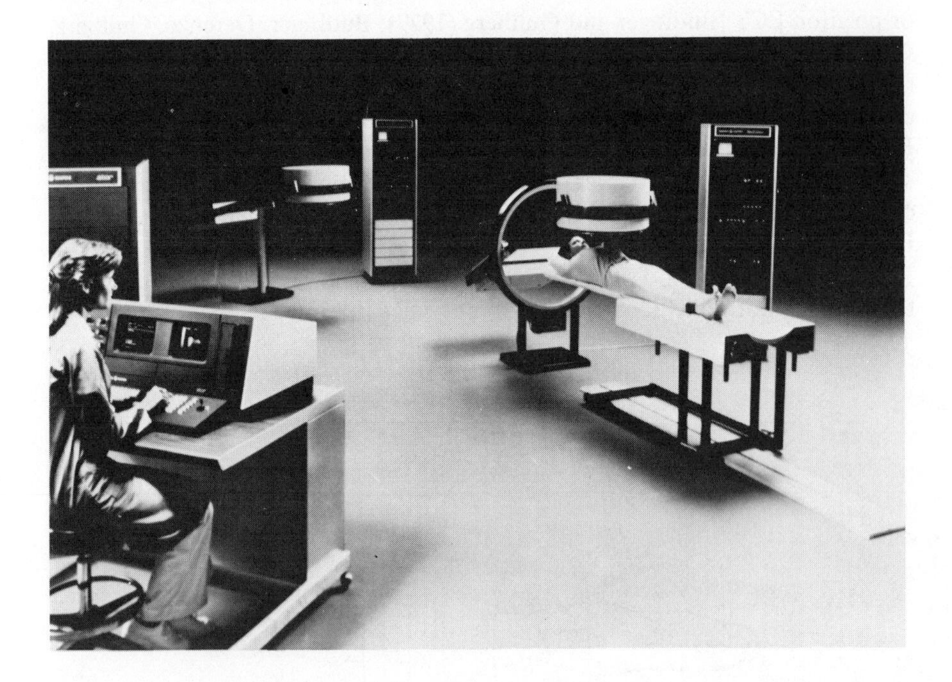

Figure 1.10 The G.E. 400T camera is a single photon ECT detector supported by a rotating gantry. The left insert shows reconstructed transaxial images of the heart. The data were obtained by rotating the camera $\pm 90°$ from the LAO position to obtain 64 angular views over 180°. The right insert shows transventricular views of the heart obtained by reformatting the transaxial images. In this orientation, the left ventricle is shown as one would view it looking from the apex to the base of the heart. A normal heart would show a complete "donut" but because of a transmural infarct in this particular patient, a cold area is shown due to a diminished uptake of activity. Courtesy of G. T. Gullberg, and General Electric Company.

exponential Radon transform [Gullberg (1979); Tretiak and Metz (1980)]. In some cases, such as heart imaging with a rotating gamma camera following an intravenous injection of ^{201}Tl, good tomographic results are obtained without attenuation correction (Fig. 1.10). However, imaging larger organs, such as the liver, attenuation correction is necessary in order to have better definition of the internal structure. Further discussion of the attenuated Radon transform is given in [8.8].

In addition to the need to compensate for attenuation, there is another important difference between ECT and TCT. In ECT, several limitations lead

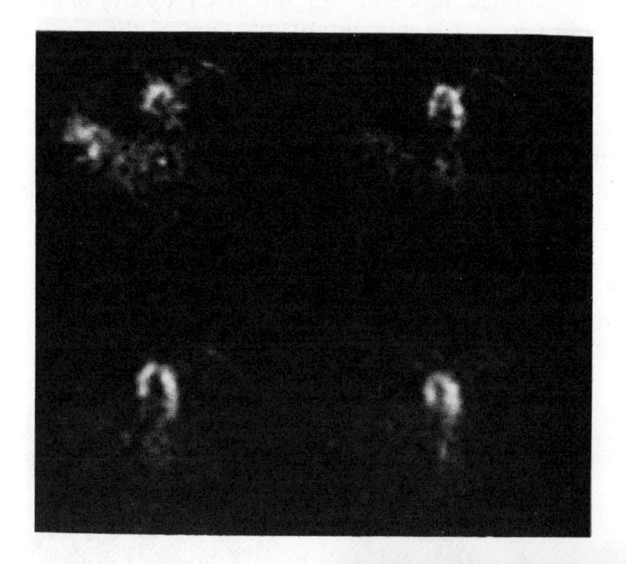

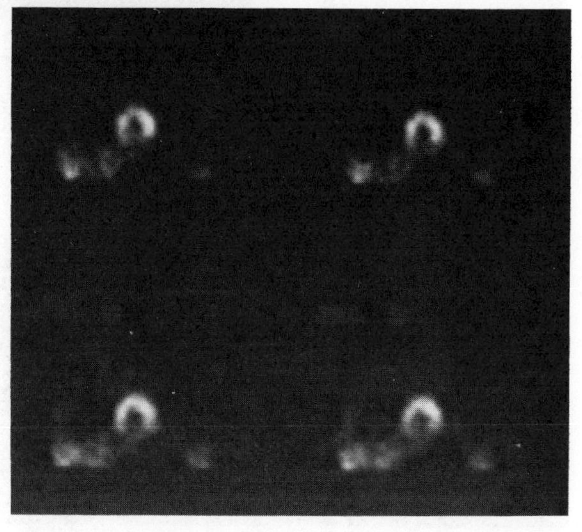

Figure 1.10 (*Continued*).

to a factor of over 10^3 less statistics than in TCT [Budinger, Gullberg, and Huesman (1979)], thus emphasizing the need for optimization and accounting for statistical limitations.

Finally, use of ECT in medical science provides a dramatic advancement in nuclear medicine just as X-ray TCT does in radiology. The clinical value is already established both independently and in a complementary role with X-ray TCT [Kuhl, Edwards, Ricci, Yacob, Mich, and Alavi (1976); Phelps, Hoffman, Huang, and Kuhl (1978); Hill, Lovett, and McNeil (1980)]. It is now possible to use radionuclides and newly developed radiopharmaceuticals to measure metabolism and physiologic function. As discussed by Budinger (1979) and by Budinger, Gullberg, and Huesman (1979), some of these new possibilities include investigations of brain and heart metabolism, such as neuroreceptor distribution studies; studies on the metabolic basis of schizophrenia; and heart metabolism of sugar, amino acids, and fats. Other possibilities include new methods of cancer detection, determination of myocardial infarction volume, and detection of pulmonary embolism, cerebral aneurysms, and strokes. There are several operating ECT machines and designs for new systems. One of these, the Donner ring array, is shown in Fig. 1.11. These ECT systems are discussed by Kuhl, Edwards, Ricci, Yacob, Mich, and Alavi (1976); Phelps (1977); Brownell, Correia, and Zamenhof (1978); Phelps, Hoffman, Huang, and Kuhl (1978); Budinger, Gullberg and Huesman (1979).

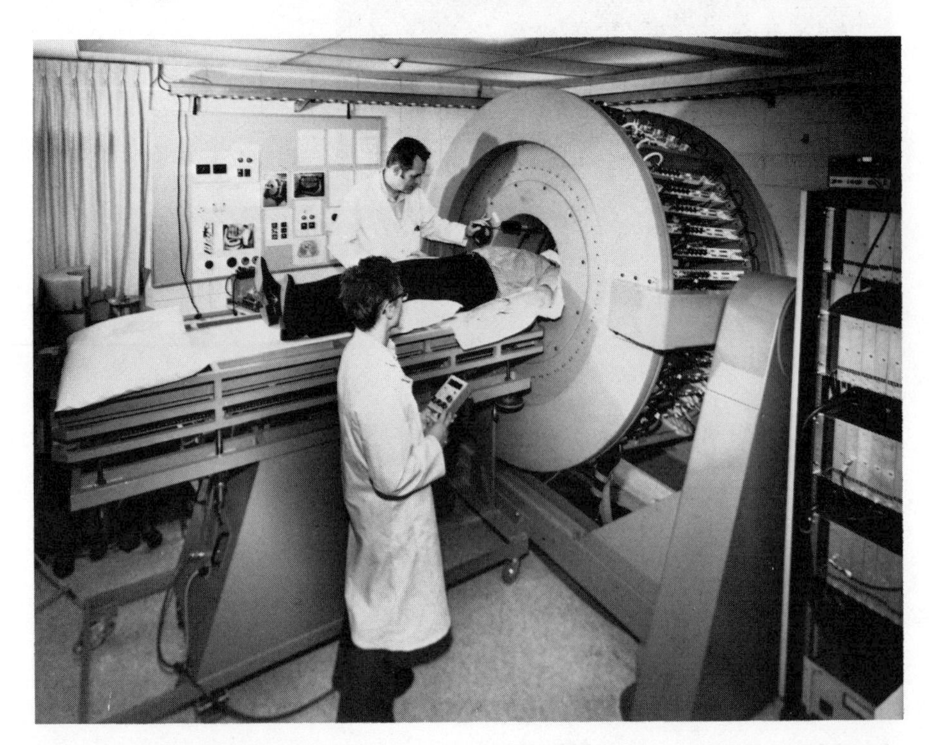

Figure 1.11 The Donner ring array at Lawrence Berkeley Laboratory for positron ECT work. Courtesy of T. F. Budinger.

[1.4] MEDICINE; ULTRASOUND CT

As in X-ray CT, the aim in UCT is to reconstruct transverse cross sectional images from projection data obtained when the probe, ultrasound in this case, passes through the object. Typical central acoustic probe frequencies are 3.5 MHz [Dick, Carson, Bayly, Oughton, Kubicheck, and Kitson (1977)] and 5.0 MHz [Glover and Sharp (1977)]. Under appropriate conditions, the ultrasound probe may be used to determine acoustic attenuation and acoustic velocity distributions in tissues. Investigators at the Mayo Clinic have demonstrated the method for two-dimensional acoustic attenuation distributions from amplitude data [Greenleaf, Johnson, Lee, Herman, and Wood (1974)] and for two-dimensional acoustic velocity distributions from propagation time data [Greenleaf, Johnson, Samayoa, and Duck (1975)]. The UCT work at Mayo is a continuing effort [see Johnson and Greenleaf (1979) for a brief history] that is producing considerable success in the area of breast imaging [Greenleaf, Kenue, Rajagopalan, Bahn, and Johnson (1980)]. Other clinical investigations include those of Colorado groups [Carson, Dick, Thieme, Dick, Bayly, Oughton, Dubuque, and Bay (1978); Dick, Elliott, Metz, and Rojohn (1979)] and work associated with efforts at General Electric [Glover and Sharp (1977)]. For a representative list of laboratories where UCT work is in progress, see Mueller, Kaveh, and Wade (1979), and for general reviews see Kak (1979); Greenleaf, Johnson, Bahn, Rajagopalan, and Kenue (1979); Mueller, Kaveh, and Wade (1979); Wade, Mueller, and Kaveh (1979); Greenleaf and Bahn (1981).

In order to cast an image reconstruction problem where ultrasound is the probe into the framework covered by the Radon transform, it is necessary to make a very restrictive assumption: the straight path approximation. In X-ray CT, the path of the probe from the source to the detector is a straight line, but when ultrasound propagates through tissue it undergoes a deflection at every interface between tissues of different refractive index [Morse and Ingard (1968)]. To see precisely how this is important, consider the way in which amplitude and propagation time data may be identified with the Radon transform.

Amplitude Data

Under the most ideal conditions, the attenuation coefficient $\mu(x, y)$ for an ultrasonic pulse propagating along the s direction in Fig. 1.12 may be related to the pressure amplitude A_0 at the transmitting transducer and the pressure amplitude $A(p, \phi)$ at the receiving transducer by the same type of equation as used in [1.2] [Wells (1975)]

$$A(p, \phi) = A_0 \exp\left[-\int_L \mu(x, y)\, ds\right].$$

By arguments similar to those in [1.2], the Radon transform $\check{f}$ may be identified with $-\log(A/A_0)$ and f identified with $\mu(x, y)$. [A derivation based on the

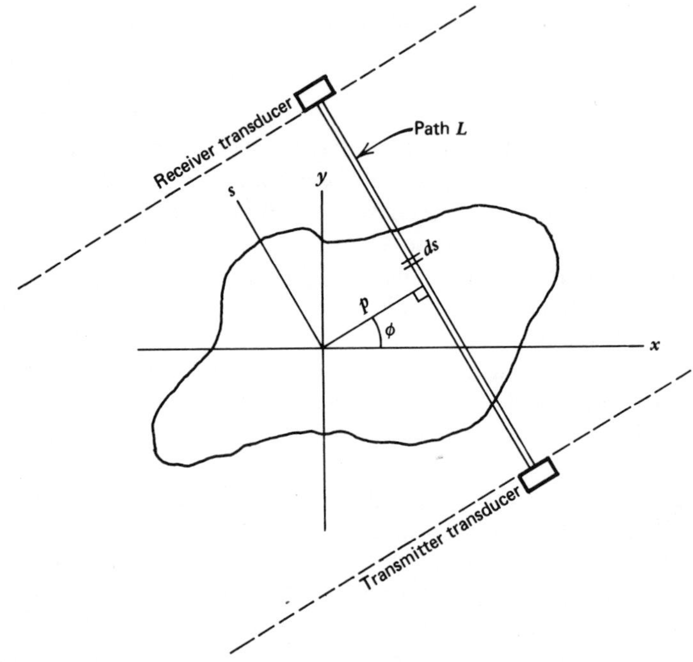

Figure 1.12 The pressure amplitude is A_0 at the transmitting transducer and $A(p, \phi)$ at the receiving transducer.

acoustic wave equation is given by Greenleaf, Johnson, and Lent (1978).] This approach using only absorption data has been only moderately successful even for soft tissue since reflection and refraction losses cause distortions that severely limit the accuracy of reconstructions [Greenleaf, Johnson, Samayoa, and Duck (1975); Glover and Sharp (1977); Klepper, Brandenburger, Busse, and Miller (1977); Jones, Kitsen, Carson, and Bayly (1979); Johnson and Greenleaf (1979)]. By measuring projections at several frequencies, Klepper, Brandenburger, Mimbs, Sobel, and Miller (1981) have developed methods for attenuation reconstruction based on the slope of the attenuation coefficient as a function of frequency. Artifacts arising from reflection and refraction may be reduced by this method.

Other methods for measuring the integrated attenuation coefficient have been developed by Kak and Dines (1978) and Dines and Kak (1979). Their frequency shift method is especially interesting since it is independent of transmittances at tissue-tissue and tissue-couplant interfaces. The integrated attenuation is simply proportional to the difference between the center frequencies ν_i and ν_t of the incident and transmitted pulses,

$$\int_L \mu(x, y)\, ds \propto \nu_i - \nu_t.$$

Further details and comparisons with other methods are given by Kak (1979) and Dines and Kak (1979).

Propagation Time Data

Another approach that is meeting with some success is the use of propagation speed (time-of-flight) data to reconstruct acoustic velocity distributions or, equivalently, index of refraction distributions. Suppose an ultrasonic pulse travels along the path L in Fig. 1.12. The time required for the pulse to go from the transmitting transducer to the receiving transducer is given by the line integral

$$T_\phi(p) = \int_L \frac{ds}{V},$$

where $V = V(x, y)$ is the velocity along the path L. A normalized set of time-of-flight projections may be defined by [Glover and Sharp (1977)]

$$\tau_\phi(p) = \int_L \frac{ds}{V_w} - \int_L \frac{ds}{V},$$

where V_w is the velocity in water. Since the relative index of refraction n is given by [Kak (1979)]

$$n(x, y) = \frac{V_w}{V(x, y)},$$

it follows that

$$V_w \tau_\phi(p) = \int_L [1 - n(x, y)]\, ds.$$

If $V_w \tau_\phi(p)$ is measured for arbitrary ϕ and p and identified with $\check{f}(p, \phi)$ and $1 - n(x, y)$ is identified with $f(x, y)$, then it is clear how the Radon transform emerges once again.

Although the time-of-flight data yield generally better results than the amplitude data, several problems still remain [Johnson and Greenleaf (1979)]. One problem is that the resolution is diffraction limited; that is, the wavelength of the probe is of the same order of magnitude as the details to be resolved. This has led to the use of diffraction methods using the wave equation and the Born and Rytov approximations [Iwata and Nagata (1975); Mueller, Kaveh and Wade (1979); Kaveh, Mueller, and Iverson (1979); Vezzetti and Aks (1979)]. Other methods have been used to correct for the straight-ray approximation [Johnson, Greenleaf, Samayoa, Duck, and Sjostrand (1975); Schomberg (1978); Bates and McKinnon (1979)]. However, interesting arguments by McKinnon and Bates (1980) lead to the conclusion that such corrections are unnecessary and even misleading in many practical situations.

Finally, Wade, Elliott, Khogeer, Flesher, Eisler, Mensa, Ramesh, and Heidbreder (1980) have presented principles for obtaining projectionlike data using reflections of acoustic waves. However, results of preliminary pulse-echo studies demonstrate the need for further development of specific algorithms for processing this type of acoustic information. Other approaches are being developed at the National Bureau of Standards and at the Mayo Foundation employing both circular and spherical transducer arrays to analyze pulse-echo data and reconstruct two- and three-dimensional images of acoustic reflectivity [Norton and Linzer (1981), 1979a, b); Johnson, Greenleaf, Rajagopalan, and Tanaka (1980)].

[1.5] ASTRONOMY

Strip Scans in Radio Astronomy

The first solution to a practical image reconstruction problem was given by Bracewell (1956a). The field was radio astronomy and the problem was to *map* the regions of emitted microwave radiation from the sun's disk. The term *map* is used here in the same sense as a two-dimensional distribution identified with $f(x, y)$.

From solar eclipse observations, it was clear by the early 1950's that microwaves were emitted from rather compact regions of the chromosphere and lower corona.* But due to the inadequate resolving power of microwave antennas, it was not possible to study these emissions in much detail during the periods of time between eclipses. There was also a more general question regarding the distribution of the residual background radiation over the sun's disk. The situation is described vividly by Bracewell (1974, 1979).

As seen from Earth, the angular diameter of the sun is about 30 minutes of arc. To map the emitted microwave radiation over the solar disk, an antenna with an angular resolution small compared to 30 minutes of arc is needed. Adequate spot resolution using a single antenna would require an antenna of enormous dimensions, impractical to construct. The way around this is to use several antennas along a line, thus achieving good resolution in one direction and poor resolution in the perpendicular direction. Thus it proved practical to construct arrays of antennas (several individual antennas along a line) with a

*The study of microwave emission from the sun, which is a currently active area of radio astronomy [Kundu and Gergely (1980)] was also an active yet top secret project during World War II. Over the period February 26–28, 1942 British radars were being jammed and it was discovered that the sun was responsible! [See Hey (1971).] These great storms of radio emission from the sun, which Hey (1946) was able to report after the war ended, marked the beginning of the modern development of radio astronomy.

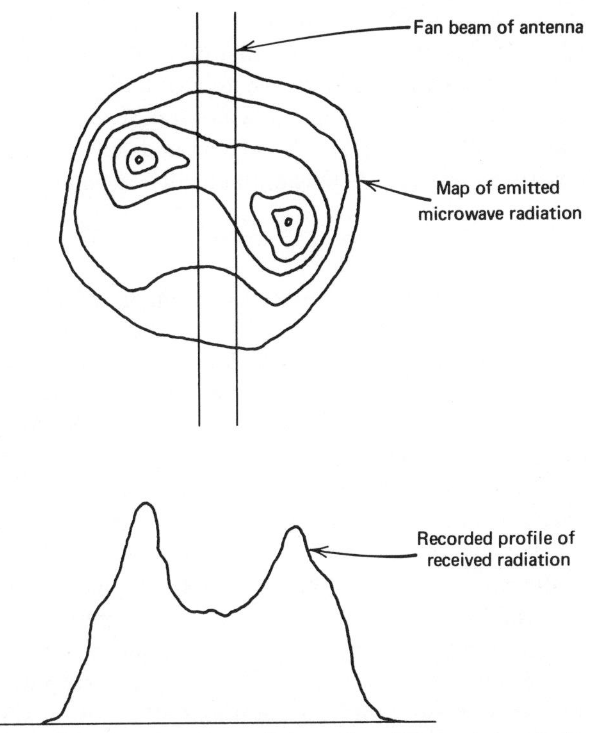

Figure 1.13 Typical map of solar microwave emission and profile recorded as the fan beam of the antenna passes over the solar disk.

beamwidth* of 3 minutes of arc, provided only one-dimensional resolution was demanded. Such an array of antennas would be sensitive to microwave radiation from sources inside a narrow strip of the sky 3 minutes of arc wide and several degrees long. In radio astronomy, a reception pattern of this type is known as a *fan beam*. When the fan beam system is aimed toward the sun, the received signal approximates a line integral of microwave intensity, as illustrated in Fig. 1.13.

There is a clear similarity with the X-ray CT and ECT problems discussed earlier, but there are major differences too. This is an emission (passive) problem, where the emitted probe is microwave radiation. Furthermore, the line of integration is perpendicular to the direction in which the radiation travels, rather than parallel as in the previous cases.

Since the antenna array is attached to the earth, the sensitive strip region sweeps across the sun as the earth rotates, and the received power rises and

*Beamwidth describes the angular resolution of a radiotelescope. A good approximation to the beamwidth in radians is given by λ/D, where λ is the wavelength of the radiation and D is the diameter of the antenna [Bracewell (1979)]. The effective value of D for a linear array is the length of the array along one direction and the width of a single antenna along the perpendicular direction. This provides sensitivity along a strip.

falls to produce a profile or fan beam scan as indicated in Fig. 1.13. Thus, insofar as this strip integration approximates a line integration, we see that if $f(x, y)$ represents the map of microwave intensity, then $\check{f}(p, \phi)$ is the profile or Radon transform of f for fixed angle ϕ. To obtain scans for other angles ϕ, it is only necessary to allow some time to elapse and do the scan again. As time passes, the position angle ϕ varies automatically and may be computed from

$$\cot \phi = \sin \delta \tan \mathfrak{h},$$

where δ is the declination and $\mathfrak{h}$ is the hour angle of the source [Bracewell (1979)].

The observational aspects of the solar microwave distribution problem were solved in a series of papers by Christiansen and Warburton (1953a, b, 1955) and the theoretical analysis using Fourier methods soon followed [Bracewell (1956a), Bracewell and Riddle (1967)]. Smerd and Wild (1957), building on suggestions of O'Brien (1953), also gave a theoretical analysis that involved use of the Fourier transform to invert the projection data.

Earlier, it was mentioned that the strip integrals should approximate line integrals if the recorded information is to be properly identified with the Radon transform. This problem has been studied extensively by Bracewell in connection with aerial smoothing in radio astronomy [Bracewell (1956a, b); Bracewell and Roberts (1954); Bracewell (1979)]. The strip distribution profile obtained by strip integration can be thought of as a smoothing of the corresponding line integration. Specifically [Bracewell (1974)],

$$\left\{ \begin{aligned} \text{strip integrate} &= \text{line integrate, then smooth} \\ &= \text{smooth, then line integrate} \end{aligned} \right\}$$

or, in terms of convolution $*$ [Bracewell (1979)],

$$\{\text{strip integral} = \text{line integral} * \text{beam cross-section shape}\}.$$

Whereas the pure reconstruction problem arises from integration along lines, when a strip integral is involved, the smoothing or blurring may cause enough loss of detail to create a need for resolution corrections. This is called *restoration* and involves error statistics, properties of recording devices, and judgment [Bracewell (1958, 1977, 1979)].

The methods described for determining solar brightness maps may be used to study other celestial regions, and emissions from other parts of the electromagnetic spectrum may serve as the probe. For example, in X-ray astronomy, strip scans have been used to reconstruct the X-ray emission structure of the Vela supernova remnant [Moore and Garmire (1975)]. Related areas of astronomy and radio astronomy also utilize reconstruction techniques. These include aperture synthesis methods in both radio and radar astronomy and occultation studies in radio and optical astronomy, where the power recorded as a function of time during the occultation may be converted to a profile curve similar to that which would have been obtained by fan beam scans [Scheuer (1965)].

A solar physics reconstruction problem of a fully three-dimensional nature has been reviewed in detail by Altschuler (1979). Data on the magnetic field of the solar surface and of the polarization brightness of the corona have been used to reconstruct the three-dimensional magnetic field and electron density distributions of the solar corona.

Aperture Synthesis

In aperture synthesis studies [Ryle and Hewish (1960); Swenson and Mathur (1968), Ryle (1972); Fomalont (1979)], an interferometer* is used to measure what is called *complex visibility* $\mathscr{V}(u, v)$ [Bracewell (1979)], a function of the vector $\mathbf{S} = (s_x, s_y)$ spacing between the two antennas; $u = s_x/\lambda$, $v = s_y/\lambda$, where λ is the wavelength to which the antennas are tuned. Complex visibility is determined by monitoring the electric field as a function of time at each of the antennas [Brouw (1975)]. The relationship between the desired radio brightness $\mathfrak{b}(l, m)$ of the extended source and the measured quantity $\mathscr{V}(u, v)$ is embodied in the van Citert–Zernike theorem [Born and Wolf (1975), section 10.4.2]. The complex visibility is the two-dimensional Fourier transform of the normalized source brightness distribution,

$$\mathscr{V}(u, v) = \int_{-\infty}^{\infty}\int_{-\infty}^{\infty} \hat{\mathfrak{b}}(l, m) e^{-i2\pi(lu+mv)}\, dl\, dm$$

where

$$\hat{\mathfrak{b}}(l, m) = \frac{\mathfrak{b}(l, m)}{\int_{-\infty}^{\infty}\int_{-\infty}^{\infty} \mathfrak{b}(l, m)\, dl\, dm},$$

and l and m are the direction cosines of the source element $dl\, dm$ with respect to the x and y axes of a plane transverse to the direction of the source. Thus, in terms of Fig. 1.3, $\check{f}$ is the measured quantity and f is to be computed. It has been suggested [D'Addario (1979)] that *Fourier synthesis* is a better term to describe what is commonly known as aperture synthesis.

The Radon transform emerges in aperture synthesis because in some instances it may be easier, or even necessary, to extract the strip distribution profile rather than the complex visibility from the observational data. Hence it

*In radio astronomy, the term interferometer signifies two (or more) separated aerials supplying signals to a common detector. The radiotelescopes may be electronically connected as with the Stanford five-element array [Bracewell, Colvin, D'Addario, Grebenkemper, Price, and Thompson (1973)] or operated separately with the respective signals related by a very accurate standard timing device [Cohen (1973)]. A comprehensive discussion and review of the use of a two-element radio astronomy interferometer to map source brightness is given by Swenson and Mathur (1968).

Figure 1.14 Photograph of the VLA. The National Radio Astronomy Observatory VLA Program is operated by Associated Universities Inc., under contract with the National Science Foundation. Courtesy of D. L. Swann.

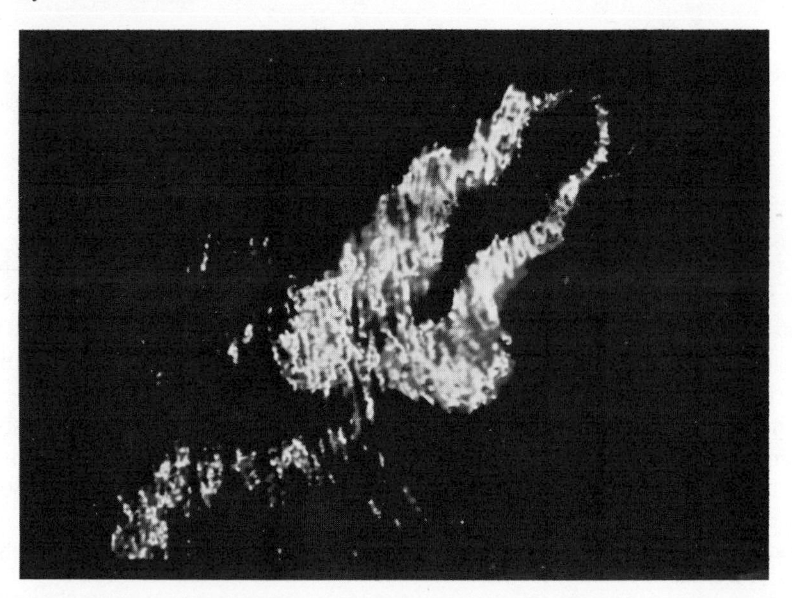

Figure 1.15 Radio "map" or "picture" of radio galaxy 3C129 prepared by L. Rudnick and J. Burns, using data from the VLA (Very Large Array) of the National Radio Astronomy Observatory (NRAO). The NRAO is operated by Associated Universities Inc., under contract with the National Science Foundation. Courtesy of L. Rudnick.

is useful to be able to go directly from strip distributions to brightness distributions [Hagfors, Nanni, and Stone (1968)].

Figure 1.14 is a photograph of the VLA (very large array) near Socorro, New Mexico, which is now operational for aperture synthesis work. The VLA map of radio galaxy 3C129 is shown in Fig. 1.15.

Radar Astronomy

Radar astronomy may be used to obtain radar maps of the moon and planetary surfaces in various ways [Thomson and Ponsonby (1968); Hagfors, Nanni, and Stone (1968); Hagfors and Campbell (1973); Pettengill, Zisk, and Thompson (1974); Pettengill (1978)]. Illustrated results are contained in these papers; see also Ponsonby, Morison, Birks, and Landon (1972), Zisk, Pettengill, and Catuna (1974), Thompson (1974).

An especially interesting method that leads directly to an interpretation in terms of the Radon transform is described by Thomson and Ponsonby (1968). In the passive aperture synthesis problem described earlier using either a two-element or multiple element interferometer, the basic idea is to measure separately enough components of the Fourier transform $\tilde{f}$ or the Radon transform $\check{f}$ to be able to reconstruct the desired brightness distribution f. In principle, all antenna spacings for every base-line direction are needed. Of course, only a finite number of spacings and directions will be available and this often involves physically moving one of the antennas several times. The (active) radar astronomy aperture synthesis problem is in some respects simpler because all antenna spacings may be obtained effectively by using an appropriate range of delays between the times the signals are correlated.

Suppose a Cartesian coordinate system is associated with the lunar disk, as indicated in Fig. 1.16. Furthermore, let the Earth-based radar be in motion relative to this coordinate frame with transverse velocity V. The reflected signal received by the radar from points on the lunar surface along a line of constant p has a Doppler shifted frequency exceeding that from points along the line with $p = 0$ by approximately

$$\nu = \frac{2Vp}{\lambda},$$

where λ is the wavelength. Let $f(x, y)$ be the radar brightness distribution normalized so that

$$\iint_{\text{moon}} f(x, y)\,dx\,dy = 1,$$

and let $\check{f}(p, \phi)$ be the line integral (Radon transform) of $f(x, y)$ along the line

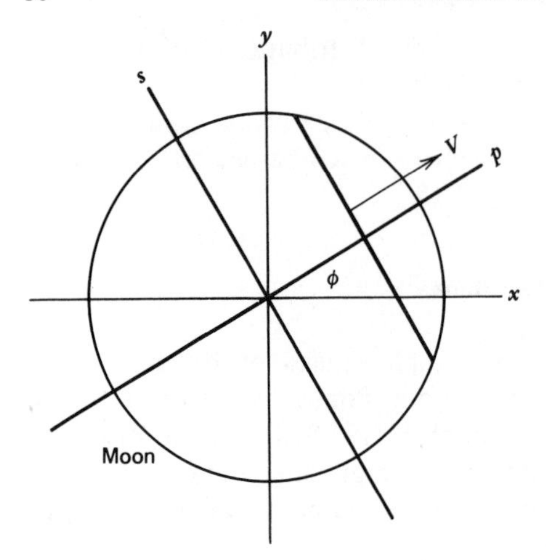

Figure 1.16 Coordinate system associated with lunar disk. The transverse velocity V is along the p axis.

defined by

$$x\cos\phi + y\sin\phi = p.$$

Furthermore, suppose the measured power spectrum of the received signals is normalized so that

$$\int_{\text{all }\nu} S(\nu,\phi)\,d\nu = 1.$$

The important result obtained by Thomson and Ponsonby (1968) is that

$$\check{f}(p,\phi) = \frac{2V}{\lambda} S\left(\frac{2V}{\lambda}p,\phi\right).$$

Using the appropriate scaling, $\check{f}$ can be computed from the measured power spectrum.

Doppler techniques may also prove useful in imaging where ultrasound is the probe [Mueller, Kaveh, and Wade (1979); Johnson, Greenleaf, Rajagopalan, Bahn, Baxter, and Christensen (1979)].

Occultation Studies

An occultation occurs when one body (say the moon) blocks our *view* of another, such as a star, a planet, or perhaps a moon belonging to one of the planets.* Here, the term *view* refers to the radiation being studied regardless of

*It may be useful to point out that the distinction between an occultation and eclipse is that an eclipse occurs when a body disappears because light from the sun has been prevented from reaching it by a second body, while an occultation occurs when one body physically blocks our view of another. The figure at the beginning of the article by Brinkmann and Millis (1973) nicely illustrates the two effects. Note that a "solar eclipse" is actually an occultation of the sun by the moon.

which part of the electromagnetic spectrum it comes from; γ-ray, X-ray, ultraviolet, visible, infrared, or radio region. There are two basic ways to think of the relation between the occultation curve (light curve, observed power) and the strip distribution or profile discussed earlier. In occultation experiments, the strip distribution profile is obtained by computing the time derivative of a filtered occultation curve [Scheuer (1965)]. Alternatively, when some object is occulted, the observed occultation curve is an integral of the object's strip brightness distribution [Elliot, Veverka, and Goguen (1975)]. Examples of the way this result may be used to extract basic properties of the occulted body include lunar occultation studies of stars [Barnes, Evans, and Moffett (1978)], planets and their moons [Elliot, Veverka, and Goguen (1975)], planetary atmospheres [Phinney and Anderson (1968)], and various radio sources [Scheuer (1962); Hazard (1962); DeJong (1966); Taylor (1967); Taylor and DeJong (1968)], including a recent study of the Crab nebula [Maloney and Gottesman (1979)]. For a review of principles, observational procedures, and results relating to occultations of stars by solar system bodies other than the moon, see Elliot (1979). Still other types of occultation studies include Jupiter and Saturn and their moons [Aksnes and Franklin (1976, 1978a, b); Smith, Green, and Shorthill (1977)], mutual occultations of planets [Albers (1979)], and occultations of spacecraft signals by planets [Kliore and Woiceshyn (1976); Fjeldbo, Kliore, Seidel, Sweetnam, and Woiceshyn (1976); Hunten and Veverka (1976)].

Use of Pulsars

One final application may be mentioned although it will take many years before enough data are available for reconstruction techniques to be applied. That is the use of pulsars as probes of the interstellar medium. Already, the first direct measurements of mean electron density and magnetic field strength in interstellar space can be made using pulsars as sources [Guélin (1973)].

[1.6] ELECTRON MICROSCOPY

Three-dimensional electron microscopy started in 1968 with the publication of several independent papers [DeRosier and Klug (1968); Hart (1968); Hoppe, Langer, Knesch, and Poppe (1968); Vainshtein, Barynin, and Gurskaya (1968)]. The method suggested by Hart (1968), called *polytropic montage*, is comparable to conventional focal-plane tomography [Brooks and Di Chiro (1976a)], which dates back to French patents in the 1920s and to the work in medical radiology in the 1930s [Gordon and Herman (1974)]. The other papers were motivated by general ideas from the study of X-ray crystallography, where Fourier synthesis of electron density is a well-known tool in crystal structure analysis dating back at least as far as early work by Bragg (1915).

The basic ideas of analysis are contained in the very nice review by DeRosier (1971). If $\mu(x, y, z)$ is the effective mass absorption coefficient for a given specimen, then for an electron beam along the x direction, the reduction (due to scattering) of the incident beam intensity I_0 is given by

$$I(y, z) = I_0 e^{-\sigma(y, z)}, \tag{6.1}$$

where $\sigma(y, z)$ is the projected density,

$$\begin{aligned} \sigma(y, z) &= \int_{-\infty}^{\infty} \mu(x, y, z)\, dx \\ &= \int_{-\infty}^{\infty} \rho(x, y, z)\, dx. \end{aligned} \tag{6.2}$$

Following DeRosier (1971), we simply equate $\mu(x, y, z)$ to the density $\rho(x, y, z)$. The infinite limits are for convenience since there is no contribution outside the specimen. The optical density $D(y, z)$ on the photographic film (micrograph) is proportional to the natural logarithm of $I(y, z)$. To a good approximation,

$$\begin{aligned} D(y, z) &= C \log I(y, z) \\ &= C \log I_0 - C\sigma(y, z). \end{aligned} \tag{6.3}$$

The scaling constant C can be determined experimentally so that knowledge of $D(y, z)$ yields $\sigma(y, z)$. Equivalently, the Fourier transform of an appropriately scaled D yields the Fourier transform of σ. This transform is given by

$$\tilde{\sigma}(k, l) = \iint_{-\infty}^{\infty} \sigma(y, z) e^{-i2\pi(ky + lz)}\, dy\, dz, \tag{6.4}$$

and the Fourier transform of ρ is given by

$$\tilde{\rho}(h, k, l) = \iiint_{-\infty}^{\infty} \rho(x, y, z) e^{-i2\pi(hx + ky + lz)}\, dx\, dy\, dz. \tag{6.5}$$

If $\rho(h, k, l)$ is evaluated at $h = 0$, then

$$\tilde{\rho}(0, k, l) = \iiint_{-\infty}^{\infty} \rho(x, y, z) e^{-i2\pi(ky + lz)}\, dx\, dy\, dz, \tag{6.6}$$

and in view of (6.2) the x integration yields

$$\tilde{\rho}(0, k, l) = \iint_{-\infty}^{\infty} \sigma(y, z) e^{-i2\pi(ky + lz)}\, dy\, dz. \tag{6.7}$$

From (6.4), this is just

$$\tilde{\rho}(0, k, l) = \tilde{\sigma}(k, l). \tag{6.8}$$

It follows that the Fourier coefficients of a projection form a central section of the Fourier coefficients of the three-dimensional density distribution. In principle, the set of Fourier coefficients $\tilde{\rho}(h, k, l)$ may be built up by collecting many different projections of the specimen, and the density $\rho(x, y, z)$ computed as an inverse Fourier transform of $\tilde{\rho}(h, k, l)$. In practice, this is by no means a trivial task. For a discussion of the experimental situation, see Crowther and Klug (1975); Vainshtein (1978); Hoppe and Hegerl (1980).

Thus, if enough projections are available, the entire problem can be solved by use of Fourier transforms. Often, it is desirable to obtain the projections by rotating the specimen about a fixed axis, say the z axis. These *coaxial* projections are samples of the Radon transform. To see this, consider (6.2) with z equal to some constant c,

$$\sigma(y, c) = \int_{-\infty}^{\infty} \rho(x, y, c)\, dx. \tag{6.9}$$

This corresponds to a line $z = c$ on the micrograph, which is coordinatized by y and z as indicated in Fig. 1.17. If the beam direction is the x direction, $\sigma(y, c)$ may be identified with the $\phi = 90°$ sample of the Radon transform of ρ for the transverse section $z = c$.

The corresponding result when the specimen is rotated about the z axis by some angle ϕ is the sample

$$\sigma(x', c) = \int_{-\infty}^{\infty} \rho(x, y, c)\, dy'. \tag{6.10}$$

For the geometry, see Fig. 1.18. Note that for $\phi = 90°$, the x' direction is just the y direction and (6.10) reduces to (6.9). Equivalently, when σ is identified with $\check{\rho} = \mathscr{R}\rho$ and polar coordinates p, ϕ are introduced (6.10) may be written

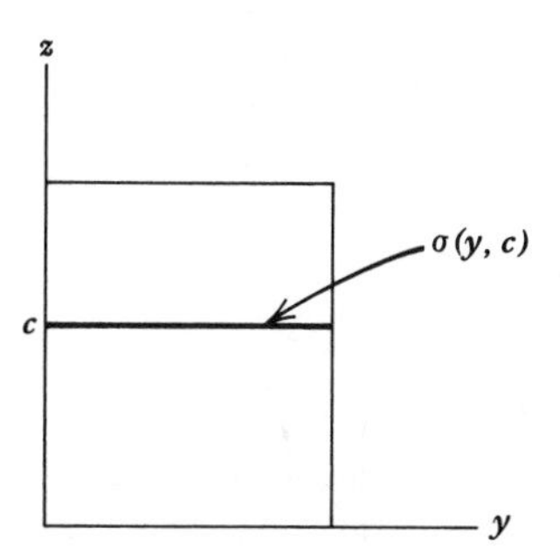

Figure 1.17 The density along the line $z = c$ corresponds to a sample of the Radon transform.

in the standard form

$$\check{\rho}(p, \phi, c) = \int_{-\infty}^{\infty} \int_{-\infty}^{\infty} \rho(x, y, c)\delta(p - x\cos\phi - y\sin\phi)\,dx\,dy \quad (6.11)$$

developed in [2.2]. Given that $\check{\rho}$ can be measured on the micrograph using a microdensitometer, the inverse Radon transform yields the $z = c$ transverse section. By changing c or, equivalently, measuring along lines parallel to the $z = c$ line of Fig. 1.18, various transverse sections may be obtained and the three-dimensional structure reconstructed by stacking these transverse sections.

Several reconstruction methods have been attempted in electron microscopy work. A discussion of the relevance of various methods to specific situations is contained in the reviews of Hoppe and Typke (1978) and Hoppe and Hegerl (1980). Special attention to the connections between the approaches of X-ray crystallography and electron microscopy is given by Hoppe (1979) using the single-unit-cell approach; see also Stroke and Halioua (1976). It is especially interesting that the very important iterative algebraic methods pioneered by Gordon, Bender, and Herman (1970) emerged as a direct application in electron microscopy. Other iterative algorithms soon appeared, for example, Goitein (1972), Gilbert (1972b), Crowther and Klug (1974). These iterative approaches, based on the Kaczmarz (1937) method for solving linear equations [Hamaker, Smith, Solmon and Wagner (1980)] are especially adaptable to the incorporation of certain a priori information, and they have been used in many other areas of image reconstruction since the original introduction in electron microscopy. It may be interesting to note that reconstructions algorithms for

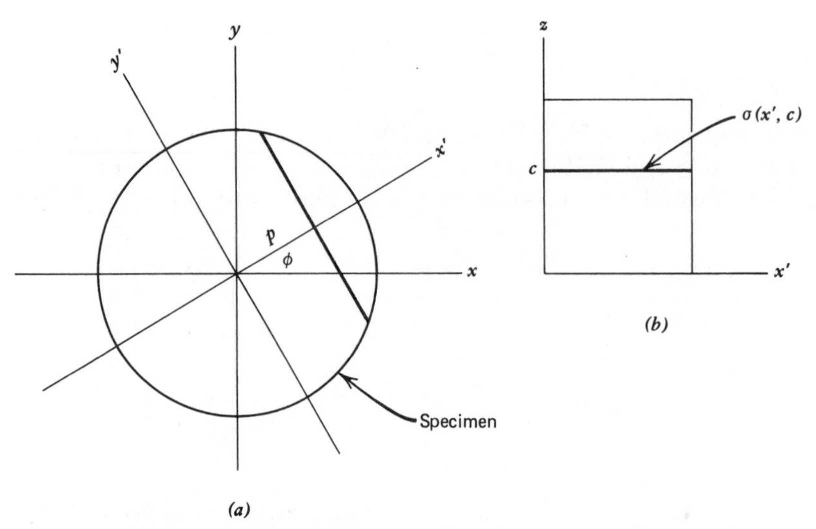

Figure 1.18 (a) Transverse section through $z = c$ with z direction perpendicular to the page. The beam direction is along y'. As p varies, a line is produced on the micrograph, as indicated in (b).

the original EMI scanner [Hounsfield (1973)] also employed iterative techniques of the Kaczmarz (1937) type, but Hounsfield and EMI were careful not to disclose that information during the early days of CT work.

[1.7] NUCLEAR MAGNETIC RESONANCE

Basic Principles

The basic principles of NMR imaging were presented by Lauterbur (1973, 1974) who called the technique *zeugmatography* for reasons that will be clear later. Before getting into current NMR imaging techniques, it may be useful to delve ever so briefly into certain basic NMR principles, although these and many others are discussed in standard texts, such as Andrew (1969), Farrar and Becker (1971), Schumacher (1970), and Slichter (1978). A brief but very useful summary of pertinent principles is given by Andrew (1980a), and a discussion with a view toward medical applications of NMR imaging is provided by Hoult (1980c).

The ground state of nuclei with an even number of protons and an even number of neutrons (even-even nuclei) is nearly always a zero intrinsic (spin) angular momentum state. Even-odd or odd-even nuclei always have a spin angular momentum which is an odd integer divided by two; that is, $I = \frac{1}{2}, \frac{3}{2}, \ldots$. The nuclear spin quantum number I may be converted to units of angular momentum by multiplying by $\hbar = h/2\pi$ where h is Planck's constant. A given nucleus with spin I in an applied magnetic field B_0 along the z direction has $2I + 1$ energy states. These levels are equally spaced with separation

$$\Delta E = \frac{\mu B_0}{I}, \tag{7.1}$$

where μ is the nuclear magnetic moment. In terms of $\hbar$ and the magnetogyric ratio γ, which is constant for a given nucleus,

$$\mu = \gamma \hbar I. \tag{7.2}$$

Thus, regardless of the value of I, it follows that

$$\Delta E = \gamma \hbar B_0. \tag{7.3}$$

The frequency of radiation that induces a transition between adjacent levels is given by

$$\nu_0 = \frac{\Delta E}{h} = \frac{\gamma B_0}{2\pi}. \tag{7.4}$$

The proton has been used for most NMR imaging due to high concentrations of hydrogen in biological tissue plus their favorable magnetic resonance properties. Since the proton has spin $\frac{1}{2}$, there are only two levels, as indicated in Fig. 1.19. The value of γ is approximately 0.0268 MHz/gauss. Thus, for a 1000-gauss magnetic field, the resonant frequency is 4.26 MHz, which is in the radio frequency (RF) region. These are typical values for NMR imaging work [Moore, Holland, and Kreel (1980)].

In conventional NMR spectroscopy, the specimen is homogeneous from a macroscopic point of view. It may be a pure crystal, powder, liquid, or gas. A small sample of the specimen is placed in a very uniform magnetic field. The uniformity may be one part in 10^9 over the sample. The excitation is achieved by a short RF pulse, at the resonant frequency, which is applied to a coil surrounding the sample in such a way as to create a magnetic field perpendicular to the constant field B_0. When the deexcitation (decay) occurs, the emitted radiation can be detected by the voltage induced in the receiver coil, which incidentally may be the same coil used to cause the initial excitation. Various properties of the specimen may then be inferred, such as relaxation times, diffusion constants, and chemical bonding information. It is worth emphasizing that the pulse must be exactly at the resonant frequency to produce a significant effect. If the specimen is not uniform and has heterogeneous internal structure, the NMR spectrometer gives a superposition of the properties of the various materials, but yields little or no information about spatial distributions of internal structure.

NMR Zeugmatography

In NMR zeugmatography, the specimen is deliberately placed in a nonuniform magnetic field. Andrew (1976) has described a two-dimensional version of Lauterbur's 1973 experiment to illustrate the idea. We shall see later that the two-dimensional version is fully adequate, because there are ways to selectively choose any plane desired through a three-dimensional object.

Consider two identical tubes of water with their axes (z direction) perpendicular to the page. If these tubes are placed in a uniform magnetic field B_0 directed along the y axis, then the NMR proton spectrum consists of a single

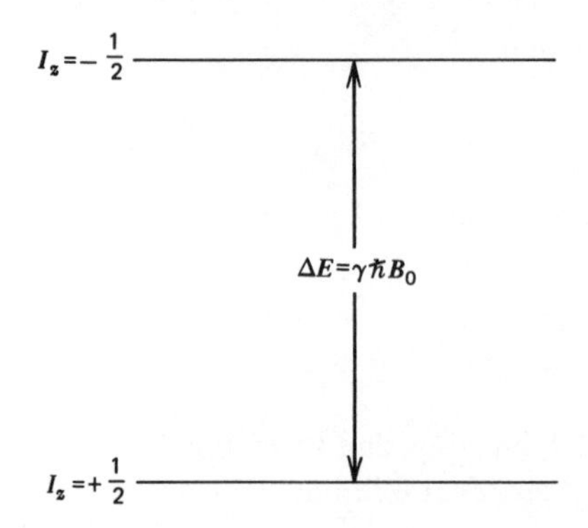

Figure 1.19 Energy levels for a proton in a magnetic field. I_z is the projection of I along the direction of the magnetic field.

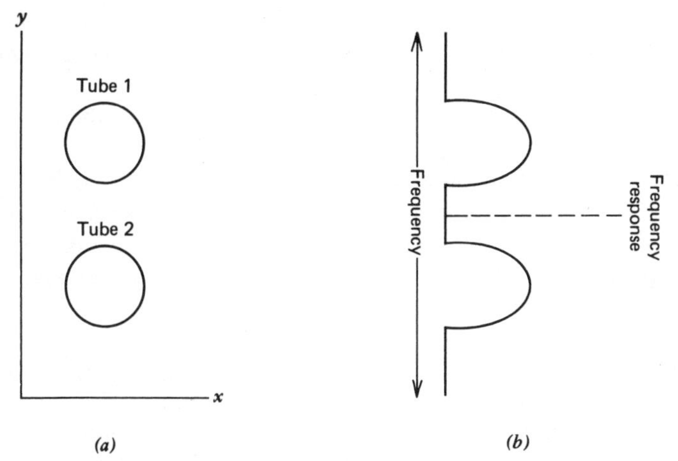

Figure 1.20 (*a*) Tubes with centers along y axis, which is the direction of the static magnetic field gradient. (*b*) Profile as function of frequency.

sharp line since all protons in the tubes resonate at the same frequency given by (7.4). On the other hand, if a linearly varying field is applied,

$$\frac{\partial B_0}{\partial y} = \text{constant}, \tag{7.5}$$

then as the frequency is changed only those protons that satisfy (7.4) resonate, and the strength of the signal is proportional to a line integral (along the x direction), which essentially counts the number of protons (hydrogen nuclei) along lines perpendicular to the y axis, as indicated in Fig. 1.20. Thus the NMR frequency response is the one-dimensional projection at right angles to the gradient direction.

Now, suppose the static gradient field (7.5) is maintained and the tubes of the homogeneous substance (water) are replaced by some arbitrary heterogeneous object. Then, the NMR spectrum constitutes a one-dimensional projection (or profile) of nuclear density;* hence the contact with the Radon transform. The Radon transform $\check{f}(p, \phi)$ for fixed ϕ is identified with the frequency response where p may be identified with frequency in this case. Thus Fig. 1.20*b* may be regarded as a plot of $\check{f}(\nu, \phi)$ as a function of ν. Such one-dimensional projections were studied by NMR spectroscopists long before

*More precisely, what is being imaged is a spatial representation of the NMR signal rather than just the proton density. This signal is directly proportional to the proton density but in general it also depends on other parameters, such as the longitudinal and transverse relaxation times (T_1 and T_2, respectively) and fluid velocity [Andrew (1980b)]. For example, the images obtained by Hawkes, Holland, Moore, and Worthington (1980) represent proton density multiplied by the ratio T_2/T_1. For fluids and soft tissue, T_2 and T_1 are approximately equal, so that the NMR signal is a measure of proton density.

the idea was incorporated into imaging. The basic principle of using an inhomogeneous magnetic field was discovered by Gooden (1950) and further investigated by Gabillard (1951, 1952). The fundamental ideas are nicely illustrated in Fig. 1.21, which was kindly provided by Dr. Katherine N. Scott.

Remark. Observe that ordinarily if x and y represent length coordinates, then p also is a distance. Here, however, since frequency is related to magnetic field strength (7.4) and the magnetic field is proportional to a distance (7.5), it follows that there is a direct proportionality between distance and frequency which justifies identifying p with frequency. □

Thus the fundamental idea underlying zeugmatography is that the gradient in the laboratory magnetic field gives spatial discrimination of those nuclear magnetic moments that interact with (or couple to) the radiation field with frequency ν_0 (the RF field). It was, in fact, this coupling of the RF field to the spatially variable magnetic field by limited regions of the sample that led Lauterbur (1973) to the name zeugmatography that comes from the Greek zeugma ($\zeta\varepsilon\upsilon\gamma\mu\alpha$) "that which is used for joining." In other words, limited regions of the sample cause the joining or coupling of the fields.

It is interesting to note that in zeugmatography the RF probe field wavelength is large compared to the size of the object to be imaged. However, due to the special nature of this technique, the wavelength of the radiation does not restrict imaging ability. For example, at 4.26 MHz, the wavelength is a little over 70 meters. Holland, Moore, and Hawkes (1980) point out that there is a very close analogy between NMR imaging and radionuclide emission tomography ECT. In both cases emitted photons (radiation) from the object (or patient) are detected. In NMR, the "isotope" is hydrogen which is rendered "radioactive" (emissive) by application of the RF pulse. The "half-life" for decay corresponds to certain characteristic times, the spin-lattice relaxation time T_1, which is a measure of the connection between the spin and its molecular environment and the spin-spin or transverse relaxation time T_2, which is a measure of the connections among the spins themselves. For a brief discussion of T_1 and T_2, see House (1980) or Andrew (1980a, b).

Oscillatory Magnetic Field Gradient

Earlier, it was stated that the NMR response could be restricted to a given plane through the specimen. The method is described by Hinshaw (1974, 1976). The basic idea is to use an oscillatory magnetic field gradient in the z direction to isolate a plane perpendicular to the z axis. To see how this works, first consider a static gradient field similar to the one discussed earlier (7.5) but this time with the gradient field in the z direction. This field can be turned into an oscillatory gradient field by using a pair of separated coils driven equally and in opposite directions by low-frequency alternating current. In the midplane parallel to the coils (and thus perpendicular to the z axis), the oscillatory

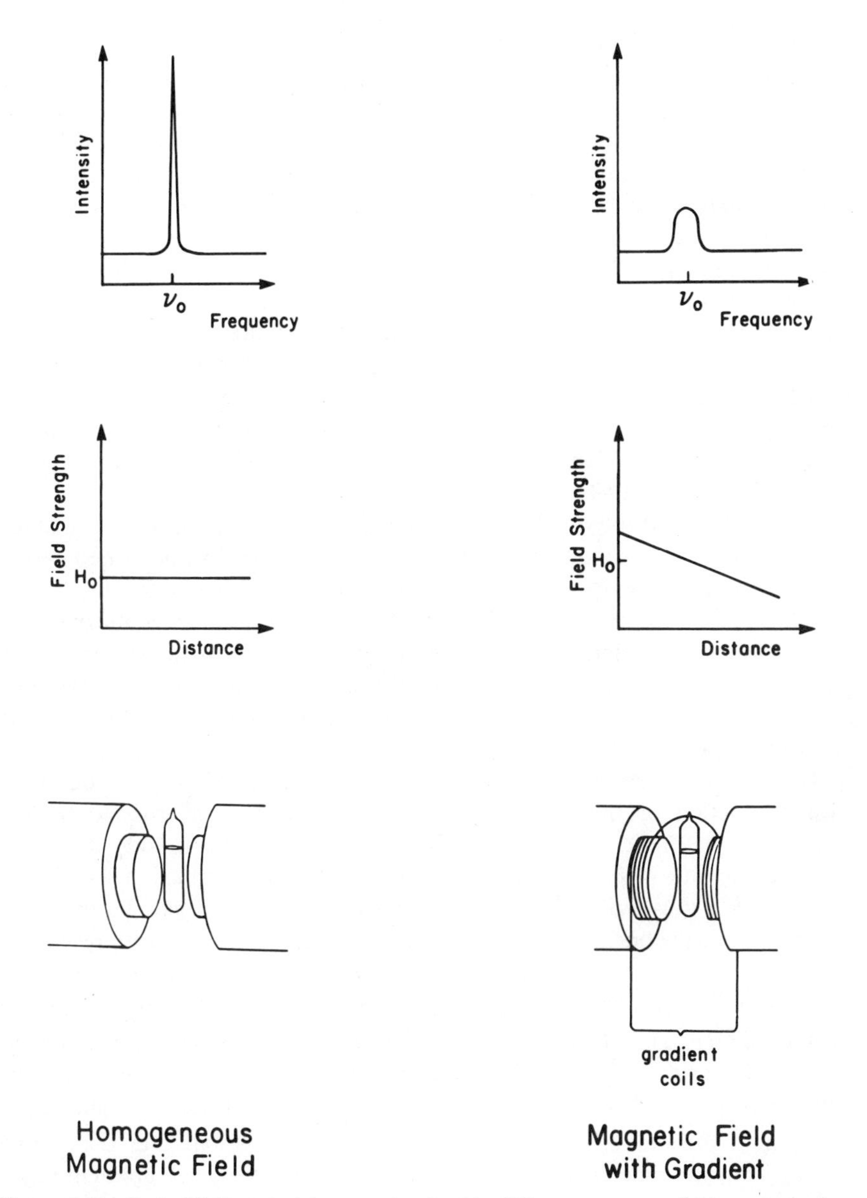

Figure 1.21 Basic NMR principles associated with different magnetic field configurations. Courtesy of Katherine N. Scott.

fields cancel and produce a null region, with no oscillatory component, known as the null plane or sensitive plane. The effect of the oscillatory field is to almost destroy the average NMR response signal picked up by the receiver coil *except* for the response from the sensitive plane. Observe that this response is proportional to an integral over the sensitive plane. The thickness of the plane can be controlled by the strength of the oscillating gradient, and the position of the plane can be moved by alternating the ratio of the currents fed into the two coils [Moore and Holland (1980); Moore, Holland and Kreel (1980)].

Thus NMR measurements can be used to approximate the *plane* integrals of a density $f(x, y, z)$ of hydrogen nuclei. These plane integrals may be identified with the Radon transform of a function on $\mathbb{R}^3$. Shepp (1980) has discussed the NMR imaging problem from this viewpoint and has provided reconstruction algorithms based on Radon's inversion formula for fully three-dimensional reconstruction. A different three-dimensional approach has been suggested by Lai and Lauterbur (1980).

If in addition to the oscillatory gradient in the z direction, there is a static gradient in the y direction, then the situation is the same as described earlier (Fig. 1.20), and the spectrometer output can be converted to the Radon transform of a two-dimensional density $f(x, y)$ or the time domain signal may be used directly [Kumar, Welti, and Ernst (1975a, b); Mansfield and Maudsley (1976), Brooker and Hinshaw (1978)]. Since the idea of directly using the time domain data is relevant to current approaches, it may be useful to consider this in more detail.

The Radon transform of the proton spin density $f(x, y)$ may be identified with a function of frequency and angle. The angle is determined by the direction of the static gradient field. (If the static gradient is in the y direction, then the line integrals are in the x direction and this corresponds to $\phi = 90°$ in Fig. 2.3.) We thus write $\mathscr{R}f$ as $\check{f}(\nu, \phi)$. The usual way to begin the inversion process is take the one-dimensional Fourier transform of $\check{f}(\nu, \phi)$ with respect to the radial variable, which in this case is frequency. Hence the Fourier transform of $\check{f}(\nu, \phi)$ just converts over to the time domain from the frequency domain, $\mathscr{F}\check{f}(\nu, \phi) = \tilde{\check{f}}(t, \phi)$. In terms of Fig. 1.3, that would correspond to being in the lower box on the right side. The beautiful part of this is that if the pulse sequences are selected properly, the NMR spectrometer signal as a function of time may be directly related to $\tilde{\check{f}}(t, \phi)$ without ever computing $\check{f}(\nu, \phi)$. Thus the time domain NMR signal obtained for a given gradient angle may be transferred to the central processing unit, stored on disk, filtered, inverse Fourier transformed, and back projected into the image array [Holland, Hawkes, and Moore (1980)]. More about this inversion procedure appears in Chapter 6.

The experimental details related to the NMR imaging system used by the Nottingham group has been discussed by Moore and Holland (1980).* Build-

*Also, see the imaging apparatus used by the Aberdeen group. Their whole-body NMR imaging machine is described in detail by Hutchison, Edelstein, and Johnson (1980).

ing mainly on the ideas of Hinshaw (1974a, 1974b, 1976), they have developed a very versatile system. The sensitive plane can be selected perpendicular to the z axis to yield transverse sections, perpendicular to the x axis to yield sagittal sections, or perpendicular to the y axis to give coronal sections [Holland, Hawkes, and Moore (1980)]. Here it is assumed that the patient is lying down in the horizontal xz plane with the feet to head direction parallel to the z axis. The y axis is up. By using two oscillatory gradients and one static gradient, the sensitive region is reduced from a plane to a line. Internal structure information obtained this way is often called the *multiple sensitive point method* [Andrew (1980a); Moore and Holland (1980)]. By using three oscillatory gradients, the sensitive region may be reduced to a point [Hinshaw (1974, 1976)]. The internal structure may be determined by moving the point around raster fashion.

Neither the sensitive point nor the multiple sensitive point methods employ reconstruction from projections. Various other ingeneous selective irradiation pulsing methods are being investigated [Lai, Shook, and Lauterbur (1979); Hoult (1979, 1980a); Maudsley (1980)]. At the present time, there is no general consensus as to which method ultimately will be optimum.

Other Work

In this section, we have concentrated mostly on the recent work of the Nottingham group. Many other workers have made important contributions to NMR imaging. See the review by Hoult (1980c), the report of the Aberdeen group [Hutchison, Edelstein, and Johnson (1980)], and the NMR references at the end of this chapter. For further discussion and comparison of methods, see Bottomley (1979) and Brunner and Ernst (1979). Reviews of ongoing efforts in the NMR imaging field are contained in papers presented at the IEEE Short Course on Nuclear Magnetic Resonance Imaging for Physicians and Engineers [House (1980); Lauterbur and Lai (1980); Andrew (1980a), Crooks (1980); Singer (1980); Hollis (1980)] and a two-day discussion on Nuclear Magnetic Resonance of Intact Biological Systems organized by R. J. P. Williams, E. R. Andrew, and G. K. Radda and sponsored by the Royal Society of London [Andrew (1980b); Lauterbur (1980a); Damadian (1980); Béné, Borcard, Hiltbrand and Magnin (1980); Mansfield, Morris, Ordidge, Pykett, Bangert, and Coupland (1980); Moore and Holland (1980); Mallard, Hutchison, Edelstein, Ling, Foster, and Johnson (1980); Bovée, Creyghton, Getreuer, Korbee, Lobregt, Smidt, Wind, Lindeman, Smid, and Posthuma (1980); Locher (1980); Hoult (1980b)].

Also, we have concentrated on imaging a function proportional to the proton density; however, other nuclei may be imaged. Using test objects and reconstruction from projections techniques, Bendel, Lai, and Lauterbur (1980) demonstrated the feasibility of imaging phosphorus-31 spin densities of inorganic phosphate, adenosine triphosphate, and creatin phosphate. Hoult (1979) has also worked with ^{31}P images and ^{19}F has been used by Holland,

Bottomley, and Hinshaw (1977). Other nuclei such as ^{2}H, ^{13}C, ^{14}N, ^{15}N, and ^{23}Na may be of interest eventually [Andrew (1980b)].

[1.8] OPTICS

Interferometry

Optical interferometry provides a method for measuring optical pathlength differences in optically transparent objects or media. *Holographic interferometry*, which is the interferometric comparison of two or more waves, at least one of which is holographically reconstructed, provides a convenient and very accurate means for measuring optical pathlength differences [Vest (1979)]. Such measurements may be related to the refractive index of the media and used to obtain information about internal properties of the media. These properties and media include mass density of fluids and gases, electron density of plasmas, temperature of fluids and gases, chemical species concentrations in reacting gases, and state of stress of transparent solids [Vest (1979)].

Our primary concern here is with the way in which the Radon transform emerges in the problem of determining the refractive index rather than the way in which the refractive index relates to the various properties and media mentioned earlier. For a comprehensive discussion of this latter connection, see Vest (1979).

In general, the optical pathlength Φ of a light ray through some medium is the path integral

$$\Phi = \int_{\text{path}} n \, ds, \tag{8.1}$$

where $n = n(x, y, z)$ is the index of refraction as a function of position and ds is the length element along the path of the ray [Born and Wolf (1975)]. Usually, the path will be curved and will have sharp corners at boundaries between media of different refractive index. Also, if the light is not monochromatic, the path will depend on the wavelength leading to dispersion as when white light incident on a prism is dispersed into its spectrum. Consequently, we shall consider *phase objects* exposed to monochromatic light of wavelength λ. These are objects (or media) in which refraction is negligible. [For extension to strongly refracting axisymmetric phase objects, see Vest (1975), and for further extension to asymmetric refractive-index fields, see Cha and Vest (1979).] This means that for all practical purposes, the probes (light rays) pass through the object in straight lines. In this so-called *refractionless limit* the optical pathlength may be written as an integral along a line rather than along some curved path. For a ray parallel to the z axis (or, equivalently, a wave moving in the $\pm z$ direction),

$$\Phi(x, y) = \int_{\text{line}} n(x, y, z) \, dz. \tag{8.2}$$

It is understood that this integral will be evaluated for all x and y. Already, it should be apparent how the Radon transform might be involved, but at this point it may not be clear how $\Phi(x, y)$ is measured.

Here we concentrate on making use of holographic interferometry for determining Φ. However, prior to the development of holography, interferometric methods developed late in the nineteenth century by Zehnder (1891) and Mach (1892) were used. Such methods have been and still are especially important in the areas of aerodynamics and flow visualization [Ladenburg, Winkler, and vanVoorhis (1948); Bennett, Carter, and Bergdolt (1952); Ladenburg and Bershader (1954), Merzkirch (1974)].

Consider an optical wave propagating in an arbitrary direction. The complex amplitude in the $z = 0$ plane may be expressed as [Goodman (1968), Vest (1979)]

$$\mathfrak{U}(x, y) = a(x, y)e^{-ik\Phi(x, y)}, \tag{8.3}$$

where $a(x, y)$ is the real amplitude and

$$k = \frac{2\pi}{\lambda}. \tag{8.4}$$

In a typical holography problem, there are two waves, one that passes through the object

$$\mathfrak{U}_0 = a(x, y)e^{-ik\Phi_0(x, y)}, \tag{8.5}$$

and the reference wave

$$\mathfrak{U}_r = a(x, y)e^{-ik\Phi_r(x, y)}. \tag{8.6}$$

[In the following discussion, it will be convenient to normalize to uniform unit real amplitudes; thus $a(x, y) = 1$.] The usual experimental arrangement is to have two waves of laser light simultaneously arrive from different directions at the $z = 0$ plane as illustrated in Fig. 1.22, where a photographic plate is placed to record the interference pattern. When this photographic plate is exposed and developed in a special way, it forms a *hologram*. If the hologram is then illuminated by the original reference wave, an image of the object may be reproduced. For details and illustrations, see Goodman (1968), Leith and Upatnieks (1965), Pennington (1968), and Leith (1976).

In holographic interferometry, the film (hologram) is exposed twice but with the same reference wave each time: first, with the phase object in a quiescent state with a uniform refractive index n_0 (this might correspond to an undisturbed gas or a solid not under stress) and second, with the object in some disturbed state with variable refractive index $n(x, y)$ (the gas might be turbulent due to heating or the solid under stress). The complex amplitude for

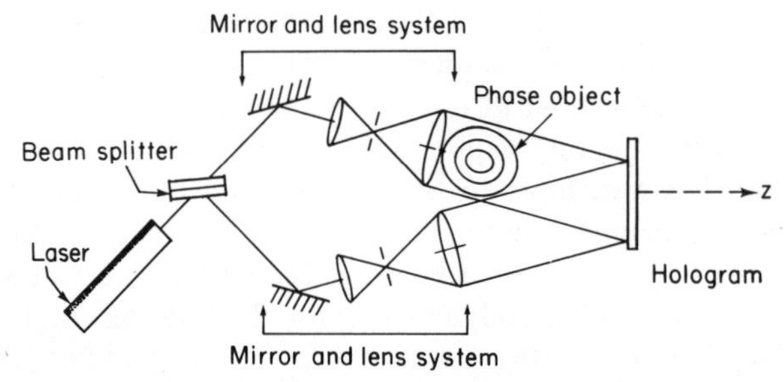

Figure 1.22 Experimental arrangement for producing a hologram. Redrawn with permission [Vest (1979), p. 265].

the first case is just $\mathfrak{U}_0$ from (8.5) with

$$\Phi_0 = \int_{\text{line}} n_0 \, dz. \tag{8.7}$$

In the second case, $\mathfrak{U}$ may be written as in (8.3) and the path length modified to

$$\Phi = \Phi_0 + \Delta\Phi, \tag{8.8}$$

with Φ given by (8.2). From (8.8), (8.7), and (8.2), it follows that the path difference $\Delta\Phi(x, y)$ is given by

$$\Delta\Phi(x, y) = \int_{\text{line}} [n(x, y, z) - n_0] \, dz. \tag{8.9}$$

This *two-exposure* hologram may then be illuminated by the reference wave $\mathfrak{U}_r$, and the waves $\mathfrak{U}_0$ and $\mathfrak{U}$ are reproduced simultaneously. The irradiance (time-average energy flux) pattern of these waves is given by [Vest (1979)]

$$I(x, y) = |\mathfrak{U}_0 + \mathfrak{U}|^2. \tag{8.10}$$

It is straightforward to show that (8.10), when expanded, yields

$$I(x, y) = 2\{1 + \cos[k \, \Delta\Phi(x, y)]\}, \tag{8.11}$$

thus providing a relationship between $I(x, y)$ and $\Delta\Phi(x, y)$. The important point is that the pattern of interference fringes $I(x, y)$ may be observed directly or photographed to yield what is called an *interferogram*. By making use of the fact that bright fringes occur whenever $\Delta\Phi$ is an integral multiple of the wavelength λ,

$$\Delta\Phi(x, y) = \lambda \mathcal{N}(x, y), \tag{8.12}$$

the interferogram can be evaluated to give $\Delta\Phi$. (Bright fringes occur for $\mathcal{N} = 0, 1, 2, \ldots,$ and dark fringes for $\mathcal{N} = .5, 1.5, \ldots$. Interpolated values may be used for regions between bright and dark.)

Connection with the Radon Transform

To make connection with the Radon transform in standard form, suppose the beam is parallel to the y axis rather than the z axis, then from (8.9) and (8.12)

$$\lambda\mathcal{N}(x, z) = \int_{\text{line}} [n(x, y, z) - n_0]\, dy. \tag{8.13}$$

Consider a transverse section plane of the phase object, say a plane perpendicular to the z axis. The corresponding region on the interferogram is a line parallel to the x axis, as indicated in Fig. 1.23*a*. By suppressing the z dependence, since z is constant on the transverse plane, (8.13) may be written as

$$\lambda\mathcal{N}(x) = \int_{\text{line}} [n(x, y) - n_0]\, dy. \tag{8.14}$$

Finally, for rays at an arbitrary angle ϕ with respect to the x axis as in Fig. 1.23*b*,

$$\check{f}(p, \phi) = \lambda\mathcal{N}(p, \phi) = \iint_{\text{plane}} f(x, y)\delta(p - x\cos\phi - y\sin\phi)\, dx\, dy, \tag{8.15}$$

where

$$f(x, y) = n(x, y) - n_0. \tag{8.16}$$

Note that for $\phi = 0°$, (8.15) reduces to (8.14). Thus, by evaluating the fringe patterns produced by interferometric or holographic interferometric methods as ϕ varies over 180°, samples of the Radon transform of the refraction function (8.16) may be determined. Once again the problem reduces to that of inversion of the transform.

Areas of Application

In optics, as in several other fields, inversion methods were developed without prior knowledge of Radon's 1917 paper. Especially notable are the inversions with a view toward optical applications by Rowley (1969), who rediscovered the Fourier method, and Berry and Gibbs (1970) and Junginger and van Haeringen (1972), who rediscovered Radon's original formula. [Also, see the

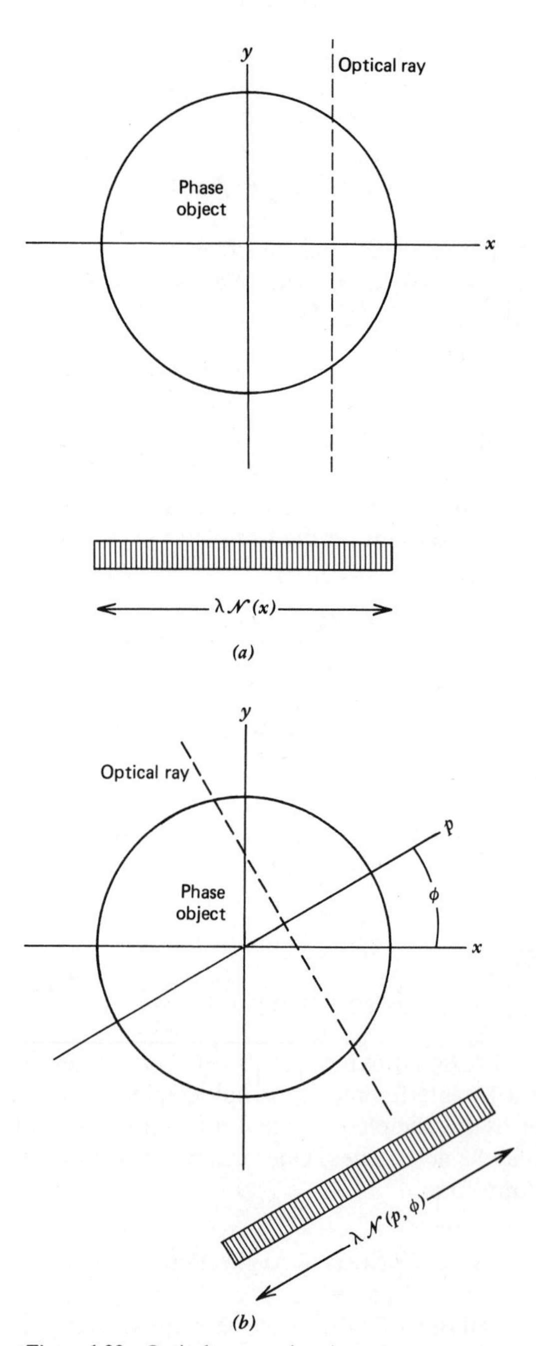

Figure 1.23 Optical ray passing through phase object.

paper by Good (1970).] About the same time in the Soviet Union, Shtein (1972) with full knowledge of Radon's early work, suggested the interpretation of interferograms in terms of the Radon transform [Presnyakov (1976)]. A comprehensive review of the various techniques as they existed early in 1973 for the reconstruction of three-dimensional refractive index fields from multidirectional interferometric data was given by Sweeney and Vest (1973), and in March of that same year, Vest (1973) pointed out the proper interpretation in terms of the Radon transform.

Prior to the development of holographic interferometry in 1965, the important optical applications were primarily for radially symmetric media and thus involved inversion of Abel's integral equation [Barakat (1964), Hochstadt (1973)]; however, Abel inversions can be avoided under certain conditions [Vest and Steel (1978)]. As we shall see in Chapter 4, the Radon transform reduces to the Abel transform under appropriate symmetry conditions, [Vest (1974)]. These applications included an *emission* problem in plasma physics; the determination of the spatial distribution of the emission coefficient from measured spectral intensity distributions emitted by optically thin sources of radiation [the physical and numerical methods have been described by several authors: Maecker (1953); Maecker and Peters (1954); Friedrich (1959); Nestor and Olsen (1960); Pearce (1961), Bockasten (1961); Barr (1962); Herlitz (1963); Maldonado, Caron, and Olsen (1965)] and a *transmission* problem in aerodynamics: the determination of three-dimensional density fields in airflow and gasflow problems [Ladenburg, Winckler, and vanVoorhis (1948); Bennett, Carter, and Bergdolt (1952); Ladenburg and Bershader (1954)]. For additional references and an evaluation of methods for solving Abel's integral equation, see Kulagin, Sorokin, and Dubrovskaya (1972) and Vest (1979).

Since 1965, methods of holographic interferometry and reconstruction techniques have been applied to symmetric and asymmetric phase objects in the areas of aerodynamics, heat and mass transfer, combustion processes, stress analysis, and plasma diagnostics. These methods and areas are discussed in detail by Vest (1979), who also gives numerous references to original papers. Here, we mention only a few of these. In the area of aerodynamics and flow visualization, Bradley (1968) and South (1970) studied axisymmetric cases, and Zien, Ragsdale, and Spring (1975) used the more general methods of Rowley (1969) and Junginger and van Haeringen (1972) to determine unsymmetrical density fields in a large-scale wind-tunnel experiment. Matulka and Collins (1971), Jagota and Collins (1972), and Kosakoski and Collins (1974) made use of an inversion procedure* developed for plasma emission coefficient studies to reconstruct various asymmetric aerodynamic density fields. Also, see Merzkirch (1974), Trolinger (1975), and Asanuma (1979). Three-dimensional temperature distributions in asymmetric natural convection plumes have been measured by use of multidirectional holographic interferometry [Sweeney and Vest (1974); Vest and Radulovic (1977)]. For applications in the area of

*For further discussion of this approach, see Chapter 7.

combustion processes, see Trolinger (1976) and Varde (1974), and for optical applications in stress analysis, see Vest (1979) and references contained therein.

In an important and interesting series of papers not involving holographic or interferometric techniques [Maldonado (1965); Maldonado, Caron and Olsen (1965); Maldonado and Olsen (1966); Olsen, Maldonado, and Duckworth (1968)] a new inversion method based on orthogonal function expansions* was developed for determining emission coefficients from emitted spectral intensities in plasma diagnostic studies. Their method has proved useful in investigating the internal properties of generally asymmetrical experimental plasmas distorted by external magnetic fields. Other work in plasma diagnostics, especially that related to using interferometric holography, is reviewed by Vest (1979); also, see Radley (1975) and Sweeney, Attwood, and Coleman (1976).

Several approaches (which do not make use of Radon transforms) to reconstruction from multidirectional interferograms have been considered. Wolf (1969, 1970) and Dändliker and Weiss (1970) used diffraction theory and the Born approximation. In a similar context, both the Born and Rytov approximation have been studied by Iwata and Nagata (1975); see also Iwata and Nagata (1970), Cha and Vest (1979). For a discussion of still other approaches to image reconstruction utilizing holographic principles, see Stroke and Halioua (1976); Stroke, Halioua, Sarma, and Srinivasan (1977); Stroke, Halioua, Thon, and Willasch (1977) and references contained therein.

[1.9] STRESS ANALYSIS, GEOPHYSICS, AND OTHER AREAS

Other areas in which the Radon transform and reconstruction methods have been found useful include stress analysis, geophysics, air pollutant studies, nondestructive testing, pattern recognition, communications, and applications of a purely mathematical nature. Some of these, especially those of a more technical nature, will be discussed in greater detail in following chapters. For now, mentioning a few of the more physical cases will suffice.

Willis (1971) made extensive use of the Radon transform of the traction vector in developing a method for determining the stresses in a composite body consisting of dissimilar isotropic elastic halfspaces bonded over a circular region. This method was extended by Willis (1972) to the dual problem of the stress analysis of two dissimilar elastic halfspaces that are perfectly bonded together at all points of their interface except over a circular region which delimits a crack. The basic idea was to formulate an integral equation for the Radon transform of the relative displacement of the faces of the crack. These elastostatic investigations were generalized by Willis (1973) to the corresponding elastodynamic cases, thus enabling the study of dynamically expanding cracks on interfaces. Here, as in the static cases, the Radon transform was

*See Chapter 7 for more detail.

fundamental to the formulation and solution of the problem. Other work in the area of stress analysis is discussed by Hildebrand and Hufferd (1976).

Recent research in theoretical seismology has resulted in effective techniques for computing synthetic seismograms. For a review, see Richards (1979). Good agreement between the actual seismic data and the synthetic record implies an understanding of the seismic source and some aspects of Earth structure. Chapman (1978, 1979) has developed a new way to compute these synthetic seismograms that makes use of the Radon transform. Standard methods are used for the computation up to the point of evaluating certain integrals over frequency and wave slowness (reciprocal velocity). Chapman's technique of evaluating the frequency integral first and keeping the final wave slowness integral real is new and gives rise to an interpretation of each seismogram as an integrated cross section of a "density" function at a given angle. The inverse of the Radon transform leads directly to the concept of the velocity stack (slant stack) of the seismograms [Eq. (45) in Chapman (1978)]. In a different, but related, geophysical application, McMechan and Ottolini (1980) made extensive use of the observation that the Radon transform of the slant stack produces a seismic profile (or, equivalently, the inverse Radon transform produces a slant stack from a seismic profile) in their study of the inversion of refraction profiles. Other applications, in the area of geophysical exploration, have been discussed by Dines and Lytle (1979) under the heading computerized geophysical tomography (CGT). For additional recent references in geophysics, see [1.10] and [8.5].

Still other applications include: an ingenious proposal for remote air-pollution monitoring using a tunable laser and reconstruction techniques [Byer and Shepp (1979)]; the estimation of the spatial concentration of certain types of air pollutants [Stuck (1977)]; use of the projection transform in image transmission [Wee and Hsieh (1976); see also Gordon and Herman (1971) and Gordon (1979)]; line and curve detection in the area of pattern analysis [Deans (1981)]; projection methods in chromosome studies [Groen, Verbeek, vanZee, and Oosterlinck (1976)]; a tomographic extension of Doppler processing [Mensa, Heidbreder, and Wade (1980)]; mapping of ultrasonic fields with conventional light diffraction [Cook (1975); Cook and Arnoult (1976)]; and use of reconstruction techniques in nondestructive testing (NDT) [Ellingson and Berger (1980); Tittmann (1980)]. In connection with NDT work, see also Barton (1978) and Sanderson (1979); multiple neutron radiographs are used to determine the condition of nuclear fuel bundles. Das and Boerner (1978) have demonstrated how the Radon transform can be used to estimate the radar target shape of perfectly conducting convex objects. These authors show that previous work primarily by Lewis (1969) and Young (1976) on the identification of a target by a radar system can be cast naturally into the framework of the theory of the Radon transform of characteristic functions. This application falls under the broad category of physical optics far field inverse scattering (POFFIS). Boerner (1980) gives a general discussion of POFFIS using Radon transform theory that can be compared with Bleistein and Cohen (1979), who use the Fourier transform.

Further reference to new and imaginative applications is contained in [1.10]. Also, see Chapter 8.

In closing, let us contemplate the "farsighted" application suggested by Hal Clement; see Swindell and Barrett (1977). Consider mapping the interior of planets by detecting the flux of solar neutrinos passing through them! We leave other such imaginative suggestions to the ingenuity of the reader.

[1.10] REFERENCES

The references for this chapter are arranged according to major area of application. The lists are not intended to be exhaustive, but they do represent important representative contributions. Most of the articles that have been omitted here are referenced in the review papers and in the references cited. Several papers that appear under the TCT section have more to do with algorithms and reconstruction in general than with TCT, but they are listed in that section because of the central role of TCT in the general development of reconstruction from projections.

TCT and Related Work on Reconstruction

Altschuler, Herman, and Lent (1978); Alvarez and Macovski (1976); Baba and Murata (1977); Baker and Sullivan (1980); Barrett, Gordon and Hershell (1976); Barrett and Swindell (1977, 1981); Bates and Peters (1971); Bracewell (1977); Brooks (1977); Brooks and Di Chiro (1975, 1976a, b, c); Brooks, Glover, Talbert, Eisner and DiBianca (1979); Brooks, Keller, O'Connor, and Sheridan (1980); Brooks and Weiss (1976); Brooks, Weiss, and Talbert (1978); Buoncore, Brody, and Macovski (1981); Chesler and Riederer (1975); Chesler, Riederer, and Pelc (1977); Chiu, Barrett, Simpson, Chou, Arendt, and Gindi (1979); Chiu, Barrett and Simpson (1980); Cho (1974); Cho, Ahn, Bohm, and Huth (1974); Cho, Chan, Hall, Kruger, and McCaughey (1975); Colsher (1977); Cormack (1963, 1964, 1973, 1978, 1980a, b); Crawford and Kak (1979); Davison and Grünbaum (1979, 1981); Denton, Friedlander, and Rockmore (1979); Dreike and Boyd (1976); Duerinckx and Macovski (1978, 1979a, b, 1980); Durrani and Goutis (1980); Edelheit, Herman, Lakshminarayanan (1977); Ein-Gal (1974); Ellingson and Berger (1980); Forgues, Goldberg, Smith, and Stuchly (1980); Frei (1978); Friedman, Beattie, and Laughlin (1974); Gilbert, Kenue, Robb, Chu, Lent, and Swartzlander (1981); Glover and Eisner (1979a, b); Goitein (1972); Good (1970); Gordon, Bender and Herman (1970); Gordon and Herman (1971); Gordon, Herman, Johnson (1975); Gordon (1976, 1979); Gore and Tofts (1978); Gullberg (1979a); Hall (1979); Hanson (1977, 1979a, 1980); Hanson, Bradbury, Cannon, Hutson, Laubacher, Macek, Paciotti, and Taylor (1978); Henrich (1980); Herman (1972, 1979, 1980a, b); Herman, Hurwitz, and Lent (1977); Herman, Hurwitz, Lent, and Lung (1979); Herman, Lakshminarayanan, and Naparstek (1976, 1977); Herman, Lakshminarayanan, Naparstek, Ritman, Robb, and Wood (1976); Herman and Lent (1976a, b, c); Herman, Lent, and Lutz (1978); Herman and Naparstek (1977); Herman and Rowland (1973, 1977); Herman, Rowland, and Yau (1979); Holden and Ip (1978); Holley, Henke, Gauger, Jones, Benton, Fabrikant, and Tobias (1979); Horn (1978, 1979); Hounsfield (1972, 1973, 1980); Huang and Wu (1976); Huesman (1977); Huesman, Gullberg, Greenberg, and Budinger (1977); Inouye (1979); Joseph, Hial, Schulz, and Kelcz (1980); Joseph and Schulz (1980); Joseph and Spital (1978); Joseph, Spital, and Stockham (1980); Judy, Swensson, and Szulc (1981); Kak (1979); Kak, Jakowatz, Baily, and Keller (1977); Kashyap and

Mittal (1975); Katz (1978); Kenue and Greenleaf (1979a, b); Kijewski and Bjarngard (1978); Knutsson, Edholm, Granlund, and Petersson (1980); Koeppe, Brugger, Schlapper, Larsen, and Jost (1981); Kowalski (1977a, b, 1978, 1979); Kreel (1977); Kwoh, Reed, and Truong (1977a, b); Lakshminarayanan (1975); Ledley (1976); Levitan (1979); Levitan, Degani, and Zak (1979); Lewitt (1979); Lewitt and Bates (1978a, b, c); Lewitt, Bates, and Peters (1978); Logan and Shepp (1975); McCullough (1975, 1980); McCullough, Baker, Houser, and Reese (1974); McDavid, Waggener, Payne, and Dennis (1975); McKinnon and Bates (1981); Mersereau (1973, 1976); Mersereau and Oppenheim (1974); Minerbo (1979a, b); Minerbo and Sanderson (1977); Morgenthaler, Brooks, and Talbert (1980); Nahamoo, Crawford, and Kak (1981); Nalcioglu and Cho (1978); Naparstek (1980); Nassi and Brody (1981); Nassi, Brody, Cipriano, and Macovski (1981); Oldendorf (1961, 1980); Oppenheim (1974); Orlov (1975a, b); Peters (1974); Peters and Lewitt (1977); Peters, Smith, and Gibson (1973); Petersson, Edholm, Granlund, and Knutsson (1980); Phelps, Hoffman, and Ter-Pogossian (1975); Post (1980); Preston, Taylor, Johnson, and Ayers (1979); Price (1979a, b); Ra and Cho (1981); Ramachandran and Lakshminarayanan (1971); Reed, Kwoh, Truong, and Hall (1977); Reed, Truong, Chang, and Kwoh (1978); Reed, Glenn, Chang, Truong, and Kwoh (1979); Reed, Glenn, Kwoh, and Truong (1980); Reed, Glenn, Truong, Kwoh, and Chang (1980); Reiderer, Pelc, and Chesler (1978); Robb, Ritman, Gilbert, Kinsey, Harris, and Wood (1979); Rosenfeld and Kak (1982); Rowland (1976, 1979); Rutt and Fenster (1980); Schlindwein (1978); Schulz, Olson, and Han (1977); Scudder (1978); Shepp, Hilal, and Schulz (1979); Shepp and Kruskal (1978); Shepp and Logan (1974); Shepp and Stein (1977); Sheridan, Keller, O'Connor, Brooks, and Hanson (1980); Smith, Peters, and Bates (1973); Snyder and Cox (1977); Southon (1981); Stark (1979a, b); Stark, Paul, and Sarna (1979); Stark, Woods, Paul, and Hingorani (1980); Stockham (1979); Stonestron, Alvarez, and Macovski (1981); Strohbehn, Yates, Curran, and Sternick (1979); Swartzlander and Gilbert (1980); Swindell and Barrett (1977); Takahaski (1969); Talbert, Brooks, and Morgenthaler (1980); Tam, Perez-Mendez, and Macdonald (1979); Tanaka (1979); Tanaka and Iinuma (1975, 1976); Tasto (1976, 1977); Tretiak (1978); Tretiak, Eden, and Simon (1969); Tretiak, Ozonoff, Klopping, and Eden (1971); Tsui and Budinger (1979); Verly and Bracewell (1979); Wade, Mueller, and Kaveh (1979); Waggener and McDavid (1979); Wagner (1976, 1979); Wang (1977); Weaver and Goodenough (1979); Wee and Prakash (1979); Weinstein (1978, 1980); Wood, Kinsey, Robb, Gilbert, Harris, and Ritman (1979); Wood and Morf (1981); Zeitler (1974); Zwick and Zeitler (1973).

ECT and Nuclear Medicine

Beattie (1975); Bohm, Eriksson, Bergstrom, Litton, Sundman, and Singh (1978); Brooks, Sank, Friauf, Leighton, Cascio, and Di Chiro (1981); Brooks, Sank, Talbert, and Di Chiro (1979); Brownell and Cochavi (1978); Brownell, Correia, and Zamenhof (1978); Budinger (1979a); Budinger, Derenzo, Gullberg, Greenberg, and Huesman (1977); Budinger, Derenzo, Gullberg, and Huesman (1979); Budinger and Gullberg (1974, 1975, 1977); Budinger, Gullberg, and Huesman (1979); Censor, Gustafson, Lent, and Tuy (1979); Chang (1978, 1979); Chesler (1973); Cho, Ahn, Bohm, and Huth (1974); Cho, Eriksson, and Chan (1977); Chu and Tam (1977); Colsher (1980); Derenzo (1977); Derenzo, Budinger, Cahoon, Huesman, and Jackson (1977); Gullberg (1975, 1976, 1979a, b, 1980); Gullberg and Budinger (1980, 1981); Gustafson, Berggren, Singh, and Dewanjec (1978); Hill, Lovett, and McNeil (1980); Hsieh and Wee (1976); Huesman (1977); Huesman, Gullberg, Greenberg, and Budinger (1977); Jaszczak, Coleman, and Lim (1980); Kak (1979); Kay, Keyes, and Simon (1974); Keys (1979); Koral and Rogers (1979); Kuhl and Edwards (1963, 1968a, b); Kuhl, Edwards, Ricci, Yacob, Mich, and Alavi (1976); Muehllehner (1971); Muehllehner and Hashmi (1972); Muehllehner and Wetzel (1971); Phelps (1977); Phelps, Hoffman, Huang, and Kuhl (1978); Pincus (1964); Ra and Cho (1981); Riederer (1981); Rowe, Undrill, and Keyes (1980); Schorr and Townsend (1981); Tam, Perez-Mendez, and Macdonald (1979); Tasto (1976); Ter-Pogossian, Raichle, and Sobel (1980); Tretiak (1980); Tretiak and Metz (1980); Tsui and Budinger (1978, 1979); Walters, Simon, Chesler, and Correia (1981).

Optics and Use of Optical Methods

Barr (1962); Bennett, Carter, and Bergdolt (1952); Berry and Gibbs (1970); Bockasten (1961); Bradley (1968); Cha and Vest (1979); Cook (1975); Dändliker and Weiss (1970); Fitzgerald and Hörster (1971); Friedrich (1959); Gabor (1948, 1949, 1951); Herlitz (1963); Horman (1965); Iwata and Nagata (1970, 1975); Jagota and Collins (1972); Junginger and van Haeringen (1972); Kosakoski and Collins (1974); Ladenburg and Bershader (1954); Ladenburg, Winckler, and vanVoorhis (1948); Leith (1976); Leith and Upatnieks (1962, 1964, 1965); Maecker (1953); Maecker and Peters (1954); Maldonado, Caron, and Olsen (1965); Maldonado and Olsen (1966); Matulka and Collins (1971); Nestor and Olsen (1960); Olsen, Maldonado, and Duckworth (1968); Pearce (1961); Pennington (1968); Powell and Stetson (1965); Presnyakov (1976); Radley (1975); Rowley (1969); Shtein (1972); South (1970); Stroke and Halioua (1976); Stroke, Halioua, Sarma, and Srinivasan (1977); Stroke, Halioua, Thon, and Willasch (1977); Stuck (1977); Sweeney (1972); Sweeney and Vest (1973, 1974); Sweeney, Attwood, and Coleman (1976); Trolinger (1975, 1976); Varde (1974); Vest (1973, 1974, 1975, 1979); Vest and Radulovic (1977); Vest and Steel (1978); Wolf (1969, 1970); Zien, Ragsdale, and Spring (1975).

Electron Microscopy and Molecular Biology

Bender, Bellman, and Gordon (1970); Crowther (1971); Crowther, Amos, Finch, DeRosier, and Klug (1970); Crowther, DeRosier, and Klug (1970); Crowther and Amos (1971); Crowther, Amos, and Klug (1972); Crowther and Klug (1974, 1975); DeRosier (1971); DeRosier and Klug (1968); DeRosier and Moore (1970); Gilbert (1972a, b); Gordon, Bender, and Herman (1970); Hart (1968); Hawkes (1978); Hoppe, Langer, Knesch, and Poppe (1968); Hoppe and Typke (1978); Hoppe (1979); Hoppe and Hegerl (1980); Klug (1971); Klug, Crick, and Wyckoff (1958); Klug and Crowther (1972); Lake (1972); Mikhailov and Vainshtein (1971); Orlov (1975a, b); Ramachandran and Lakshminarayanan (1971); Smith, Aebi, Josephs, and Kessel (1976); Vainshtein (1970, 1971, 1973, 1978); Vainshtein, Barynin, and Gurskaya (1968); Vainshtein and Mikhailou (1972); Vainshtein and Orlov (1972).

Ultrasound and Acoustic Methods

Carson, Dick, Thieme, Dick, Bayly, Oughton, Dubuque, and Bay (1978); Dick, Carson, Bayly, Oughton, Kubichek, and Kitson (1977); Dick, Elliott, Metz, and Rojohn (1979); Dines and Kak (1979); Duck and Hill (1979); Farrell (1978); Glover and Sharp (1977); Greenleaf, Johnson, Lee, Herman, and Wood (1974); Greenleaf, Johnson, Samayoa, and Duck (1975); Greenleaf, Johnson, and Lent (1978); Greenleaf, Johnson, Bahn, Rajagopalan, and Kenue (1979); Greenleaf, Kenue, Rajagopalan, Bahn, and Johnson (1980); Greenleaf and Bahn (1981); Havlice and Taenzer (1979); Hildebrand and Hufferd (1976); Iwata and Nagata (1975); Johnson and Greenleaf (1979); Johnson, Greenleaf, Rajagopalan, Bahn, Baxter, and Christensen (1979); Johnson, Greenleaf, Rajagopalan, and Tanaka (1980); Johnson, Greenleaf, Samayoa, Duck, and Sjostrand (1975); Jones, Kitsen, Carson, and Bayly (1979); Kak (1979); Kak and Dines (1978); Kaveh, Mueller, and Iverson (1979); Kenue and Greenleaf (1979b); Klepper, Bradenburger, Busse, and Miller (1977); Klepper, Brandenburger, Mimbs, Sobel, and Miller (1981); Maginness (1979); McKinnon and Bates (1980); Mensa, Heidbreder, and Wade (1980); Mueller, Kaveh, and Wade (1979); Mueller, Kaveh, and Iverson (1980); Norton and Linzer (1979a, b, 1981); Preston, Taylor, Johnson, and Ayers (1979); Vezzetti and Aks (1979); Wade (1980); Wade, Elliott, Khogeer, Flesher, Eisler, Mensa, Ramesh, and Heidbreder (1980); Wade, Mueller, and Kaveh (1979); Wells (1975, 1977a, b).

Nuclear Magnetic Resonance

Andrew (1976, 1980a, b); Bendel, Lai, and Lauterbur (1980); Béné, Borcard, Hiltbrand, and Magnin (1980); Bloch, Hansen, and Packard (1946); Bottomley (1979); Bovée, Creyghton, Getreuer, Korbee, Lobregt, Smidt, Wind, Lindeman, Smid, and Posthuma (1980); Brooker and Hinshaw (1978); Brunner and Ernst (1979); Budinger (1979b); Crooks (1980); Crooks, Hoenninger, Arakawa, Kaufman, McRee, Watts, and Singer (1980); Damadian (1971, 1972, 1980); Evens (1980); Farrar and Becker (1971); Gabillard (1951a, b, c, 1952); Gooden (1950); Hansen, Crooks, Davis, DeGroot, Herfkens, Margulis, Gooding, Kaufman, Hoenninger, Arakawa, McRee, and Watts (1980); Hawkes, Holland, Moore, and Worthington (1980); Hinshaw (1974a, b, 1976); Holland, Bottomley, and Hinshaw (1977); Holland, Hawkes, and Moore (1980); Holland, Moore, and Hawkes (1980); Hollis (1980); Hoult (1979, 1980a, b, c); House (1980); Hutchison, Edelstein, and Johnson (1980); Kumar, Welti, and Ernst (1975a, b); Lai and Lauterbur (1980); Lai, Shook, and Lauterbur (1979); Lauterbur (1973, 1974, 1979, 1980a, b); Lauterbur, Kramer, House, and Chen (1975); Lauterbur and Lai (1980); Locher (1980); Mallard, Hutchison, Edelstein, Ling, Foster, and Johnson (1980); Mansfield and Maudsley (1976); Mansfield and Morris (1982); Mansfield, Morris, Ordidge, Pykett, Bangert, and Coupland (1980); Maudsley (1980); Moore and Holland (1980); Moore, Holland, and Kreel (1980); Oldendorf (1982); Partain, James, Watson, Price, Coulam, and Rollo (1980); Shepp (1980); Singer (1980); Witcofski, Karstaedt, and Partain (1981); Wolfe, Crooks, Brown, Howard, and Painter (1980).

Astronomy and Astrophysics

Aksnes and Franklin (1976, 1978a, b); Albers (1979); Altschuler (1979); Baars, van der Brugge, Casse, Hamaker, Sondaar, Visser, and Wellington (1973); Barnes, Evans and Moffett (1978); Bracewell (1956a, b, 1958, 1974, 1979); Bracewell, Colvin, D'Addario, Grebenkemper, Price, and Thompson (1973); Bracewell, Colvin, Price, and Thompson (1971); Bracewell and Riddle (1967); Bracewell and Roberts (1954); Brinkmann and Millis (1973); Brouw (1975); Buhl (1973); Christiansen and Warburton (1953a, b, 1955); Christiansen (1973); Clark and Erickson (1973); Cohen (1973); Counselman (1973); DeJong (1966); Delannoy, Lacroix, and Blum (1973); Elliot (1979); Elliot, Veverka, and Goguen (1975); Erickson (1973); Fjeldbo, Kliore, Seidel, Sweetnam, and Woiceshyn (1976); Fomalont (1973, 1979); Gordon (1979); Guelin (1973); Hachenberg, Grahl, and Wielebinski (1973); Hagfors, Nanni, and Stone (1968); Hagfors and Campbell (1973); Hazard (1962); Hazard, Mackey, and Shimmins (1963); Hills, Janssen, Thornton, and Welch (1973); Högbom (1974); Högbom and Brouw (1974); Hunten and Veverka (1976); Kellerman and Pauliny-Toth (1973); Kerr (1973); Kliore and Woiceshyn (1976); Kundu and Gergely (1980); Maloney and Gottesman (1979); Manchester (1973); Moore and Garmire (1975); Pasachoff and Kutner (1978); Pettengill (1978); Phinney and Anderson (1968); Ponsonby, Morison, Birks, and Landon (1972); Reid, Clauss, Bathker, and Stelzried (1973); Riddle (1968); Roger, Costain, Lacey, Landecker, and Bowers (1973); Ryle and Hewish (1960); Ryle (1972); Scheuer (1962, 1965); Smerd and Wild (1957); Smith, Green, and Shorthill (1977); Swarup and Bagri (1973); Swenson and Mathur (1968); Taylor (1967); Taylor and DeJong (1968); Thompson (1974); Thomson and Ponsonby (1968); van Schooneveld (1979); Zisk, Pettengill, and Catuna (1974).

Geophysics

Chapman (1977, 1978, 1979, 1981a, b); Clayton and McMechan (1981); Coen (1981); Dey-Sarkar and Chapman (1978); Dines and Lytle (1979); Frazier and Phinney (1980); Garmany, Orcutt, and Parker (1979); Helmberger and Brudick (1979); Lager and Lytle (1977); Levin (1980); Lytle, Laine, Lager, and Davis (1979); Mason (1981); McMechan and Ottolini (1980); McMechan and

Yedlin (1981); Newton (1981); Phinney and Anderson (1968); Phinney, Chowdhury, and Frazer (1981); Richards (1979); Robinson (1982); Schultz and Claerbout (1978); Stoffa, Buhl, Diebold, and Wenzel (1981); Yilmaz and Claerbout (1980).

Mathematical and Generalizations

Artzy, Elfving, and Herman (1979); Ball, Johnson, and Stenger (1980); Bang (1951); Bhatia and Wolf (1954); Birkhoff (1940); Bockwinkel (1906); Bureau (1955); Cavaretta, Micchelli, and Sharma (1980a, b); Chapman (1981b); Chen (1978); Cormack and Quinto (1980); Cramér and Wold (1936); Crowther, Amos, and Klug (1972); Davison (1981a, b); Davison and Grünbaum (1979, 1981); Deans (1977, 1978, 1979); Eggermont (1975); Ein-Gal (1974); Falconer (1979); Funk (1913, 1916); Gel'fand, Graev, and Vilenkin (1966); Gilbert (1955); Green (1958); Guenther, Kerber, Killian, Smith, and Wagner (1974); Guillemin and Sternberg (1977); Hamaker and Solmon (1978); Hamaker, Smith, Solmon, and Wagner (1980); Helgason (1965, 1973, 1980); Herman and Lent (1976c); Herman, Hurwitz, and Lent (1977); Herman and Naparstek (1977); Herman, Hurwitz, Lent, and Lung (1979); Herman and Natterer (1981); Hurwitz (1975); John (1934, 1955); Kaczmarz (1937); Katz (1978, 1979); Klug and Crowther (1972); Koornwinder (1982); Lax and Phillips (1970, 1971, 1979); Leahy, Smith, and Solmon (1982); Lerche and Zeitler (1976); Lindgren and Rattey (1981); Logan and Shepp (1975); Louis (1980, 1981); Ludwig (1966); Mader (1927); Marr (1974, 1982); Miller (1978); Minkowski (1904–06); Montaldi (1979); Müller and Richberg (1980); Naparstek (1980); Nashed (1982); Natterer (1977, 1978, 1979, 1980); Perry (1975); Peters (1980a, b, c); Petersen, Smith, and Solmon (1979); Quinto (1980a, b, 1981a, b); Radon (1917); Rattey and Lindgren (1981); Rényi (1952); Shepp and Kruskal (1978); Smith, Solmon, and Wagner (1977); Smith, Solmon, Wagner, and Hamaker (1978); Solmon (1976, 1979); Tanabe (1971); Tewarson and Narain (1974a, b); Tretiak and Metz (1980); Uhlenbeck (1925); Wilkins (1948); Zeitler (1974); Zernike (1934).

Other Areas and Unique Approaches

Baily (1979); Ballard (1981); Barton (1978); Bleistein and Cohen (1979); Boerner (1980); Byer and Shepp (1979); Censor, Gustafson, Lent, and Tuy (1979); Cohen and Toussaint (1977); Cook (1975); Cook and Arnoult (1976); Das and Boerner (1978); Deans (1981); Duda and Hard (1972); Duda, Nitzan, and Barrett (1979); Dudani and Luk (1978); Ellingson and Berger (1980); Forgues, Goldberg, Smith, and Stuchly (1980); Groen, Verbeek, van Zee, and Oosterlinck (1976); Hildebrand and Hufferd (1976); Iannino and Shapiro (1978); Johnson and Greenleaf (1979); Kak, Jakowatz, Baily, and Keller (1977); Kramer and Lauterbur (1979); Price (1979a, b); Sanderson (1979); Shapiro (1975, 1978, 1979a, b); Shapiro and Iannino (1979); Sklansky (1978); Stuck (1977); Tewarson (1972); Tittman (1980); Wee and Hsieh (1976); Willis (1971, 1972, 1973); Zabele and Koplowitz (1979).

Chapter 2

Definition of the Radon Transform

[2.1] INTRODUCTION

In this chapter, the Radon transform of functions f on Euclidean space is defined, first for functions on $\mathbb{R}^2$, then on $\mathbb{R}^3$, and finally on $\mathbb{R}^n$. More general definitions can be given [Helgason (1980)]. However, our purposes are served adequately by restricting the discussion to $\mathbb{R}^n$ and often $\mathbb{R}^2$ or $\mathbb{R}^3$ will suffice.

It will be convenient to select the functions f from some nice class such as the class $\mathscr{S}$ of rapidly decreasing C^∞ functions or from the class $\mathscr{D}$ of C^∞ functions with compact support [Schwartz (1966)]. These classes of functions are discussed briefly in Appendix B. There are important advantages in working with functions that have such nice properties: The Radon transform of f designated by $\check{f} = \mathscr{R}f$ may be differentiated as often as desired, and changes in the order of various integrations may be made with full confidence. A broader class of functions and distributions will be utilized in later chapters; however, for the present, unless stated otherwise, the functions f are assumed to have the properties of $\mathscr{S}$ or $\mathscr{D}$.

The approach here will be to give several forms of the definition for the two-dimensional case. Since many of the applications are from $\mathbb{R}^2$, it will be convenient to tabulate some of the various forms that have appeared in the literature in connection with different applications. Hopefully, this will lead to a better understanding of the transform itself and make the generalization to higher dimensions plausible.

Finally, as an illustration, the Radon transform of a very important function will be computed for both the two- and three-dimensional cases. Further examples of the explicit computation of transforms appear in Chapter 3.

General References

General references for this chapter include the first part of Chapter 1 in Gel'fand, Graev, and Vilenkin (1966) and the paper by Ludwig (1966).

A Convention

Vectors in the transform space *will not* be written in boldface type. Unless specifically stated otherwise the symbols ξ, ζ, and ω are used to represent such vectors. Vectors in $\mathbb{R}^2$, $\mathbb{R}^3$, and $\mathbb{R}^n$ *will* be written in the conventional boldface type. Typical symbols for such vectors are **x**, **y**, **z**, **a**, and **b**. This is more than just a notational convenience. It is a great help in the immediate identification of the space on which various functions are defined and of special use when integrating over various spaces.

[2.2] TWO DIMENSIONS

Let (x, y) designate coordinates of points in the plane, and consider some arbitrary function f defined on some domain D of $\mathbb{R}^2$. If L is any line in the plane, then the mapping defined by the *projection* or line integral of f along all possible lines L is the (two-dimensional) Radon transform of f provided the integral exists. Explicitly,

$$\check{f} = \mathscr{R}f = \int_L f(x, y)\, ds, \tag{2.1}$$

where ds is an increment of length along L. The domain D may include the entire plane or some region of the plane, as indicated in Fig. 2.1.

The mapping defined by (2.1), along with its inverse and certain generalizations, was studied first by Johann Radon (1917). A translation of the original paper, which was in German, appears in Appendix A. Radon showed that if f is continuous and has compact support, then $\mathscr{R}f$ is uniquely determined by integrating along all lines L.

It will be useful to set up some coordinates and be somewhat more precise about the integration along all lines L. Consider Fig. 2.2, where the equation of

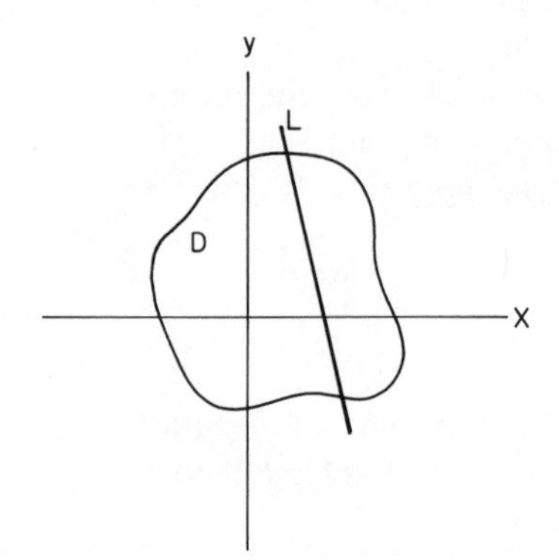

Figure 2.1 Line L through domain D.

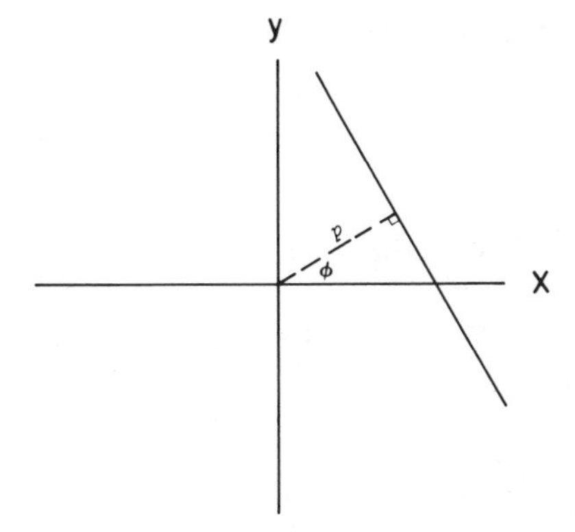

Figure 2.2 Coordinates to describe the line in Fig. 2.1.

the line L is given (in normal form) by

$$p = x \cos \phi + y \sin \phi. \tag{2.2}$$

The line integral (2.1) depends on the values of p and ϕ. This is indicated explicitly by writing

$$\check{f}(p, \phi) = \mathscr{R}f = \int_L f(x, y)\, ds. \tag{2.3}$$

If $\check{f}(p, \phi)$ is known for all p and ϕ, then $\check{f}(p, \phi)$ is the two-dimensional Radon transform of $f(x, y)$. (When $\check{f}$ is known for only certain values of p and ϕ, we say we have a *sample* of the Radon transform.)

Now suppose a new coordinate system is introduced with axes rotated by the angle ϕ. If the new axes are labeled by p and s as in Fig. 2.3, then

$$x = p \cos \phi - s \sin \phi$$

$$y = p \sin \phi + s \cos \phi,$$

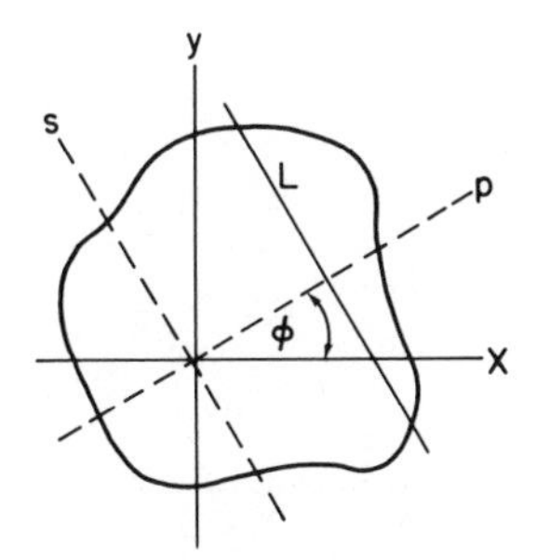

Figure 2.3 The line in Fig. 2.1 relative to original and rotated coordinates.

and we have a somewhat more explicit form for the transform

$$\check{f}(p,\phi) = \int_{-\infty}^{\infty} f(p\cos\phi - s\sin\phi,\ p\sin\phi + s\cos\phi)\,ds. \tag{2.4}$$

Of course, the limits may be finite if f vanishes outside some domain D.

Other explicit forms may be obtained by resorting to a vector notation. Let $\mathbf{x} = (x, y)$ be a vector with components x and y, then $f(\mathbf{x})$ means the same thing as $f(x, y)$. Furthermore, introduce unit vectors

$$\xi = (\cos\phi, \sin\phi),$$

and

$$\xi^{\perp} = (-\sin\phi, \cos\phi),$$

as indicated in Fig. 2.4. Observe that a scalar parameter t can be found such that $\mathbf{x} = p\xi + t\xi^{\perp}$. In terms of these new variables, (2.4) becomes

$$\check{f}(p,\xi) = \int_{-\infty}^{\infty} f(p\xi + t\xi^{\perp})\,dt. \tag{2.5}$$

The notation $\check{f}(p,\xi)$ and $\check{f}(p,\phi)$ may be used interchangeably depending on whether it is desired to emphasize the dependence on the vector ξ or the scalar ϕ.

Finally, one more form will be introduced. This form is selected because its generalization to higher dimensions is both natural and simple. First, observe that the equation of the line (2.2) may be written as

$$p = \xi \cdot \mathbf{x} = x\cos\phi + y\sin\phi.$$

Then the transform may be written as an integral over $\mathbb{R}^2$ by allowing the

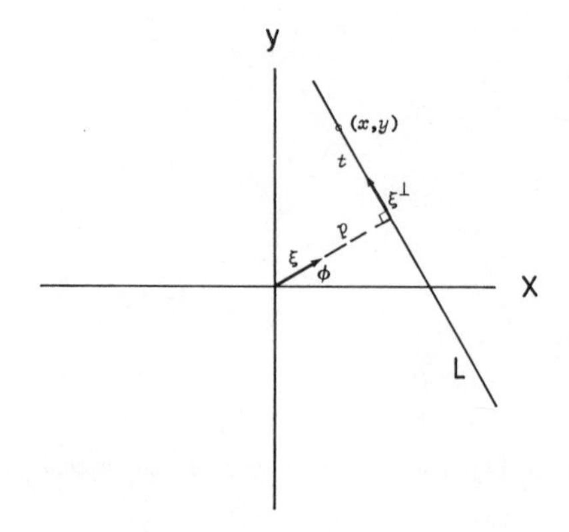

Figure 2.4 Another way to describe the line in Fig. 2.1.

Dirac delta function (see Appendix B) to select the line $p = \xi \cdot \mathbf{x}$ from $\mathbb{R}^2$,

$$\check{f}(p, \xi) = \iint_{\mathbb{R}^2} f(\mathbf{x})\, \delta(p - \xi \cdot \mathbf{x})\, dx\, dy. \tag{2.6}$$

Although this is the desired form, a further modification of notation is in order. In place of $dx\, dy$, simply write $d\mathbf{x}$ and indicate the integral over the entire space by a single symbol $\int$ rather than $\iint_{\mathbb{R}^2}$. With these modifications, (2.6) becomes

$$\check{f}(p, \xi) = \int f(\mathbf{x})\, \delta(p - \xi \cdot \mathbf{x})\, d\mathbf{x}. \tag{2.7}$$

Keep in mind that the unit vector ξ defines direction in terms of the angle ϕ.

Reference to Fig. 2.5 may be helpful in visualizing the space over which $\check{f}$ is defined. Given that ξ is a unit vector characterized by ϕ, $\check{f}$ is defined on a half-infinite cylinder. The cylinder need be only half-infinite since the value of $\check{f}$ is the same at $(-p, \phi)$ and $(p, \phi + \pi)$. Alternatively, negative values of p may be used and ϕ restricted to vary through 180°.

It is important to observe that in xy space (Fig. 2.4) for a fixed angle ϕ the variable p changes along the direction defined by ξ. In the transform space (Fig. 2.5), ξ defines the direction associated with a given profile. Suppose ϕ is held constant as p varies, then the points (p, ϕ) define a line on the cylinder. A real number is associated with every point along this line, and this set of numbers defines the profile $\check{f}_\phi(p)$.

[2.3] THREE DIMENSIONS

The generalization of (2.7) to three dimensions is accomplished by letting $\mathbf{x}$ be a vector in $\mathbb{R}^3$, $\mathbf{x} = (x, y, z)$ and $d\mathbf{x} = dx\, dy\, dz$. The vector ξ is still a unit vector, now in $\mathbb{R}^3$, and the equation

$$p = \xi \cdot \mathbf{x} = \xi_1 x + \xi_2 y + \xi_3 z$$

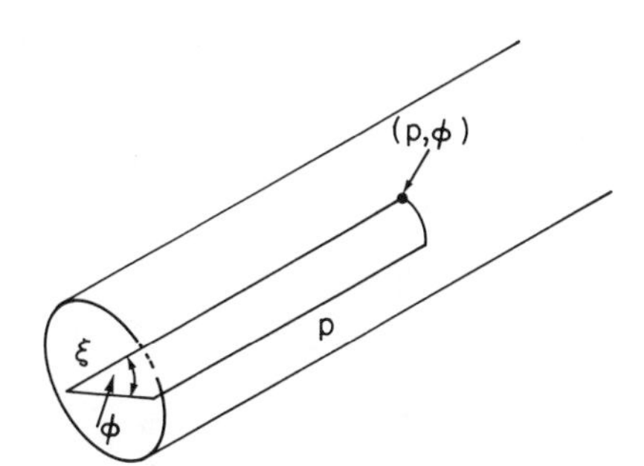

Figure 2.5 Space over which the Radon transform is defined.

defines a plane rather than a line. The Radon transform is still given by

$$\check{f}(p, \xi) = \int f(\mathbf{x})\, \delta(p - \xi \cdot \mathbf{x})\, d\mathbf{x}, \tag{3.1}$$

but the integrations are over planes rather than lines. Geometrically, as indicated in Fig. 2.6, p is the (perpendicular) distance from the origin to the plane and ξ is a unit vector along p that defines the orientation of the plane. As before, in order to fully define the transform, $\check{f}(p, \xi)$ must be known for all p and ξ.

[2.4] EXTENSION TO HIGHER DIMENSIONS

Points in $\mathbb{R}^n$ are designated by $\mathbf{x} = (x_1, x_2, \ldots, x_n)$ and functions defined on $\mathbb{R}^n$ by $f(\mathbf{x}) = f(x_1, x_2, \ldots, x_n)$. The "volume" element is designated by $d\mathbf{x} = dx_1\, dx_2 \cdots dx_n$, and ξ is a unit vector that defines the orientation of a hyperplane with equation

$$p = \xi \cdot \mathbf{x} = \xi_1 x_1 + \xi_2 x_2 + \cdots + \xi_n x_n.$$

Once again, by using the δ function, the Radon transform of f may be written in the convenient form used by Gel'fand, Graev, and Vilenkin (1966)

$$\check{f}(p, \xi) = \mathscr{R}f = \int f(\mathbf{x})\, \delta(p - \xi \cdot \mathbf{x})\, d\mathbf{x}. \tag{4.1}$$

Thus, in the general case, given a function f defined on $\mathbb{R}^n$, the Radon transform of f, designated by $\check{f}$, is determined by integrating over each

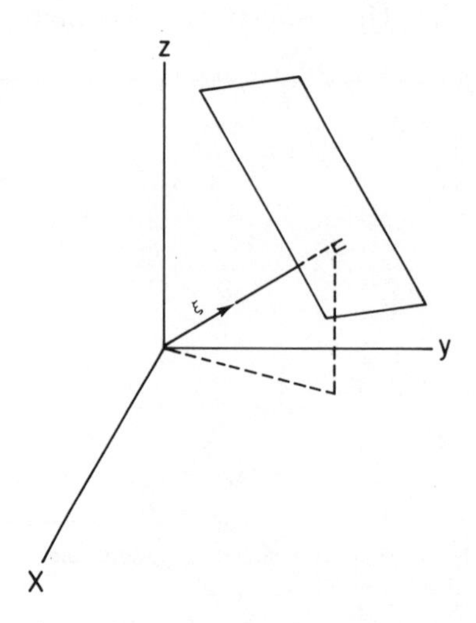

Figure 2.6 Geometry for the Radon transform in three dimensions.

hyperplane in the space. If S^{n-1} is a generalized unit sphere, then we may say that $\check{f}(p, \xi)$ is defined on $\mathbb{R}^1 \times S^{n-1}$. Even more specifically, if $\mathscr{P}^n$ represents the space of all hyperplanes in $\mathbb{R}^n$ then the Radon transform of $f \in \mathscr{S}(\mathbb{R}^n)$ is a function $\check{f} \in \mathscr{S}(\mathscr{P}^n)$, which can be computed using (4.1). Furthermore, in view of the fact that (p, ξ) and $(-p, -\xi)$ define the same hyperplane in $\mathbb{R}^n[\xi \cdot \mathbf{x} = p$ and $-\xi \cdot \mathbf{x} = -p]$, the mapping

$$(p, \xi) \to \text{hyperplane} \in \mathscr{P}^n$$

is a double covering of $\mathbb{R}^1 \times S^{n-1}$ onto $\mathscr{P}^n$.

[2.5] SOME IMPORTANT EXAMPLES

The explicit calculation of the Radon transform of a Gaussian distribution function is presented, first for the two-dimensional case and then for the three-dimensional case. The general case follows immediately.

Example 1. Let $f(x, y) = e^{-x^2-y^2}$. Then

$$\check{f} = \mathscr{R}f = \int_{-\infty}^{\infty}\int_{-\infty}^{\infty} e^{-x^2-y^2}\,\delta(p - \xi_1 x - \xi_2 y)\,dx\,dy.$$

Now make the orthogonal linear transformation (in matrix notation)

$$\begin{pmatrix} u \\ v \end{pmatrix} = \begin{pmatrix} \xi_1 & \xi_2 \\ -\xi_2 & \xi_1 \end{pmatrix}\begin{pmatrix} x \\ y \end{pmatrix}.$$

The vector $\xi = (\xi_1, \xi_2)$ is still a unit vector. Following the change of variables,

$$\begin{aligned}\check{f}(p, \xi) &= \int_{-\infty}^{\infty}\int_{-\infty}^{\infty} e^{-u^2-v^2}\,\delta(p - u)\,du\,dv \\ &= e^{-p^2}\int_{-\infty}^{\infty} e^{-v^2}\,dv \\ &= \sqrt{\pi}\,e^{-p^2}.\end{aligned}$$

Hence we have the important result

$$\mathscr{R}\{e^{-x^2-y^2}\} = \sqrt{\pi}\,e^{-p^2}. \qquad \square \quad (5.1)$$

Example 2. The result obtained in Example 1 can be extended to higher values of n. If $n = 3$, an appropriate transformation is

$$\begin{pmatrix} u \\ v \\ w \end{pmatrix} = \begin{pmatrix} \xi_1 & \xi_2 & \xi_3 \\ -\xi_1\xi_2/q & q & -\xi_2\xi_3/q \\ -\xi_3/q & 0 & \xi_1/q \end{pmatrix}\begin{pmatrix} x \\ y \\ z \end{pmatrix},$$

where $q = (\xi_1^2 + \xi_3^2)^{1/2}$ and $|\xi| = (\xi_1^2 + \xi_2^2 + \xi_3^2)^{1/2} = 1$. This leads to

$$\mathcal{R}\{e^{-x^2-y^2-z^2}\} = \pi e^{-p^2}. \qquad \square \quad (5.2)$$

For arbitrary n, the first row of the orthogonal transformation matrix will consist of the components of ξ, just as for the $n = 2$ and $n = 3$ cases. Thus, for the general case,

$$\mathcal{R}\{\exp(-x_1^2 - x_1^2 - \cdots - x_n^2)\} = (\sqrt{\pi})^{n-1}\exp(-p^2). \qquad (5.3)$$

Two more examples will be presented, one involving a square and the other a circle.

Example 3. Here we consider the transform over the unit square. Let $f(x, y)$ be unity for $0 < x < 1$, $0 < y < 1$ and zero otherwise, as illustrated in Fig. 2.7. The appropriate geometry for performing the integration is shown in Fig. 2.8. By symmetry, it is adequate to consider ϕ in the region $0 < \phi < \pi/4$. The transform is given by

$$\check{f} = \begin{cases} \dfrac{p}{\sin\phi\cos\phi}, & 0 < p < \sin\phi, \quad \text{region 1} \\[2ex] \dfrac{1}{\cos\phi}, & \sin\phi \leqslant p \leqslant \cos\phi, \quad \text{region 2} \\[2ex] \dfrac{\sin\phi + \cos\phi - p}{\sin\phi\cos\phi}, & \cos\phi < p < \sin\phi + \cos\phi, \quad \text{region 3.} \end{cases}$$

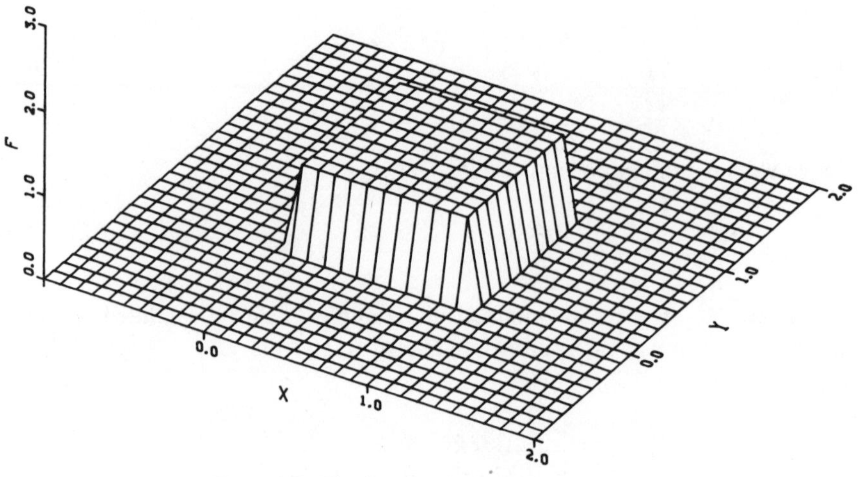

Figure 2.7 The function $f(x, y)$ of Example 3.

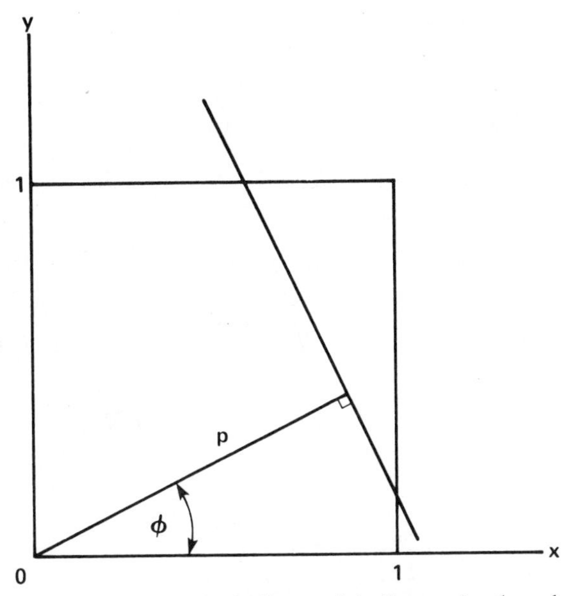

Figure 2.8 The unit square and coordinates of the line passing through the square.

This surface is shown in Fig. 2.9 for ϕ between 0 and 45° and for ϕ between 0 and 90°. Details of the computation follow for region 3. The other two regions may be treated in a similar fashion.

$$\check{f}(p,\phi) = \iint_{\text{region 3}} f(x,y)\,\delta(p - x\cos\phi - y\sin\phi)\,dx\,dy$$

$$= \frac{1}{\cos\phi}\iint_{\text{region 3}} f(x,y)\,\delta(p\sec\phi - x - y\tan\phi)\,dx\,dy$$

$$= \frac{1}{\cos\phi}\int_{p\csc\phi-\cot\phi}^{1} f(p\sec\phi - y\tan\phi,\, y)\,dy$$

$$= \frac{1 - p\csc\phi + \cot\phi}{\cos\phi}$$

$$= \frac{\sin\phi + \cos\phi - p}{\sin\phi\cos\phi}, \qquad \text{region 3 with } 0 < \phi < \frac{\pi}{4}. \qquad \square$$

Example 4. In this example, the transform is over the unit circle. Let $f(x, y)$ be defined by

$$f(x,y) = \begin{cases} (1 - x^2 - y^2)^{\lambda-1} & \text{for } x^2 + y^2 < 1 \\ 0 & \text{for } x^2 + y^2 \geqslant 1 \end{cases}$$

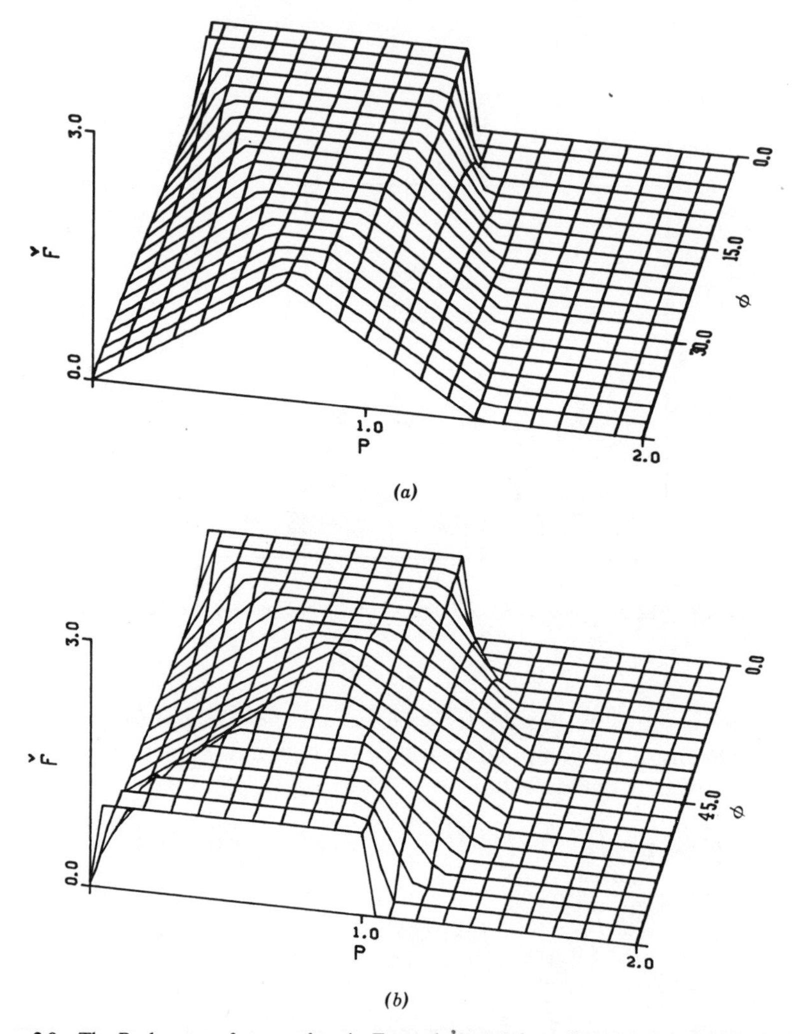

Figure 2.9 The Radon transform surface in Example 3: (a) for ϕ between 0 and 45°. (b) for ϕ between 0 and 90°.

with $\lambda > 1$. From (2.4), it follows that

$$\check{f}(p, \phi) = \int_{-\sqrt{1-p^2}}^{\sqrt{1-p^2}} \left[1 - p^2 - s^2\right]^{\lambda - 1} ds.$$

The integral can be evaluated in terms of beta or gamma functions [Erdélyi,

Magnus, Oberhettinger, and Tricomi (1953a)]. From the formula

$$\int_{-a}^{a} (a^2 - t^2)^{\lambda-1}\, dt = a^{2\lambda-1} B\left(\tfrac{1}{2}, \lambda\right) = \frac{a^{2\lambda-1}\sqrt{\pi}\,\Gamma(\lambda)}{\Gamma\left(\lambda + \frac{1}{2}\right)}$$

the transform follows:

$$\check{f}(p, \phi) = \frac{\sqrt{\pi}\,\Gamma(\lambda)}{\Gamma\left(\lambda + \frac{1}{2}\right)} (1 - p^2)^{\lambda - 1/2}, \qquad -1 \leqslant p \leqslant 1$$

□

Chapter 3

Basic Properties

[3.1] INTRODUCTION

Several elementary properties of the Radon transform of a function f defined on $\mathbb{R}^n$ follow directly from the definition

$$\check{f}(p, \boldsymbol{\xi}) = \mathscr{R}f = \int f(\mathbf{x})\, \delta(p - \boldsymbol{\xi} \cdot \mathbf{x})\, d\mathbf{x}$$

discussed in [2.4]. In this chapter, these properties are developed for $\mathbb{R}^n$ and illustrated by examples from $\mathbb{R}^2$ and $\mathbb{R}^3$. As usual, the notation $x_1 = x$, $x_2 = y$, $x_3 = z$ is used whenever convenient.

General Reference

A brief discussion of many of the properties treated here is given by Gel'fand, Graev, and Vilenkin (1966).

[3.2] HOMOGENEITY

The function $\check{f}(p, \boldsymbol{\xi})$ is an even homogeneous function of degree -1. To show this, observe that

$$\check{f}(sp, s\boldsymbol{\xi}) = \int f(\mathbf{x})\, \delta(sp - s\boldsymbol{\xi} \cdot \mathbf{x})\, d\mathbf{x}$$

$$= |s|^{-1} \int f(\mathbf{x})\, \delta(p - \boldsymbol{\xi} \cdot \mathbf{x})\, d\mathbf{x},$$

for arbitrary real number $s \neq 0$. Thus

$$\check{f}(sp, s\boldsymbol{\xi}) = |s|^{-1} \check{f}(p, \boldsymbol{\xi}). \tag{2.1}$$

Note that if $s = -1$, we have the important symmetry property

$$\check{f}(-p, -\boldsymbol{\xi}) = \check{f}(p, \boldsymbol{\xi}). \tag{2.2}$$

Another important observation follows from (2.1): Knowledge of $\check{f}(p, \xi)$ for fixed p and for all possible values of the vector ξ serves to determine $\check{f}$ completely. An immediate consequence of this observation is that the Radon transform of f depends on n independent variables if f is a function of n independent variables. The n independent variables of $\check{f}$ *include* the parameter p and the n components of ξ subject to the *condition* $|\xi| = 1$. As mentioned in [2.4], $\check{f}$ lives on $\mathbb{R}^1 \times \mathbf{S}^{n-1}$.

Example 1. Given that $\zeta = s\xi$, $|\zeta| = s > 0$, it follows that

$$\check{f}(p, \zeta) = \check{f}(p, s\xi) = |s|^{-1}\check{f}\left(\frac{p}{s}, \xi\right),$$

or, equivalently,

$$\check{f}(p, \zeta) = |\zeta|^{-1}\check{f}\left(\frac{p}{|\zeta|}, \frac{\zeta}{|\zeta|}\right). \qquad \square \quad (2.3)$$

Example 2. If as in (5.3) of [2.5]

$$\check{f}(p, \xi) = (\sqrt{\pi})^{n-1}e^{-p^2},$$

then

$$\check{f}(p, \zeta) = \frac{(\sqrt{\pi})^{n-1}}{|\zeta|}\exp\left(\frac{-p^2}{|\zeta|^2}\right).$$

And if

$$\check{f}(p, \xi) = (\sqrt{\pi})^{n-1}\xi_1 pe^{-p^2},$$

then

$$\check{f}(p, \zeta) = \frac{(\sqrt{\pi})^{n-1}\zeta_1 p}{|\zeta|^3}\exp\left(\frac{-p^2}{|\zeta|^2}\right),$$

where

$$|\zeta| = (\zeta_1^2 + \zeta_2^2 + \cdots + \zeta_n^2)^{1/2}. \qquad \square$$

[3.3] LINEARITY

Given two functions f and g and two constants c_1 and c_2,

$$\mathscr{R}\{c_1 f + c_2 g\} = \int [c_1 f(\mathbf{x}) + c_2 g(\mathbf{x})]\, \delta(p - \xi \cdot \mathbf{x})\, d\mathbf{x}$$

$$= c_1\check{f} + c_2\check{g}.$$

Thus

$$\mathscr{R}\{c_1 f + c_2 g\} = c_1 \mathscr{R} f + c_2 \mathscr{R} g, \qquad (3.1)$$

and the Radon transform is a linear transformation.

[3.4] TRANSFORM OF A LINEAR TRANSFORMATION

By making the appropriate change of integration variables, it is possible to find the Radon transform of a function of a linear transformation of coordinates. A brief digression about notation is in order first, however.

For convenience we are using the notation

$$\xi \cdot \mathbf{x} = \xi_1 x_1 + \xi_2 x_2 + \cdots + \xi_n x_n$$

for inner products of vectors in $\mathbb{R}^n$. In a matrix notation, this would appear as

$$\xi^{\mathsf{T}} \mathbf{x} = (\xi_1 \xi_2 \ldots \xi_n) \begin{pmatrix} x_1 \\ x_2 \\ \vdots \\ x_n \end{pmatrix} = \xi_1 x_1 + \xi_2 x_2 + \cdots + \xi_n x_n,$$

where the superscript T means transpose. In bracket notation,

$$\langle \xi, \mathbf{x} \rangle = \xi_1 x_1 + \xi_2 x_2 + \cdots + \xi_n x_n.$$

Thus, for a nonsingular matrix A with real elements such that $\mathbf{y} = \mathsf{A}\mathbf{x}$ with $\mathbf{x}, \mathbf{y} \in \mathbb{R}^n$, we may write the identity

$$\langle \xi, \mathbf{y} \rangle = \langle \xi, \mathsf{A}\mathbf{x} \rangle = \langle \mathsf{A}^{\mathsf{T}} \xi, \mathbf{x} \rangle$$

in the bracket notation, or

$$\xi^{\mathsf{T}} \mathbf{y} = \xi^{\mathsf{T}} \mathsf{A} \mathbf{x} = \left(\mathsf{A}^{\mathsf{T}} \xi\right)^{\mathsf{T}} \mathbf{x}$$

in the matrix notation, or

$$\xi \cdot \mathbf{y} = \xi \cdot \mathsf{A}\mathbf{x} = \mathsf{A}^{\mathsf{T}} \xi \cdot \mathbf{x}$$

in the "dot" notation.

Now, consider $\mathscr{R}\{f(\mathsf{A}\mathbf{x})\}$, where A is a nonsingular matrix such that

$$\mathbf{y} = \mathsf{A}\mathbf{x},$$

or equivalently in terms of components

$$y_k = \sum_{l=1}^{n} \mathsf{A}_{kl} x_l, \qquad (k = 1, 2, \ldots, n),$$

where the (real) elements of A are written as A_{kl}, $(k, l = 1, 2, \ldots, n)$. Perhaps an example would be useful in understanding the notation $f(\mathsf{A}\mathbf{x})$.

Example 1. Suppose $\mathbf{x} \in \mathbb{R}^2$ and

$$f(\mathbf{x}) = f(x_1, x_2) = \exp(-x_1^2 - x_2^2),$$

then

$$\begin{aligned} f(\mathsf{A}\mathbf{x}) &= f(\mathsf{A}_{11}x_1 + \mathsf{A}_{12}x_2, \mathsf{A}_{21}x_1 + \mathsf{A}_{22}x_2) \\ &= \exp\{-(\mathsf{A}_{11}x_1 + \mathsf{A}_{12}x_2)^2 - (\mathsf{A}_{21}x_1 + \mathsf{A}_{22}x_2)^2\}. \end{aligned}$$

□

It is useful to define $\mathsf{B} = \mathsf{A}^{-1}$, then $\mathbf{x} = \mathsf{A}^{-1}\mathbf{y} = \mathsf{B}\mathbf{y}$. The desired result follows immediately:

$$\begin{aligned} \mathscr{R}\{f(\mathsf{A}\mathbf{x})\} &= \int f(\mathsf{A}\mathbf{x})\,\delta(p - \boldsymbol{\xi}\cdot\mathbf{x})\,d\mathbf{x} \\ &= |\det \mathsf{B}| \int f(\mathbf{y})\,\delta(p - \boldsymbol{\xi}\cdot\mathsf{B}\mathbf{y})\,d\mathbf{y} \\ &= |\det \mathsf{B}| \int f(\mathbf{y})\,\delta(p - \mathsf{B}^{\mathsf{T}}\boldsymbol{\xi}\cdot\mathbf{y})\,d\mathbf{y}, \end{aligned}$$

since the Jacobian of the transformation is the magnitude of the determinant of B. Hence

$$\mathscr{R}\{f(\mathsf{A}\mathbf{x})\} = |\det \mathsf{B}|\check{f}(p, \mathsf{B}^{\mathsf{T}}\boldsymbol{\xi}), \tag{4.1}$$

where $\mathsf{B} = \mathsf{A}^{-1}$. Equivalently,

$$\mathscr{R}\{f(\mathsf{B}^{-1}\mathbf{x})\} = |\det \mathsf{B}|\check{f}(p, \mathsf{B}^{\mathsf{T}}\boldsymbol{\xi}). \tag{4.2}$$

Caution. Since $\mathsf{B}^{\mathsf{T}}\boldsymbol{\xi}$ is not in general a unit vector, the results of Example 1 in [3.2] may have to be employed. In particular, see Examples 4 and 5 in this section. □

An important special case of (4.1) is for B orthogonal, then $\mathsf{B}^{-1} = \mathsf{B}^{\mathsf{T}} = \mathsf{A}$, $|\det \mathsf{B}| = 1$, and

$$\mathscr{R}\{f(\mathsf{A}\mathbf{x})\} = \check{f}(p, \mathsf{A}\boldsymbol{\xi}). \tag{4.3}$$

Another important special case occurs when A is a multiple of the identity, $\mathsf{A} = \lambda\mathsf{I}$ with λ real. Then

$$\mathscr{R}f(\lambda\mathbf{x}) = \frac{1}{\lambda^n}\check{f}\left(p, \frac{\boldsymbol{\xi}}{\lambda}\right),$$

or

$$\mathscr{R}f(\lambda\mathbf{x}) = \frac{1}{\lambda^{n-1}}\check{f}(\lambda p, \xi). \tag{4.4}$$

Example 2. Given that

$$\mathscr{R}\{e^{-x^2-y^2}\} = \sqrt{\pi}\, e^{-p^2},$$

it follows from (4.4) with $\lambda = 1/\sigma\sqrt{2}$ that

$$\mathscr{R}\left\{\exp\left(-\frac{x^2}{2\sigma^2} - \frac{y^2}{2\sigma^2}\right)\right\} = \sigma\sqrt{2\pi}\exp\left(-\frac{p^2}{2\sigma^2}\right).$$

In a more standard form,

$$\mathscr{R}\left\{\frac{1}{2\pi\sigma^2}\exp\left(-\frac{x^2}{2\sigma^2} - \frac{y^2}{2\sigma^2}\right)\right\} = \frac{1}{\sigma\sqrt{2\pi}}\exp\left(-\frac{p^2}{2\sigma^2}\right). \quad \square \tag{4.5}$$

Example 3. It will be useful to observe that the factor of $(\sqrt{\pi})^{n-1}$ in (5.3) of [2.5] may be eliminated by use of (4.4) with $\lambda = \sqrt{\pi}$. The result

$$\mathscr{R}\{\exp[-\pi(x_1^2 + x_2^2 + \cdots + x_n^2]\} = \exp(-\pi p^2).$$

In particular, in two dimensions

$$\mathscr{R}\{e^{-\pi(x^2+y^2)}\} = e^{-\pi p^2}. \qquad \square$$

Example 4. It is desired to find the transform of

$$\exp\left[-\left(\frac{x}{a}\right)^2 - \left(\frac{y}{b}\right)^2\right].$$

Start with

$$f(x, y) = e^{-x^2-y^2}$$

and use (4.2) with

$$\mathsf{B} = \begin{pmatrix} a & 0 \\ 0 & b \end{pmatrix}, \qquad \mathsf{B}^{-1} = \begin{pmatrix} 1/a & 0 \\ 0 & 1/b \end{pmatrix}.$$

Clearly, f is transformed into the desired form. Since $\zeta = \mathsf{B}^{\mathsf{T}}\xi = (a\cos\phi, b\sin\phi)$ is not a unit vector, we write the transform from (4.2) as

$$|\det \mathsf{B}|\check{f}(p, \zeta) = \frac{ab}{|\zeta|}\check{f}\left(\frac{p}{|\zeta|}, \frac{\zeta}{|\zeta|}\right)$$

prior to using the results of [2.5]. It follows immediately that

$$\mathscr{R}\left\{\exp\left[-\left(\frac{x}{a}\right)^2-\left(\frac{y}{b}\right)^2\right]\right\}=\frac{ab\sqrt{\pi}}{s}\exp\left(-\frac{p^2}{s^2}\right),$$

where $s^2 = |\zeta|^2 = a^2\cos^2\phi + b^2\sin^2\phi$. □

Example 5. It is desired to find the transform of the characteristic function of an ellipse:

$$e(x, y) = \begin{cases} 1, & \left(\frac{x}{a}\right)^2 + \left(\frac{y}{b}\right)^2 < 1 \\ 0, & \left(\frac{x}{a}\right)^2 + \left(\frac{y}{b}\right)^2 > 1. \end{cases}$$

First, observe that the transform of the characteristic function of a unit disk

$$u(x, y) = \begin{cases} 1, & x^2 + y^2 < 1 \\ 0, & x^2 + y^2 > 1 \end{cases}$$

is given by the length of a chord at a distance p from the center:

$$\check{u}(p, \phi) = \begin{cases} 2(1 - p^2)^{1/2}, & p < 1 \\ 0, & p > 1. \end{cases}$$

Now select B exactly as in Example 4. Then $u(x, y) \to e(x, y)$, and the method presented in Example 4 can be used to obtain the desired transform

$$\check{e}(p, \phi) = \begin{cases} \frac{2ab}{s}\left[1 - \left(\frac{p}{s}\right)^2\right]^{1/2}, & \frac{p}{s} < 1 \\ 0, & \frac{p}{s} > 1, \end{cases}$$

where $s = [a^2\cos^2\phi + b^2\sin^2\phi]^{1/2}$. □

[3.5] SHIFTING PROPERTY

Consider the function $f(\mathbf{x} - \mathbf{a})$, where the components of the vector $\mathbf{x} - \mathbf{a}$ are $(x_1 - a_1, x_2 - a_2, \ldots, x_n - a_n)$. The Radon transform of this function is given by

$$\mathscr{R}f(\mathbf{x} - \mathbf{a}) = \int f(\mathbf{x} - \mathbf{a})\, \delta(p - \boldsymbol{\xi} \cdot \mathbf{x})\, d\mathbf{x}$$

$$= \int f(\mathbf{y})\, \delta(p - \boldsymbol{\xi} \cdot \mathbf{a} - \boldsymbol{\xi} \cdot \mathbf{y})\, d\mathbf{y}.$$

Thus the *shift theorem* is

$$\mathscr{R}f(\mathbf{x} - \mathbf{a}) = \check{f}(p - \boldsymbol{\xi} \cdot \mathbf{a}, \boldsymbol{\xi}). \tag{5.1}$$

For later use, observe that if $\mathbf{a}$ is replaced by $-\mathbf{a}$ in (5.1), the shift theorem reads

$$\mathscr{R}f(\mathbf{x} + \mathbf{a}) = \check{f}(p + \boldsymbol{\xi} \cdot \mathbf{a}, \boldsymbol{\xi}). \tag{5.2}$$

This form will prove useful in [3.6].

Example 1. Apply (5.1) to (4.5) with $\mathbf{a} = (a, b)$ and $\boldsymbol{\xi} = (\cos\phi, \sin\phi)$. The result is

$$\mathscr{R}\left\{\frac{1}{2\pi\sigma^2} \exp\left[-\frac{(x-a)^2}{2\sigma^2} - \frac{(y-b)^2}{2\sigma^2}\right]\right\} = \frac{1}{\sigma\sqrt{2\pi}} \exp\left[-\frac{(p-p_0)^2}{2\sigma^2}\right], \tag{5.3}$$

where

$$p_0 = a\cos\phi + b\sin\phi. \qquad \square \tag{5.4}$$

Note that the equation

$$p = a\cos\phi + b\sin\phi$$

is the polar form for a circle with center at $(a/2, b/2)$ and radius $\frac{1}{2}(a^2 + b^2)^{1/2}$, where p is the radial coordinate and ϕ the angular coordinate.

Example 2. Consider the limiting form of (5.3) as $\sigma \to +0$. Both sides are δ sequences, and in the limit

$$\mathscr{R}\{\delta(x-a)\,\delta(y-b)\} = \delta(p - p_0),$$

with p_0 still given by (5.4). $\square$

A Digression

The result in Example 2 is the same result one would obtain by the formal substitution of

$$f(x, y) = \delta(x-a)\,\delta(y-b)$$

into the definition [(2.7) of Chapter 2]. Thus the Radon transform of a function concentrated at a point $\mathbb{R}^2$ yields a function concentrated on a circle in $\mathbb{R}^2$. In this sense, points transform to circles. For example, the circle associated with $a = b = 4$ is shown in Fig. 3.1. This can lead to some confusion though, since the space over which $\check{f}$ is defined is the half-infinite cylinder of Fig. 2.5. There are interesting differences that perhaps can be seen most clearly by comparing graphs of $\check{f}$.

In Fig. 3.2, we show the surface

$$f(x, y) = \exp\left[-(x-4)^2 - (y-4)^2\right],$$

and in Fig. 3.3, the Radon transform surface

$$\check{f}(p, \phi) = \sqrt{\pi} \exp\left[-(p - 4\cos\phi - 4\sin\phi)^2\right].$$

For purposes of illustration, the cylinder has been "cut" along the $\phi = 180°$

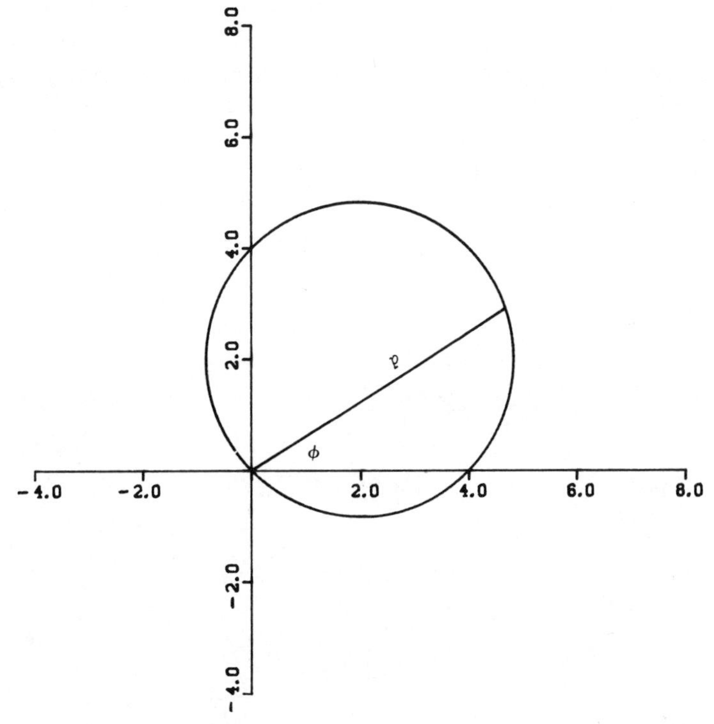

Figure 3.1 The circle $p = 4\cos\phi + 4\sin\phi$ when p and ϕ are regarded as polar coordinate variables.

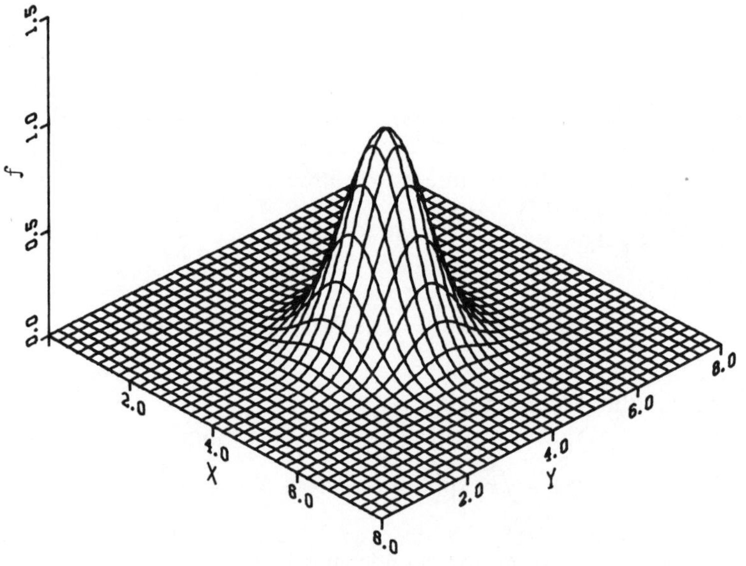

Figure 3.2 The surface $f(x, y) = \exp[-(x-4)^2 - (y-4)^2]$.

line and "unfolded" to lie in a plane designated the ϕp plane. Clearly $\check{f}$ has its maximum value of $\sqrt{\pi}$ along the sinusoidal curve $p = 4\cos\phi + 4\sin\phi$. Since the values of $\check{f}$ for negative p are obtained from the symmetry property discussed in [3.2], it is adequate to show the surface for $p \geqslant 0$, as in Fig. 3.4.

On the other hand, if $\check{f}$ is drawn with p and ϕ considered as polar coordinates, the surface shown in Fig. 3.5 emerges.

Something strange appears to be happening near the origin. After changing the view angle to reveal this more clearly, Fig. 3.6 results. The explanation lies

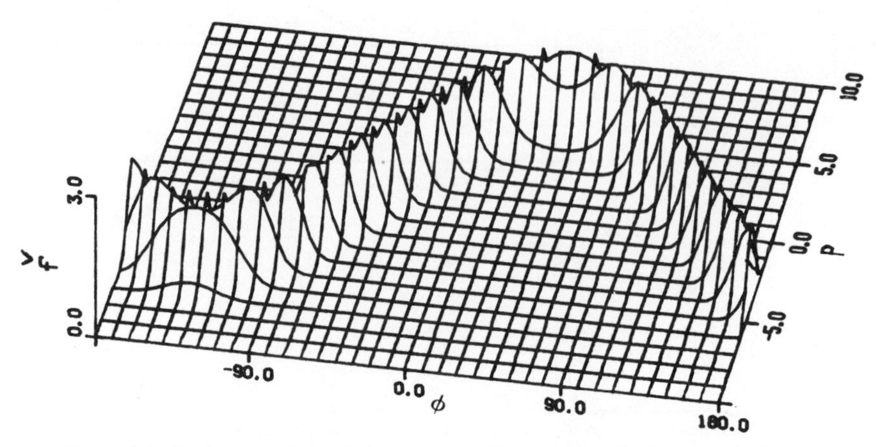

Figure 3.3 Radon transform of the surface in Fig. 3.2. Negative values of p shown.

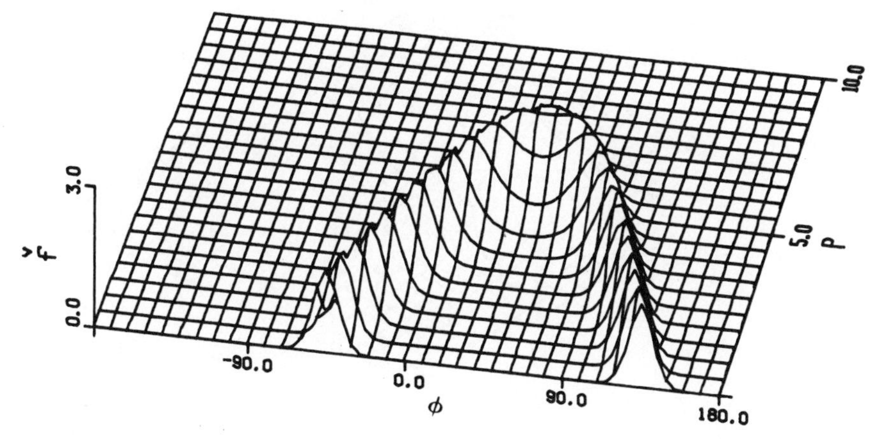

Figure 3.4 Radon transform of the surface in Fig. 3.2. Only positive values of p shown.

in the variables used. When working with the polar form,

$$p = \left(u^2 + v^2\right)^{1/2}$$

$$\cos\phi = \frac{u}{\left(u^2 + v^2\right)^{1/2}},$$

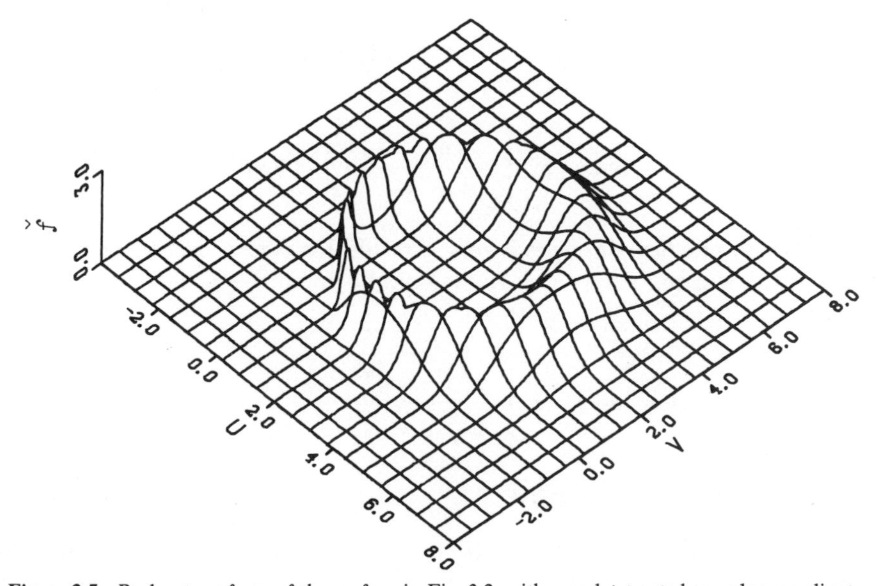

Figure 3.5 Radon transform of the surface in Fig. 3.2, with p and ϕ treated as polar coordinates.

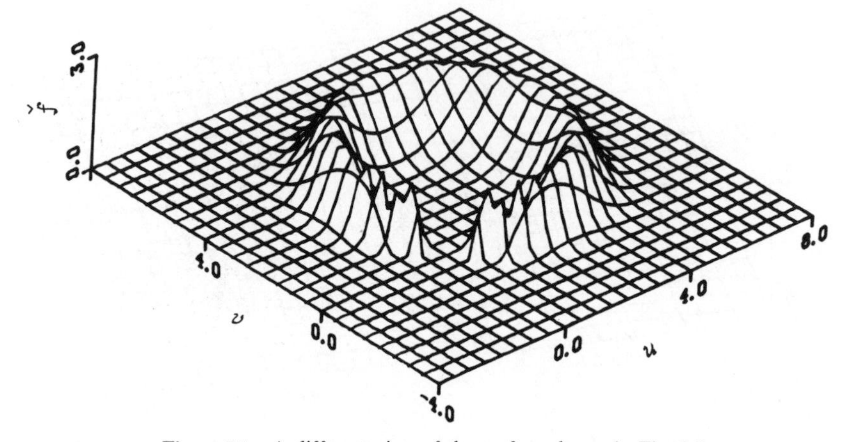

Figure 3.6 A different view of the surface shown in Fig. 3.5.

and

$$\sin \phi = \frac{v}{\left(u^2 + v^2\right)^{1/2}}.$$

The transform $\check{f}$ in terms of u and v becomes

$$\check{f}(u, v) = \sqrt{\pi} \exp\left[-\left(\frac{u^2 + v^2 - 4u - 4v}{\sqrt{u^2 + v^2}}\right)^2\right].$$

Now, as u approaches zero along the $v = 0$ line (as in the graphics display of Fig. 3.6), it is easy to see that

$$\check{f} = \lim_{u \to 0} \sqrt{\pi} \exp\left[-\left(\frac{u^2 - 4u}{u}\right)^2\right] = \sqrt{\pi}\, e^{-16},$$

which for all practical purposes is zero, a result that is also clear from Fig. 3.6. In contrast, if u and v approach zero along the line $v = -u$, the limit is quite different:

$$\begin{aligned}\check{f} &= \lim_{u \to 0} \sqrt{\pi} \exp\left[-\left(\frac{u^2 + u^2 - 4u + 4u}{\sqrt{2u^2}}\right)^2\right] \\ &= \sqrt{\pi}\, e^0 = \sqrt{\pi}.\end{aligned}$$

Although this type of discontinuity is not at all unusual when working with a function of several variables [Taylor (1955)], here it has been introduced by the choice of coordinates. The behavior of $\check{f}$ is illustrated much better by Fig. 3.4, where the behavior at the origin (along the $p = 0$ line) is clearly shown with maximum values at $\phi = -45°$ and $\phi = 135°$. Consequently, the format of Fig. 3.4 will be followed for future plots of $\check{f}$.

Rattey and Lindgren (1981) also observe the fallacy in considering p and ϕ polar coordinates in a plane. They note that for $\phi_1 \neq \phi_2$, $\check{f}(0, \phi_1) \neq \check{f}(0, \phi_2)$; this is contrary to the result that follows from considering p and ϕ as polar coordinates.

[3.6] TRANSFORM OF DERIVATIVES

Given the function $f(\mathbf{x}) = f(x_1, \ldots, x_n)$, the problem is to find the Radon transform of $\partial f / \partial x_k$. First, observe that

$$\frac{\partial f}{\partial x_k} = \lim_{\varepsilon \to 0} \frac{f\left(\mathbf{x} + \dfrac{\varepsilon}{\xi_k}\right) - f(\mathbf{x})}{\dfrac{\varepsilon}{\xi_k}}, \tag{6.1}$$

where $f(\mathbf{x} + \varepsilon/\xi_k)$ means $f(x_1, \ldots, x_k + \varepsilon/\xi_k, \ldots, x_n)$ and ξ_k is the kth component of ξ. Now take the Radon transform of (6.1) and apply (5.2) with $\mathbf{a} = (0, 0, \ldots, \varepsilon/\xi_k, \ldots, 0)$,

$$\mathscr{R}\left\{\frac{\partial f}{\partial x_k}\right\} = \xi_k \lim_{\varepsilon \to 0} \frac{\check{f}(p + \varepsilon, \xi) - \check{f}(p, \xi)}{\varepsilon}.$$

Thus

$$\mathscr{R}\left\{\frac{\partial f}{\partial x_k}\right\} = \xi_k \frac{\partial \check{f}(p, \xi)}{\partial p}. \tag{6.2}$$

By use of the linearity property, it follows that the Radon transform of

$$\sum_{k=1}^{n} a_k \frac{\partial f}{\partial x_k}$$

for arbitrary scalars a_k is given by

$$\mathscr{R}\left\{\sum_{k=1}^{n} a_k \frac{\partial f}{\partial x_k}\right\} = \mathbf{a} \cdot \xi \frac{\partial \check{f}(p, \xi)}{\partial p}. \tag{6.3}$$

For example, if $n = 3$, this equation may be interpreted as the Radon transform of a directional derivative,

$$\mathscr{R}\{\mathbf{a}\cdot\nabla f\} = \mathbf{a}\cdot\xi\frac{\partial\check{f}(p,\xi)}{\partial p}, \tag{6.4}$$

where, as usual, ∇ is the *grad* operator $(\partial/\partial x_1, \partial/\partial x_2, \partial/\partial x_3)$.

From (6.2), it is clear that the Radon transform of the derivative with respect to x_k of the function f yields ξ_k times the derivative with respect to p of the Radon transform of f. Thus we have the rather obvious generalization

$$\mathscr{R}\left\{\frac{\partial^2 f}{\partial x_l\,\partial x_k}\right\} = \xi_l\xi_k\frac{\partial^2\check{f}(p,\xi)}{\partial p^2} \tag{6.5}$$

and for arbitrary constant vectors $\mathbf{a},\mathbf{b}\in\mathbb{R}^n$,

$$\mathscr{R}\left\{\sum_{l=1}^{n}\sum_{k=1}^{n} a_l b_k\frac{\partial^2 f}{\partial x_l\,\partial x_k}\right\} = (\mathbf{a}\cdot\xi)(\mathbf{b}\cdot\xi)\frac{\partial^2\check{f}(p,\xi)}{\partial p^2}. \tag{6.6}$$

The preceding results may be generalized even further to a linear operator $\mathsf{L}(\partial/\partial x_1,\ldots,\partial/\partial x_n)$ with constant coefficients,

$$\mathscr{R}\{\mathsf{L}f\} = \mathsf{L}\left(\xi_1\frac{\partial}{\partial p},\ldots,\xi_n\frac{\partial}{\partial p}\right)\check{f}(p,\xi). \tag{6.7}$$

Example 1. Suppose $\mathsf{L} = a\,\partial/\partial x_1 + b\,\partial^3/\partial x_2\,\partial x_3^2$ is the linear operator. Then

$$\mathscr{R}\{\mathsf{L}f\} = \left(a\xi_1\frac{\partial}{\partial p} + b\xi_2\xi_3^2\frac{\partial^3}{\partial p^3}\right)\check{f}(p,\xi). \qquad \square$$

Example 2. Consider an important special case of (6.6). Suppose $n = 3$ and

$$a_l b_k = \delta_{kl} = \begin{cases} 1 & \text{if } l = k \\ 0 & \text{if } l \neq k. \end{cases}$$

In this case, (6.6) reduces to the Laplacian operator $\sum_{k=1}^{3}\partial^2/\partial x_k^2$, which will be designated by $\Delta_{\mathbf{x}}$ or simply by Δ when no confusion can arise. Now (6.6) becomes

$$\mathscr{R}\{\Delta_{\mathbf{x}} f(\mathbf{x})\} = |\xi|^2\frac{\partial^2\check{f}(p,\xi)}{\partial p^2} = \frac{\partial^2\check{f}(p,\xi)}{\partial p^2} \tag{6.8}$$

for $|\xi| = 1$. $\square$

The result obtained in the last example clearly suggests applications in the area of partial differential equations [John (1955)].

Example 3. Suppose f depends on an additional parameter, say time, for $n = 3$, $f = f(x, y, z; t)$. Consider the three-dimensional wave equation

$$\frac{\partial^2 f}{\partial t^2} = \frac{\partial^2 f}{\partial x^2} + \frac{\partial^2 f}{\partial y^2} + \frac{\partial^2 f}{\partial z^2}.$$

If the Radon transform is applied to both sides of this equation, an interesting simplification takes place. Using (6.8), it follows that

$$\frac{\partial^2 \check{f}}{\partial t^2} = \frac{\partial^2 \check{f}}{\partial p^2},$$

where $\check{f} = \check{f}(p, \xi; t) = \mathscr{R}f(x, y, z; t)$. Observe that $\mathscr{R}$ commutes with $\partial^2/\partial t^2$. □

In this example, the three-dimensional equation has been reduced to a one-dimensional equation. We shall return to this concept in Chapter 8.

Finally, observe that if the linear operator L is a homogeneous polynomial (of degree m with constant coefficients) in the derivatives, then the Radon transform of $\mathsf{L}f$ is simple,

$$\mathscr{R}\{\mathsf{L}f\} = \mathsf{L}(\xi)\frac{\partial^m \check{f}(p, \xi)}{\partial p^m}. \tag{6.9}$$

Example 4. Let $f = f(x, y)$ and suppose L is the operator

$$\mathsf{L} = \mathsf{L}\left(\frac{\partial}{\partial x}, \frac{\partial}{\partial y}\right) = \frac{\partial^3}{\partial x^3} + \frac{\partial^3}{\partial x\, \partial y^2} + 4\frac{\partial^3}{\partial y^3}.$$

Then

$$\mathscr{R}\{\mathsf{L}f\} = \left(\xi_1^3 + \xi_1\xi_2^2 + 4\xi_2^3\right)\frac{\partial^3 \check{f}(p, \xi)}{\partial p^3}.$$ □

It is useful [Helgason (1965)] to introduce the differential operator □ such that by definition

$$\Box\psi(p, \xi) = \frac{\partial^2 \psi(p, \xi)}{\partial p^2} \tag{6.10}$$

for an arbitrary function of the scalar p and vector ξ. If ψ is $\check{f}$, then from (6.8) it follows that

$$\mathcal{R}\,\Delta f = \Box \mathcal{R} f \tag{6.11}$$

or, equivalently,

$$(\Delta f)\check{} = \Box \check{f}. \tag{6.12}$$

[3.7] TRANSFORMS INVOLVING HERMITE POLYNOMIALS

In [2.5], we found that the Radon transform of $e^{-x^2-y^2}$ was

$$\mathcal{R}\{e^{-x^2-y^2}\} = \sqrt{\pi}\, e^{-p^2}. \tag{7.1}$$

This result coupled with the derivative relations of [3.6] suggests that we consider finding the Radon transform of functions of the form

$$H_k(x)H_l(y)e^{-x^2-y^2},$$

where $H_k(x)$ is the kth Hermite polynomial in the variable x and $H_l(y)$ is the lth Hermite polynomial in the variable y. A brief discussion of Hermite polynomials is contained in Appendix C.

The starting point for computing the transform $\mathcal{R}\{H_k(x)H_l(y)e^{-x^2-y^2}\}$ is to make use of the Rodrigues formula

$$e^{-t^2}H_k(t) = (-1)^k\left(\frac{\partial}{\partial t}\right)^k e^{-t^2} \tag{7.2}$$

for the Hermite polynomials [Szegö (1939); Rainville (1960)]. From this formula, it follows that

$$H_k(x)H_l(y)e^{-x^2-y^2} = (-1)^{k+l}\left(\frac{\partial}{\partial x}\right)^k\left(\frac{\partial}{\partial y}\right)^l e^{-x^2-y^2}, \tag{7.3}$$

for nonnegative integers k and l. From [3.6], we know that

$$\mathcal{R}\left\{\left(\frac{\partial}{\partial x}\right)^k\left(\frac{\partial}{\partial y}\right)^l f(x, y)\right\} = \xi_1^k\xi_2^l\left(\frac{\partial}{\partial p}\right)^{k+l}\check{f}(p, \xi). \tag{7.4}$$

Hence, with $f(x, y) = e^{-x^2-y^2}$ so that $\check{f}$ is known from (7.1), we have

$$\begin{aligned}\mathcal{R}\{H_k(x)H_l(y)e^{-x^2-y^2}\} &= (-1)^{k+l}\mathcal{R}\left\{\left(\frac{\partial}{\partial x}\right)^k\left(\frac{\partial}{\partial y}\right)^l e^{-x^2-y^2}\right\}\\ &= (-1)^{k+l}\xi_1^k\xi_2^l\left(\frac{\partial}{\partial p}\right)^{k+l}\sqrt{\pi}\, e^{-p^2}.\end{aligned}$$

However, from (7.2),

$$(-1)^{k+l}\left(\frac{\partial}{\partial p}\right)^{k+l} e^{-p^2} = e^{-p^2} H_{k+l}(p).$$

Therefore,

$$\mathscr{R}\left\{H_k(x)H_l(y)e^{-x^2-y^2}\right\} = \sqrt{\pi}\,(\cos\phi)^k(\sin\phi)^l e^{-p^2} H_{k+l}(p). \quad (7.5)$$

Here, the explicit form of the unit vector ξ has been used,

$$\xi = (\xi_1, \xi_2) = (\cos\phi, \sin\phi).$$

Remark. Results of the type in (7.5) along with corresponding results involving Laguerre polynomials in [3.8] have been utilized by several authors: Maldonado (1965); Maldonado, Caron, and Olsen (1965); Maldonado and Olsen (1966); Olsen, Maldonado, and Duckworth (1968); Matulka and Collins (1971); Jagota and Collins (1972); Kosakoski and Collins (1974); Eggermont (1975). However, the original derivations did not make use of the Radon transform and lack the elegance of this approach. □

Example 1. (*Scaled Variables*). It may be desirable to scale the variables x and y. The transform of

$$H_k(\lambda x)H_l(\lambda y)e^{-\lambda^2(x^2+y^2)}$$

must be computed. This may be accomplished by using (4.4), where the right-hand side of (7.5) is identified with $\check{f}(p, \phi)$:

$$\mathscr{R}\left\{H_k(\lambda x)H_l(\lambda y)e^{-\lambda^2(x^2+y^2)}\right\} = \frac{\sqrt{\pi}}{\lambda}(\cos\phi)^k(\sin\phi)^l e^{-\lambda^2 p^2} H_{k+l}(\lambda p). \quad □$$

For future convenience, define

$$\mathscr{H}_{kl} = \mathscr{H}_{kl}(\lambda x, \lambda y) = \frac{\lambda}{\sqrt{\pi}} H_k(\lambda x)H_l(\lambda y). \quad (7.6)$$

Then, from Example 1,

$$\mathscr{R}\left\{\mathscr{H}_{kl}(\lambda x, \lambda y)e^{-\lambda^2(x^2+y^2)}\right\} = (\cos\phi)^k(\sin\phi)^l e^{-\lambda^2 p^2} H_{k+l}(\lambda p). \quad (7.7)$$

Another useful formula follows by differentiating (7.7) with respect to p:

$$\frac{1}{\lambda^m}\left(-\frac{\partial}{\partial p}\right)^m \mathscr{R}\left\{\mathscr{H}_{kl}(\lambda x, \lambda y)e^{-\lambda^2(x^2+y^2)}\right\}$$

$$= (\cos\phi)^k(\sin\phi)^l e^{-\lambda^2 p^2} H_{k+l+m}(\lambda p). \quad (7.8)$$

Since it is always possible to express members of the sequence of monomials

$$1, x, y, x^2, xy, y^2, x^3, x^2y, \ldots$$

in terms of Hermite polynomials, it follows that Radon transform of functions of the form

$$x^k y^l e^{-x^2-y^2}$$

can be computed. The method may be illustrated by an example with $k = 2$ and $l = 1$.

Example 2. The problem is to find the Radon transform of $x^2ye^{-x^2-y^2}$. First note that

$$x^2y = \tfrac{1}{8}H_2(x)H_1(y) + \tfrac{1}{4}H_0(x)H_1(y).$$

(The information needed to derive this result can be found in Appendix C.) Consequently,

$$x^2y = \frac{\sqrt{\pi}}{8}\mathscr{H}_{21}(x, y) + \frac{\sqrt{\pi}}{4}\mathscr{H}_{01}(x, y),$$

where (7.6) with $\lambda = 1$ has been used for shorthand. The desired transform follows immediately from (7.7) and linearity:

$$\begin{aligned}\mathscr{R}\left\{x^2ye^{-x^2-y^2}\right\} &= \frac{\sqrt{\pi}}{8}\mathscr{R}\left\{(\mathscr{H}_{21} + 2\mathscr{H}_{01})e^{-x^2-y^2}\right\} \\ &= \frac{\sqrt{\pi}}{8}\cos^2\phi\sin\phi\, e^{-p^2}H_3(p) \\ &\quad + \frac{\sqrt{\pi}}{4}\sin\phi\, e^{-p^2}H_1(p).\end{aligned}$$

Or, after evaluation of $H_3(p)$ and $H_1(p)$,

$$\begin{aligned}\mathscr{R}\left\{x^2ye^{-x^2-y^2}\right\} &= \sqrt{\pi}\,p^3e^{-p^2}\cos^2\phi\sin\phi \\ &\quad + \frac{\sqrt{\pi}}{2}pe^{-p^2}\sin\phi(1 - 3\cos^2\phi).\end{aligned}$$ □

The extension of the method utilized in this example to more complicated cases is straightforward. A short table of conversion coefficients from powers of x to Hermite polynomials is included in Appendix C, and a more extensive table is available in Abramowitz and Stegun (1964).

A few additional transforms follow along with corresponding graphical results. Note that for display purposes the shifting property discussed in [3.5] has been utilized in the plots.

$$\begin{aligned}\mathcal{R}\{xe^{-x^2-y^2}\} &= \mathcal{R}\left\{\tfrac{1}{2}H_1(x)e^{-x^2-y^2}\right\} \\ &= \frac{\sqrt{\pi}}{2}\cos\phi\, e^{-p^2}H_1(p) \\ &= \sqrt{\pi}\, pe^{-p^2}\cos\phi.\end{aligned}$$

This transform is illustrated in Fig. 3.7. Note that (5.1) has been used,

$$x \to x - 4, \qquad y \to y - 4, \qquad p \to p - 4\cos\phi - 4\sin\phi.$$

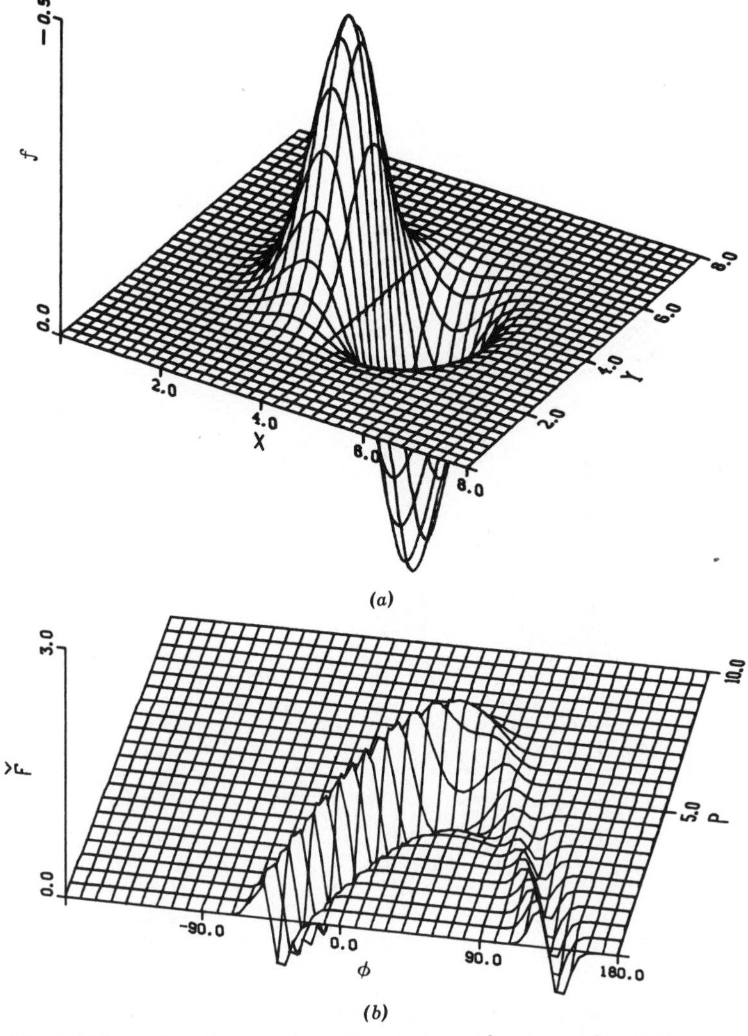

Figure 3.7 (a) The surface $f(x, y) = (x - 4)\exp[-(x - 4)^2 - (y - 4)^2]$. ($b$) Radon transform of the surface in (a).

Also, it has been required that $f < 0$ above the xy plane to better illustrate the surface. The next transform is illustrated in Fig. 3.8.

$$\mathscr{R}\{x^2 e^{-x^2-y^2}\} = \mathscr{R}\{[\tfrac{1}{4}H_2(x) + \tfrac{1}{2}H_0(x)]e^{-x^2-y^2}\}$$

$$= \frac{\sqrt{\pi}}{4}\cos^2\phi\, e^{-p^2}H_2(p) + \frac{\sqrt{\pi}}{2}H_0(p)e^{-p^2}$$

$$= \sqrt{\pi}\,(p^2\cos^2\phi + \tfrac{1}{2}\sin^2\phi)e^{-p^2}.$$

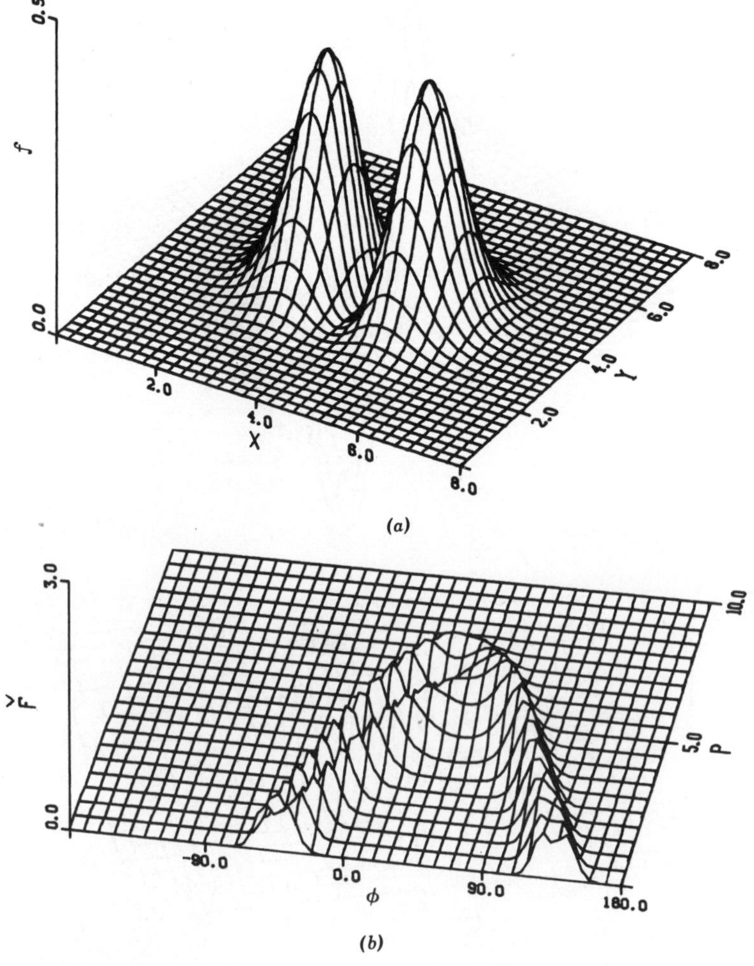

Figure 3.8 (a) The surface $f(x, y) = (x-4)^2\exp[-(x-4)^2 - (y-4)^2]$. ($b$) Radon transform of the surface in (a).

The following transform is illustrated in Fig. 3.9.

$$\begin{aligned}
\mathscr{R}\left\{x^2y^2e^{-x^2-y^2}\right\} &= \mathscr{R}\left\{\left[\tfrac{1}{16}H_2(x)H_2(y) + \tfrac{1}{8}H_2(x)H_0(y) \right.\right.\\
&\quad \left.\left. + \tfrac{1}{8}H_0(x)H_2(y) + \tfrac{1}{4}H_0(x)H_0(y)\right]e^{-x^2-y^2}\right\}\\
&= \sqrt{\pi}\left[\tfrac{1}{16}\cos^2\phi\sin^2\phi\, H_4(p) + \tfrac{1}{8}\cos^2\phi\, H_2(p)\right.\\
&\quad \left. + \tfrac{1}{8}\sin^2\phi\, H_2(p) + \tfrac{1}{4}H_0(p)\right]e^{-p^2}\\
&= \sqrt{\pi}\left[(\cos^2\phi\sin^2\phi)\left(p^4 - 3p^2 + \tfrac{3}{4}\right) + \frac{p^2}{2}\right]e^{-p^2}.
\end{aligned}$$

Finally, the results of Example 2 appear in Fig. 3.10.

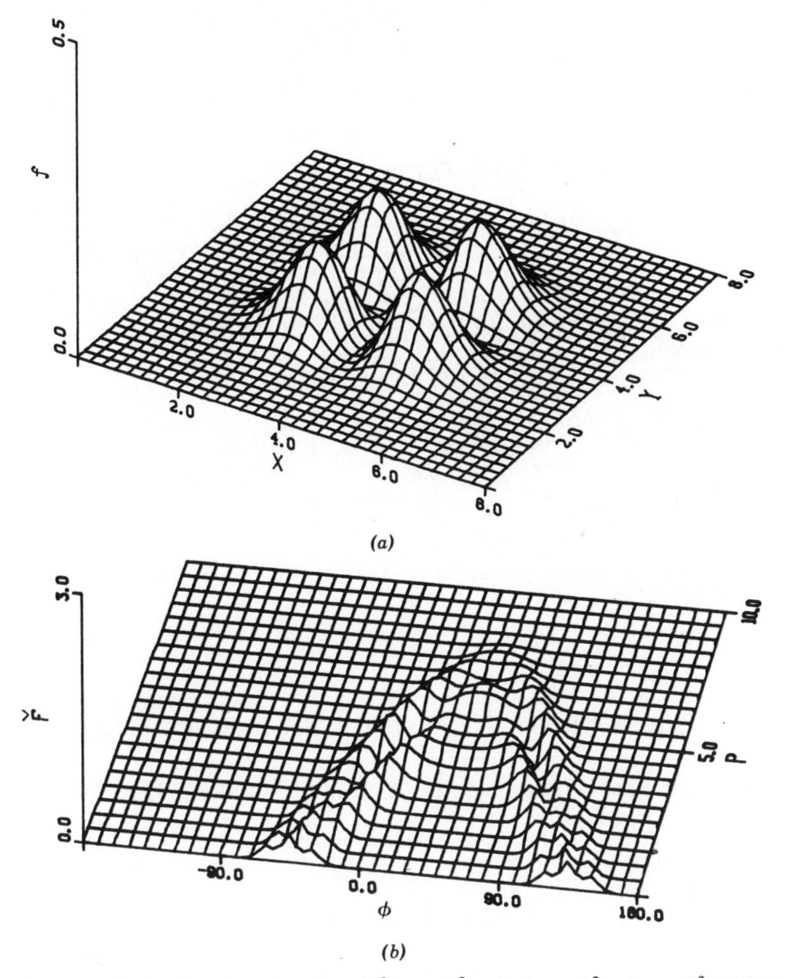

(b)

Figure 3.9 (a) The surface $f(x, y) = (x - 4)^2(y - 4)^2\exp[-(x - 4)^2 - (y - 4)^2]$. ($b$) Radon transform of the surface in (a).

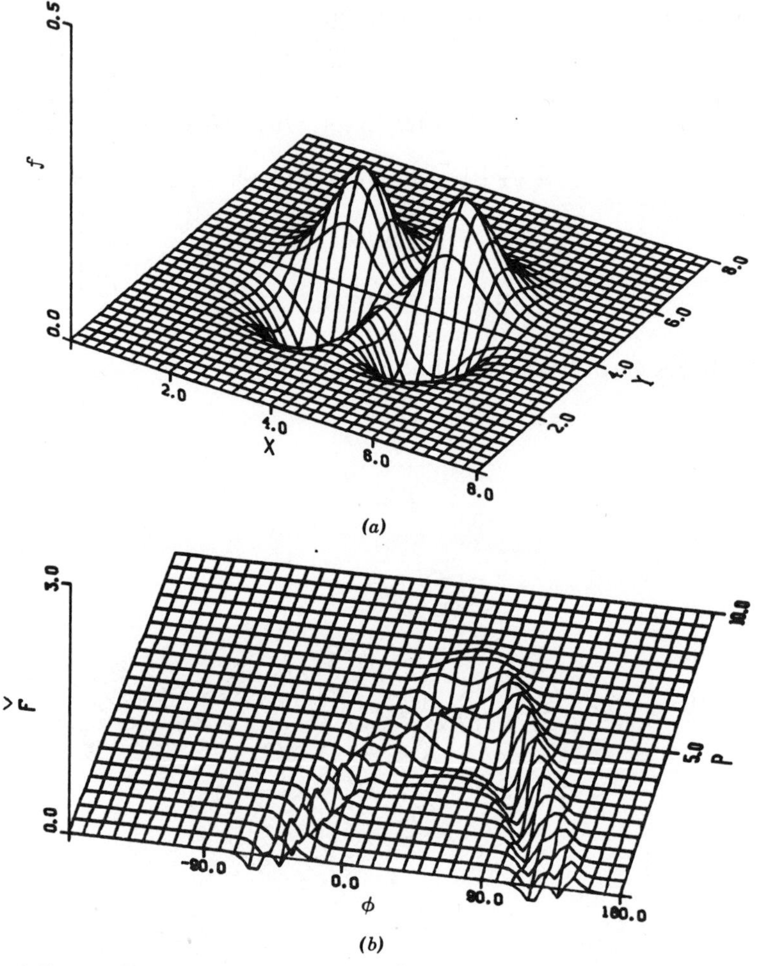

Figure 3.10 (*a*) The surface $f(x, y) = (x - 4)^2(y - 4)\exp[-(x - 4)^2(y - 4)^2]$. (*b*) Radon transform of the surface in (*a*).

[3.8] TRANSFORMS INVOLVING LAGUERRE POLYNOMIALS

The procedure here is patterned after that of [3.7], with (7.4) replaced by

$$\left(\frac{\partial}{\partial x} \pm i\frac{\partial}{\partial y}\right)^l \left(\frac{\partial^2}{\partial x^2} + \frac{\partial^2}{\partial y^2}\right)^k e^{-x^2-y^2}$$

$$= (-1)^{l+k} 2^{2k+l} k!(x \pm iy)^l e^{-x^2-y^2} L_k^l(x^2 + y^2), \tag{8.1}$$

where $L_k^l(t)$ is an associated Laguerre polynomial with both l and k nonnegative integers and $i = \sqrt{-1}$. (Appendix C contains some relevant properties of the Laguerre polynomials.) The expression (8.1) is not commonly given in discussions of Laguerre polynomials; however, it follows by induction and use of the Rodrigues-type formula [Szegö (1939); Rainville (1960)]:

$$e^{-t}t^l L_k^l(t) = \frac{1}{k!}\left(\frac{\partial}{\partial t}\right)^k (e^{-t}t^{l+k}).$$

The Radon transform of (8.1) follows from the methods developed in [3.6]:

$$\check{f}(p,\phi) = \mathscr{R}\left\{\left(\frac{\partial}{\partial x} \pm i\frac{\partial}{\partial y}\right)^l \left(\frac{\partial^2}{\partial x^2} + \frac{\partial^2}{\partial y^2}\right)^k e^{-x^2-y^2}\right\}$$

$$= (-1)^{2k+l}\sqrt{\pi}\, e^{\pm il\phi} e^{-p^2} H_{l+2k}(p), \tag{8.2}$$

where $H_{l+2k}(p)$ is a Hermite polynomial of order $l + 2k$. From (8.1) and (8.2),

$$\mathscr{R}\left\{\frac{(-1)^k}{\sqrt{\pi}}\left[\frac{k!}{(l+k)!}\right]^{1/2} (x^2+y^2)^{l/2} e^{\pm il\theta} L_k^l(x^2+y^2) e^{-x^2-y^2}\right\}$$

$$= \left[\frac{1}{k!(l+k)!}\right]^{1/2} \frac{1}{2^{2k+l}} e^{\pm il\phi} e^{-p^2} H_{l+2k}(p).$$

The $x \pm iy$ term in (8.1) has been replaced by

$$(x \pm iy)^l = (x^2+y^2)^{l/2} e^{\pm il\theta} \tag{8.3a}$$

with

$$\theta = \tan^{-1}\left(\frac{y}{x}\right). \tag{8.3b}$$

In terms of scaled variables,

$$\mathcal{R}\left\{\frac{(-1)^k\lambda}{\sqrt{\pi}}\left[\frac{k!}{(l+k)!}\right]^{1/2}[\lambda^2(x^2+y^2)]^{l/2}\right.$$

$$\left.\times\ e^{\pm il\theta}L_k^l[\lambda^2(x^2+y^2)]e^{-\lambda^2(x^2+y^2)}\right\}$$

$$=\left[\frac{1}{k!(l+k)!}\right]^{1/2}\frac{1}{2^{2k+l}}e^{\pm il\phi}e^{-\lambda^2p^2}H_{l+2k}(\lambda p).$$

If we define the complex polynomial ($l \geqslant 0$) of degree $l + 2k$ by

$$\mathcal{L}^{\pm l}_{l+2k}(\lambda x, \lambda y) = \frac{(-1)^k\lambda}{\sqrt{\pi}}\left[\frac{k!}{(l+k)!}\right]^{1/2}$$

$$\times[\lambda^2(x^2+y^2)]^{l/2}e^{\pm il\theta}L_k^l[\lambda^2(x^2+y^2)] \tag{8.4}$$

and designate the normalization by

$$N_k^l = \left[\frac{1}{k!(l+k)!}\right]^{1/2}\frac{1}{2^{l+2k}}, \tag{8.5}$$

then a more concise and manageable expression follows,

$$\mathcal{R}\left\{\mathcal{L}^{\pm l}_{l+2k}(\lambda x, \lambda y)e^{-\lambda^2(x^2+y^2)}\right\} = N_k^l e^{\pm il\phi}e^{-\lambda^2p^2}H_{l+2k}(\lambda p). \tag{8.6}$$

It is useful to observe that the complex conjugate of (8.4) yields the equation

$$[\mathcal{L}^{\pm l}_{l+2k}]^* = \mathcal{L}^{\mp l}_{l+2k}, \qquad l \geqslant 0. \tag{8.7}$$

Example 1. To evaluate $\mathcal{L}_4^{-2}$, first observe that

$$\mathcal{L}_4^2(\lambda x, \lambda y) = \frac{-\lambda^3}{\sqrt{\pi}}\frac{1}{\sqrt{6}}(x+iy)^2(3-\lambda^2x^2-\lambda^2y^2),$$

and then, by complex conjugation,

$$\mathcal{L}_4^{-2} = \frac{-\lambda^3}{\sqrt{\pi}}\frac{1}{\sqrt{6}}(x-iy)^2(3-\lambda^2x^2-\lambda^2y^2).$$

□

Example 2. By direct substitution in (8.6) with $\lambda = 1$, we find the transforms

$$\mathscr{R}\left\{\mathscr{L}_1^1 e^{-x^2-y^2}\right\} = \tfrac{1}{2} e^{i\phi} e^{-p^2} H_1(p)$$

$$\mathscr{R}\left\{\mathscr{L}_3^3 e^{-x^2-y^2}\right\} = \frac{1}{8\sqrt{6}} e^{3i\phi} e^{-p^2} H_3(p)$$

$$\mathscr{R}\left\{\mathscr{L}_3^1 e^{-x^2-y^2}\right\} = \frac{1}{8\sqrt{2}} e^{i\phi} e^{-p^2} H_3(p). \qquad \square$$

Example 3. From (8.4),

$$\mathscr{L}_1^1 = \frac{1}{\sqrt{\pi}}(x + iy),$$

$$\mathscr{L}_3^3 = \frac{1}{\sqrt{\pi}} \frac{1}{\sqrt{6}}(x + iy)^3,$$

$$\mathscr{L}_3^1 = \frac{-1}{\sqrt{\pi}} \frac{1}{\sqrt{2}}(x + iy)(2 - x^2 - y^2).$$

Note that

$$\frac{\sqrt{\pi}}{2}\mathscr{L}_1^1 + \frac{\sqrt{6\pi}}{4}\mathscr{L}_3^3 + \frac{\sqrt{2\pi}}{4}\mathscr{L}_3^1 = \frac{x^3 - xy^2}{2} + ix^2 y,$$

and the imaginary part of the left side is just $x^2 y$. □

Example 4. By combining the previous examples, it is possible to compute $\mathscr{R}\{x^2 y e^{-x^2-y^2}\}$, since from Example 3

$$x^2 y = \operatorname{Im}\left[\frac{\sqrt{\pi}}{2}\mathscr{L}_1^1 + \frac{\sqrt{6\pi}}{4}\mathscr{L}_3^3 + \frac{\sqrt{2\pi}}{4}\mathscr{L}_3^1\right]$$

$$= \operatorname{Im}[\cdot]$$

for brevity, and

$$\mathscr{R}\left\{x^2 y e^{-x^2-y^2}\right\} = \mathscr{R}\left\{\operatorname{Im}[\cdot] e^{-x^2-y^2}\right\}$$

$$= \operatorname{Im} \mathscr{R}\left\{[\cdot] e^{-x^2-y^2}\right\}$$

Thus, from Example 2,

$$\begin{aligned}
\mathscr{R}\{x^2ye^{-x^2-y^2}\} &= \frac{\sqrt{\pi}}{4}\,\mathrm{Im}(e^{i\phi})e^{-p^2}H_1(p) \\
&\quad + \frac{\sqrt{\pi}}{32}\,\mathrm{Im}(e^{3i\phi})e^{-p^2}H_3(p) \\
&\quad + \frac{\sqrt{\pi}}{32}\,\mathrm{Im}(e^{i\phi})e^{-p^2}H_3(p) \\
&= \frac{\sqrt{\pi}}{4}\sin\phi\, e^{-p^2}H_1(p) \\
&\quad + \frac{\sqrt{\pi}}{32}(\sin 3\phi + \sin\phi)e^{-p^2}H_3(p).
\end{aligned}$$

With simplification, this reduces to the result obtained in Example 2 of [3.7]. □

This last example clearly suggests the possibility of a relationship among Radon transforms of functions of the form

$$x^k y^l e^{-x^2-y^2}, \qquad \mathscr{H}_{kl}(x, y)e^{-x^2-y^2}, \qquad \text{and} \qquad \mathscr{L}^{\pm l}_{l+2k}(x, y)e^{-x^2-y^2}.$$

Furthermore, by the Weierstrass approximation theorem [Courant and Hilbert (1953); Jackson (1936)], the sets $\mathscr{H}_{kl}$ and $\mathscr{L}^{\pm l}_{l+2k}$ are complete since for fixed degree m there are exactly $\frac{1}{2}(m+1)(m+2)$ linearly independent members of each set of degree $\leqslant m$. See Tables 3.1 and 3.2; the corresponding table for the $\mathscr{H}_{kl}$ is precisely the same as that for Table 3.1 with $x^k y^l$ replaced by $\mathscr{H}_{kl}$. These functions are discussed further in Chapter 7.

Table 3.1 Linearly Independent Monomials of The Form $x^k y^l$

M	Degree $k + l \leqslant M$	Total Number
0	1	1
1	$1, x, y$	$3\ (= 1 + 2)$
2	$1, x, y, x^2, xy, y^2$	$6\ (= 1 + 2 + 3)$
3	$1, x, y, x^2, xy, y^2, x^3, x^2y, xy^2, y^3$	$10\ (= 1 + 2 + 3 + 4)$
⋮		
m	$1, \ldots, y^m$	$\frac{1}{2}(m+1)(m+2) = \sum_{1}^{m+1} n$

Table 3.2 Linearly Independent Polynomials $\mathscr{L}_{l+2k}^{\pm l}$

M	Degree $l + 2k \leqslant M$	Total Number
0	$\mathscr{L}_0^0$	1
1	$\mathscr{L}_0^0, \mathscr{L}_1^1, \mathscr{L}_1^{-1}$	3
2	$\mathscr{L}_0^0, \mathscr{L}_1^1, \mathscr{L}_1^{-1}, \mathscr{L}_2^0, \mathscr{L}_2^2, \mathscr{L}_2^{-2}$	6
3	$\mathscr{L}_0^0, \mathscr{L}_1^1, \mathscr{L}_1^{-1}, \mathscr{L}_2^0, \mathscr{L}_2^2, \mathscr{L}_2^{-2}, \mathscr{L}_3^1, \mathscr{L}_3^{-1}, \mathscr{L}_3^3, \mathscr{L}_3^{-3}$	10
⋮		
m	$\mathscr{L}_0^0, \ldots, \mathscr{L}_m^{-m}$	$\frac{1}{2}(m+1)(m+2)$

[3.9] DERIVATIVES OF THE TRANSFORM

In the preceding three sections we have been working with the Radon transform of derivatives of f. The results involve derivatives of $\check{f}$ with respect to p as in (6.2). We now investigate what happens when $\check{f}(p, \xi)$ is differentiated with respect to one of the components of ξ. Start with the definition

$$\check{f}(p, \xi) = \int f(\mathbf{x})\, \delta(p - \xi \cdot \mathbf{x})\, d\mathbf{x}$$

and differentiate:

$$\frac{\partial \check{f}}{\partial \xi_k} = \int f(\mathbf{x}) \frac{\partial}{\partial \xi_k} \delta(p - \xi \cdot \mathbf{x})\, d\mathbf{x}.$$

Use of the identity (Appendix B)

$$\frac{\partial}{\partial \xi_k} \delta(p - \xi \cdot \mathbf{x}) = -x_k \frac{\partial}{\partial p} \delta(p - \xi \cdot \mathbf{x})$$

yields

$$\frac{\partial \check{f}}{\partial \xi_k} = -\frac{\partial}{\partial p} \int x_k f(\mathbf{x})\, \delta(p - \xi \cdot \mathbf{x})\, d\mathbf{x}, \tag{9.1}$$

or, equivalently,

$$\frac{\partial}{\partial \xi_k} \mathscr{R}\{f(\mathbf{x})\} = -\frac{\partial}{\partial p} \mathscr{R}\{x_k f(\mathbf{x})\}. \tag{9.2}$$

Caution. It is important to point out that when derivatives with respect to components of ξ are taken, the vector ξ is not considered as a unit vector initially. Of course, the derivatives may then be evaluated at points where $|\xi| = 1$. The following examples in $\mathbb{R}^2$ illustrate the use of (9.1) and (9.2) and point out the need for caution when taking derivatives with respect to components of ξ. □

Example 1. Suppose $f = f(x, y)$ is

$$f(x, y) = e^{-x^2-y^2}.$$

Then (2.3) applies and

$$\check{f}(p, \xi) = \sqrt{\pi}\left(\xi_1^2 + \xi_2^2\right)^{-1/2} \exp\left[\frac{-p^2}{\left(\xi_1^2 + \xi_2^2\right)}\right].$$

For $|\xi| = 1$, this reduces to $\sqrt{\pi}\, e^{-p^2}$ and the dependence on ξ is not explicitly visible. Now, differentiate $\check{f}$ with respect to ξ_1 to obtain

$$\frac{\partial \check{f}}{\partial \xi_1} = \sqrt{\pi}\left(\xi_1^2 + \xi_2^2\right)^{-5/2} \xi_1 (2p^2 - 1) \exp\left[\frac{-p^2}{\left(\xi_1^2 + \xi_2^2\right)}\right].$$

Evaluation of this expression at $|\xi| = 1$, where $\xi = (\cos\phi, \sin\phi)$ yields

$$\frac{\partial \check{f}}{\partial \xi_1} = \sqrt{\pi}\cos\phi (2p^2 - 1) e^{-p^2}.$$ □

In preparation for the next example, note that

$$\frac{\partial \check{f}}{\partial \xi_1} = \frac{\sqrt{\pi}}{2}\cos\phi\, H_2(p) e^{-p^2}.$$

Example 2. Consider the same function $f(x, y) = e^{-x^2-y^2}$ as in Example 1. From (9.2),

$$\frac{\partial \check{f}}{\partial \xi_1} = -\frac{\partial}{\partial p}\mathscr{R}\{x e^{-x^2-y^2}\}.$$

However, from (7.6),

$$x = \tfrac{1}{2} H_1(x) = \frac{\sqrt{\pi}}{2}\mathscr{H}_{10}(x, y),$$

and from (7.7),

$$\mathscr{R}\left\{\frac{\sqrt{\pi}}{2}\mathscr{H}_{10}(x, y) e^{-x^2-y^2}\right\} = \frac{\sqrt{\pi}}{2}\cos\phi\, e^{-p^2} H_1(p).$$

Thus

$$-\frac{\partial}{\partial p}\mathscr{R}\{xe^{-x^2-y^2}\} = -\frac{\sqrt{\pi}}{2}\cos\phi\,\frac{\partial}{\partial p}\left[e^{-p^2}H_1(p)\right]$$

$$= \frac{\sqrt{\pi}}{2}\cos\phi\,(4p^2-2)e^{-p^2}$$

$$= \frac{\sqrt{\pi}}{2}\cos\phi\,H_2(p)e^{-p^2}.$$

Of course, this is in agreement with Example 1. Also, note that a special case of (7.8) yields this same result. □

Suppose $\mathbf{a} \in \mathbb{R}^n$ is some arbitrary constant vector; then it follows from (9.1), (9.2), and linearity that

$$\sum_{k=1}^{n} a_k \frac{\partial \check{f}(p,\xi)}{\partial \xi_k} = -\frac{\partial}{\partial p}\mathscr{R}\{(\mathbf{a}\cdot\mathbf{x})f(\mathbf{x})\} \tag{9.3}$$

Or, if we designate

$$\mathbf{a}\cdot\frac{\partial}{\partial \xi} = \sum_{k=1}^{n} a_k\frac{\partial}{\partial \xi_k},$$

then (9.3) may be written as

$$\left(\mathbf{a}\cdot\frac{\partial}{\partial \xi}\right)\mathscr{R}f(\mathbf{x}) = -\frac{\partial}{\partial p}\mathscr{R}\{(\mathbf{a}\cdot\mathbf{x})f(\mathbf{x})\}. \tag{9.4}$$

Higher derivatives follow from differentiating (9.1):

$$\frac{\partial^2 \check{f}(p,\xi)}{\partial \xi_k\,\partial \xi_l} = \frac{\partial^2}{\partial p^2}\mathscr{R}\{x_l x_k f(\mathbf{x})\}. \tag{9.5}$$

And for arbitrary constant vectors $\mathbf{a}, \mathbf{b} \in \mathbb{R}^n$,

$$\sum_{l=1}^{n}\sum_{k=1}^{n} a_l b_k \frac{\partial \check{f}(p,\xi)}{\partial \xi_l\,\partial \xi_k} = \frac{\partial^2}{\partial p^2}\mathscr{R}\{(\mathbf{a}\cdot\mathbf{x})(\mathbf{b}\cdot\mathbf{x})f(\mathbf{x})\}, \tag{9.6}$$

where as usual $\check{f} = \mathscr{R}f$. □

Example 3. Again consider $\mathbb{R}^2$ and let $f = f(x, y)$. Generalization of (9.5) yields

$$\frac{\partial^{k+l}}{\partial \xi_1^k\,\partial \xi_2^l}\check{f}(p,\xi) = \left(-\frac{\partial}{\partial p}\right)^{k+l}\mathscr{R}\{x^k y^l f(x,y)\}.$$ □

Example 4. Verify Example 3 for the special case

$$f(x, y) = xe^{-x^2-y^2},$$

with $k = 1$ and $l = 1$.

From (7.5),

$$\check{f}(p, \xi) = \sqrt{\pi}\, \xi_1 p e^{-p^2}$$

for ξ a unit vector. If ξ is not a unit vector, then from (2.3)

$$\check{f}(p, \xi) = \frac{\sqrt{\pi}\, \xi_1 p \exp\left[\dfrac{-p^2}{(\xi_1^2 + \xi_2^2)}\right]}{(\xi_1^2 + \xi_2^2)^{3/2}}.$$

It is a straightforward (but tedious) calculation to show that $\partial^2 \check{f}/\partial\xi_1 \partial\xi_2$ evaluated at $|\xi| = 1$ yields

$$\left.\frac{\partial^2 \check{f}}{\partial\xi_1\, \partial\xi_2}\right|_{|\xi|=1} = \xi_1^2\xi_2(15p - 20p^3 - 4p^5)e^{-p^2} + \xi_2(2p^3 - 3p)e^{-p^2}.$$

The same result is obtained by evaluating $(\partial^2/\partial p^2)\{\mathscr{R}(x^2ye^{-x^2-y^2})\}$. See Example 2 of [3.7] for $\mathscr{R}(x^2ye^{-x^2-y^2})$, where $\xi_1 = \cos\phi$ and $\xi_2 = \sin\phi$. □

These examples serve to demonstrate the overall consistency of the properties of the Radon transform and at the same time point out the necessity for proper caution when taking derivatives of $\check{f}(p, \xi)$ with respect to components of ξ, especially when $\check{f}(p, \xi)$ was originally obtained for $|\xi| = 1$. Of course, no generality is lost by working with $|\xi| = 1$, since (2.3) can always be used, as in the previous example, to convert to $\check{f}(p, \xi)$, where ξ is not a unit vector.

[3.10] TRANSFORM OF CONVOLUTION

Suppose

$$\check{g} = \mathscr{R}g \qquad \text{and} \qquad \check{h} = \mathscr{R}h.$$

Let f be the convolution of g with h,

$$f(\mathbf{x}) = g * h = \int g(\mathbf{y})h(\mathbf{x} - \mathbf{y})\, d\mathbf{y},$$

where $\mathbf{x} \in \mathbb{R}^n$ and $\mathbf{y} \in \mathbb{R}^n$. Then

$$\check{f}(p, \boldsymbol{\xi}) = \mathscr{R}f = \mathscr{R}\{g * h\}$$

$$= \int d\mathbf{x} \int d\mathbf{y}\, g(\mathbf{y})\, h(\mathbf{x} - \mathbf{y})\, \delta(p - \boldsymbol{\xi} \cdot \mathbf{x})$$

$$= \int d\mathbf{y}\, g(\mathbf{y}) \int d\mathbf{x}\, h(\mathbf{x} - \mathbf{y})\, \delta(p - \boldsymbol{\xi} \cdot \mathbf{x}).$$

Making the change of variables $\mathbf{z} = \mathbf{x} - \mathbf{y}$,

$$\check{f}(p, \boldsymbol{\xi}) = \int d\mathbf{y}\, g(\mathbf{y}) \int d\mathbf{z}\, h(\mathbf{z})\, \delta(p - \boldsymbol{\xi} \cdot \mathbf{y} - \boldsymbol{\xi} \cdot \mathbf{z})$$

$$= \int d\mathbf{y}\, g(\mathbf{y})\, \check{h}(p - \boldsymbol{\xi} \cdot \mathbf{y}, \boldsymbol{\xi})$$

$$= \int d\mathbf{y}\, g(\mathbf{y}) \int_{-\infty}^{\infty} ds\, \check{h}(p - s, \boldsymbol{\xi})\, \delta(s - \boldsymbol{\xi} \cdot \mathbf{y})$$

$$= \int_{-\infty}^{\infty} ds\, \check{h}(p - s, \boldsymbol{\xi}) \int d\mathbf{y}\, g(\mathbf{y})\, \delta(s - \boldsymbol{\xi} \cdot \mathbf{y})$$

$$= \int_{-\infty}^{\infty} \check{g}(s, \boldsymbol{\xi}) \check{h}(p - s, \boldsymbol{\xi})\, ds.$$

Thus

$$\check{f} = \check{g} * \check{h},$$

and we have the interesting result that the Radon transform of the convolution is the convolution of the Radon transforms,

$$\mathscr{R}\{g * h\} = \check{g} * \check{h}. \tag{10.1}$$

Observe the difference here as compared with the Fourier transform, where the Fourier transform of the convolution is the product of the Fourier transforms.

Chapter 4

Relation to Other Transforms

[4.1] INTRODUCTION

In the first part of this chapter, the relationship between the Radon transform and the Fourier transform is developed. The main result is that the n-dimensional Fourier transform may be obtained by a Radon transform followed by a one-dimensional Fourier transform. The second part of the chapter illustrates how a Gegenbauer transform emerges in a natural way in connection with certain types of Radon transforms. Finally, the connection between the Radon transform and the Hough transform is discussed.

Although the general aim is the eventual inversion of the Radon transform, there is no direct discussion of inversion in this chapter; however, many of the concepts developed here serve as useful background material for the following chapters in which inversion and the ways the Radon transform relates to still other types of transforms are discussed. The connections established here (with Fourier and Gegenbauer) while of fundamental importance are by no means exhaustive.

General References

For a general treatment of the Fourier transform, see Bracewell (1978) and Sneddon (1951). Gel'fand, Graev, and Vilenkin (1966) and Helgason (1965) give the fundamental connection between the Fourier transform and the Radon transform, and Ludwig (1966) makes use of both Fourier and Gegenbauer transforms in his study of the Radon transform on Euclidean space.

[4.2] RELATION TO THE FOURIER TRANSFORM

As in earlier discussions, points in $\mathbb{R}^n$ are designated by $\mathbf{x} = (x_1, x_2, \ldots, x_n)$ and in Fourier space $\mathbf{k} = (k_1, k_2, \ldots, k_n)$. The n-dimensional Fourier transform of $f(\mathbf{x})$ can be written as

$$\tilde{f}(\mathbf{k}) = \mathscr{F}_n f = \int f(\mathbf{x}) e^{-i2\pi \mathbf{k}\cdot\mathbf{x}}\, d\mathbf{x}, \tag{2.1}$$

and the inverse transform is given by

$$f(\mathbf{x}) = \mathscr{F}_n^{-1}\tilde{f} = \int \tilde{f}(\mathbf{k})e^{i2\pi\mathbf{k}\cdot\mathbf{x}}\,d\mathbf{k}. \tag{2.2}$$

As usual, the integral is understood to be over the entire space when the shorthand $\int d\mathbf{x}$ or $\int d\mathbf{k}$ appears.

To connect the Fourier transform with the Radon transform, observe that (2.1) may be rewritten in the form

$$\tilde{f}(\mathbf{k}) = \int_{-\infty}^{\infty} dt \int d\mathbf{x}\, f(\mathbf{x})e^{-i2\pi t}\,\delta(t - \mathbf{k}\cdot\mathbf{x}),$$

where t is real. Now, let $\mathbf{k} = s\xi$ and $t = sp$, with s real and ξ a unit vector in $\mathbb{R}^n$. Then

$$\begin{aligned}\tilde{f}(s\xi) &= |s|\int_{-\infty}^{\infty} dp \int d\mathbf{x}\, f(\mathbf{x})e^{-i2\pi sp}\,\delta(sp - s\xi\cdot\mathbf{x}) \\ &= \int_{-\infty}^{\infty} dp\, e^{-i2\pi sp}\int d\mathbf{x}\, f(\mathbf{x})\delta(p - \xi\cdot\mathbf{x}).\end{aligned}$$

The last integral on the right-hand side is just the Radon transform of f. Hence it follows that

$$\tilde{f}(s\xi) = \int_{-\infty}^{\infty} \check{f}(p, \xi)e^{-i2\pi sp}\,dp. \tag{2.3}$$

Note that this equation also holds for $s = 0$, since for $s = 0$, (2.3) reduces to (2.1) with the vector $\mathbf{k}$ equal to the zero vector.

This establishes the fundamental connection between the Fourier transform and the Radon transform. Observe that the right-hand side of (2.3) is the one-dimensional Fourier transform *along the radial direction* of the Radon transform. If $\mathscr{F}_1$ designates this radial transform, then symbolically

$$\tilde{f} = \mathscr{F}_1\check{f} = \mathscr{F}_1\mathscr{R}f. \tag{2.4}$$

See Fig. 4.1 for a diagrammatic interpretation.

By using the definition of the inverse Fourier transform (2.2) with $n = 1$, it follows that (2.3) can be inverted,

$$\check{f}(p, \xi) = \int_{-\infty}^{\infty} \tilde{f}(s\xi)e^{i2\pi sp}\,ds. \tag{2.5}$$

In effect, this says that $\check{f}$ may be obtained by applying the inverse Fourier

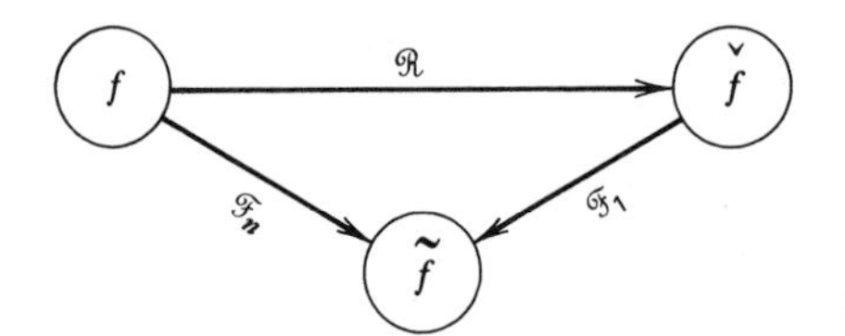

Figure 4.1 $\mathscr{F}_n f = \mathscr{F}_1 \mathscr{R} f$.

transform in the radial direction once $\tilde{f}$ is known. Symbolically,

$$\mathscr{R}f = \mathscr{F}_1^{-1} \mathscr{F}_n f. \tag{2.6}$$

See Fig. 4.2 for the corresponding diagrammatic interpretation.

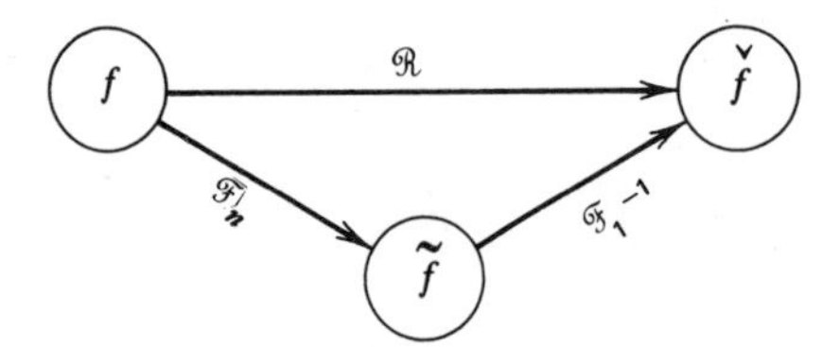

Figure 4.2 $\mathscr{R} f = \mathscr{F}_1^{-1} \mathscr{F}_n f$.

Remark. The Fourier transform was used before 1965 for studying the relation of a function to its projections [e.g., John (1934); Cramér and Wold (1936); Rényi (1952); Bracewell (1956a)]. However, important developments in radio astronomy [Bracewell (1956a); Bracewell and Riddle (1967)] provided a major impetus for further study and other applications [e.g., De Rosier and Klug (1968); Ramachandran and Lakshminarayanan (1971); Mersereau and Oppenheim (1974)]. □

Example 1. Let $f(x, y) = e^{-x^2 - y^2}$. Then

$$\tilde{f}(u, v) = \mathscr{F}_2 f = \pi e^{-\pi^2(u^2+v^2)}$$

and

$$\tilde{f}(su, sv) = \pi e^{-\pi^2 s^2(u^2+v^2)}.$$

Thus

$$\mathscr{F}_1^{-1} \tilde{f} = \pi \int_{-\infty}^{\infty} e^{-\pi^2 s^2(u^2+v^2)} e^{i2\pi sp}\, ds.$$

Evaluation of this integral yields

$$\check{f}(p; u, v) = \frac{\sqrt{\pi}}{(u^2 + v^2)^{1/2}} \exp\left(\frac{-p^2}{u^2 + v^2}\right).$$

See Example 1 of [3.9]. If one makes the identification $\xi = (u, v)$ with $u^2 + v^2 = 1$, the following familiar result is obtained:

$$\check{f}(p, \xi) = \sqrt{\pi}\, e^{-p^2}. \qquad \square$$

Example 2. Find $\mathscr{R}\{(x^2 + y^2)e^{-\pi(x^2+y^2)}\}$ by direct and by Fourier methods.

Direct method. From [3.7],

$$\mathscr{R}\{x^2 e^{-x^2-y^2}\} = \sqrt{\pi}\left(p^2\cos^2\phi + \tfrac{1}{2}\sin^2\phi\right)e^{-p^2},$$

$$\mathscr{R}\{y^2 e^{-x^2-y^2}\} = \sqrt{\pi}\left(p^2\sin^2\phi + \tfrac{1}{2}\cos^2\phi\right)e^{-p^2}.$$

Therefore,

$$\mathscr{R}\{(x^2 + y^2)e^{-x^2-y^2}\} = \sqrt{\pi}\left(p^2 + \tfrac{1}{2}\right)e^{-p^2},$$

and using (4.4) in [3.4],

$$\mathscr{R}\left\{(x^2 + y^2)e^{-\pi(x^2+y^2)}\right\} = \left(\frac{1}{2\pi} + p^2\right)e^{-\pi p^2}.$$

Fourier transform method. By direct computation or using standard tables,

$$\tilde{f}(u, v) = \left[\frac{1}{\pi} - (u^2 + v^2)\right]e^{-\pi(u^2+v^2)}.$$

Thus $\tilde{f}(su, sv)$ evaluated at $u^2 + v^2 = 1$ yields $\tilde{f}(s\xi)$ with $\xi = (u, v)$ and $|\xi| = 1$,

$$\tilde{f}(s\xi) = \left(\frac{1}{\pi} - s^2\right)e^{-\pi s^2}.$$

Now, from (2.5),

$$\begin{aligned}\check{f}(p, \xi) &= \int_{-\infty}^{\infty}\left(\frac{1}{\pi} - s^2\right)e^{-\pi s^2}e^{i2\pi sp}\,ds\\ &= \frac{1}{\pi}e^{-\pi p^2} - 2\int_0^{\infty} s^2 e^{-\pi s^2}\cos(2\pi sp)\,ds.\end{aligned}$$

The integral on the right-hand side is given on page 495 of Gradshteyn and Ryzhik (1965),

$$\check{f}(p, \xi) = \frac{1}{\pi}e^{-\pi p^2} - \left(\frac{1}{2\pi} - p^2\right)e^{-\pi p^2}.$$

After simplification, this yields

$$\check{f}(p, \xi) = \left(\frac{1}{2\pi} + p^2\right)e^{-\pi p^2},$$

the result obtained using the direct method. □

The next two examples justify the interrelationships illustrated in Fig. 1.3 of [1.1].

Example 3. Let $f = f(x, y)$ and show that $\mathscr{F}_2 f = \mathscr{F}_1\mathscr{R}f$. By definition,

$$\mathscr{F}_1\mathscr{R}f = \int_{-\infty}^{\infty} dp\, e^{-i2\pi tp} \iint_{-\infty}^{\infty} f(x, y)\delta(p - x\cos\phi - y\sin\phi)\, dx\, dy$$

$$= \iint_{-\infty}^{\infty} e^{-i2\pi t(x\cos\phi + y\sin\phi)} f(x, y)\, dx\, dy$$

$$= \iint_{-\infty}^{\infty} f(x, y) e^{-i2\pi(ux+vy)} dx\, dy$$

$$= \mathscr{F}_2 f = \tilde{f}(u, v),$$

where $u = t\cos\phi$ and $v = t\sin\phi$. □

Example 4. Given that $\tilde{f}(u, v) = \mathscr{F}_2 f(x, y)$, show that $f(x, y) = \mathscr{F}_1\mathscr{R}\tilde{f}$. By definition,

$$\mathscr{F}_1\mathscr{R}\tilde{f} = \int_{-\infty}^{\infty} dt\, e^{-i2\pi rt} \iint_{-\infty}^{\infty} \tilde{f}(u, v)\delta(t - u\cos\theta - v\sin\theta)\, du\, dv$$

$$= \iint_{-\infty}^{\infty} \tilde{f}(u, v) e^{-i2\pi(ux+vy)}\, du\, dv$$

$$= f(x, y),$$

where $x = r\cos\theta$ and $y = r\sin\theta$. □

[4.3] RELATION TO THE GEGENBAUER TRANSFORM

Special Factored Form

Consider the problem of finding the Radon transform of $f(\mathbf{x})$, where $f(\mathbf{x})$ may be written as the product of a radial part multiplied by an angular part:

$$f(\mathbf{x}) = g_l(r) S_{lm}(\omega). \tag{3.1}$$

(Since the Radon transform is linear this may be easily extended to a more general decomposition by superposition.) Here, $\mathbf{x} \in \mathbb{R}^n$, $r = |\mathbf{x}| \leqslant 0$, $\omega = \mathbf{x}/|\mathbf{x}|$, and the doubly subscripted $S_{lm}(\omega)$ is a real spherical (or surface) harmonic [Hochstadt (1971); Erdélyi et al. (1953b)] of degree l defined on $\mathbb{R}^n$.

The $S_{lm}(\omega)$ come from an orthonormal set with $\mathcal{N} = \mathcal{N}(n, l)$ members. Explicitly, members of the set

$$\{S_{l1}, S_{l2}, \ldots, S_{l\mathcal{N}}\}$$

satisfy the orthonormality condition

$$\int_{|\omega|=1} S_{lm}(\omega) S_{l'm'}(\omega)\, d\omega = \delta_{ll'}\delta_{mm'}, \tag{3.2}$$

where $d\omega$ is the surface element on a unit sphere in hyperspherical polar coordinates and the integral $\int_{|\omega|=1}$ is over this surface. More explicitly, we could write $S(\theta_1, \theta_2, \ldots, \theta_{n-2}, \phi)$ in place of $S(\omega)$, and for the surface element on the unit sphere,

$$d\omega = (\sin\theta_1)^{n-2}(\sin\theta_2)^{n-3} \cdots (\sin\theta_{n-2})\, d\theta_1\, d\theta_2 \cdots d\theta_{n-2}\, d\phi.$$

A very important property of the $S_{lm}(\omega)$ is that they satisfy the symmetry condition

$$S_{lm}(-\omega) = (-1)^l S_{lm}(\omega). \tag{3.3}$$

The integer $\mathcal{N} = \mathcal{N}(n, l)$, which depends on both the dimension n of the space and the degree l of the spherical harmonics, is the number of linearly independent spherical harmonics of degree l where each spherical harmonic depends on $n - 1$ angular coordinates $\theta_1, \ldots, \theta_{n-2}, \phi$. The value of $\mathcal{N}$ is worked out by Hochstadt (1971) and is given by

$$\mathcal{N} = \mathcal{N}(n, l) = \frac{2l + n - 2}{l}\binom{l + n - 3}{l - 1} \tag{3.4}$$

in terms of a binomial coefficient

$$\binom{p}{q} = \frac{p!}{q!(p-q)!}.$$

Transform of the Factored Form

When the Radon transform is applied to (3.1), the result is

$$\check{f}(p, \xi) = \mathcal{R}\{g_l(r) S_{lm}(\omega)\}$$

$$= \int g_l(r) S_{lm}(\omega)\, \delta(p - \xi \cdot \mathbf{x})\, d\mathbf{x}. \tag{3.5}$$

Without loss of generality, it is assumed as usual that ξ is a unit vector and $p \geqslant 0$. If for some reason $\check{f}$ is needed for negative p, the relation $\check{f}(-p, \xi) = \check{f}(p, -\xi)$ may be used. The integral (3.5) may be converted to spherical coordinates and written as

$$\check{f}(p, \xi) = \int_0^\infty dr\, r^{n-1} g_l(r) \int_{|\omega|=1} d\omega\, S_{lm}(\omega) \delta(p - r\xi \cdot \omega), \tag{3.6}$$

since $\mathbf{x} = r\omega$.

The integral from 0 to ∞ in (3.6) may be modified by observing that the δ function may be factored in various ways:

$$\delta(p - r\xi \cdot \omega) = \frac{1}{r} \delta\left(\frac{p}{r} - \xi \cdot \omega\right) \tag{3.7a}$$

or

$$\delta(p - r\xi \cdot \omega) = \frac{\delta\left(\dfrac{p}{\xi \cdot \omega} - r\right)}{|\xi \cdot \omega|}. \tag{3.7b}$$

Thus there is no contribution to the integral for $0 \leqslant r < |p|$, since $|\xi \cdot \omega| \leqslant 1$. Consequently, (3.6) may be rewritten as

$$\check{f}(p, \xi) = \int_{|p|}^\infty dr\, r^{n-2} g_l(r) \int_{|\omega|=1} d\omega\, S_{lm}(\omega) \delta\left(\frac{p}{r} - \xi \cdot \omega\right), \tag{3.8}$$

with $r \geqslant |p|$ always, so that p/r lies between -1 and $+1$.

The integration over the unit sphere can be done by the following somewhat indirect method. First, assume that the surface integral may be written in the form

$$S_{lm}(\xi) I_l(p/r) = \int_{|\omega|=1} S_{lm}(\omega) \delta\left(\frac{p}{r} - \xi \cdot \omega\right) d\omega, \tag{3.9}$$

which is quite general since $I_l(p/r)$ is unknown and the appearance of $S_{lm}(\xi)$ is to expected from considerations of symmetry. Next, multiply (3.9) by $S_{lm}(\xi)$, sum over m, and make use of the addition theorem [Hochstadt (1971); Erdélyi et al. (1955b)]

$$C_l^\nu(\xi \cdot \omega) = \frac{\mathfrak{a}_n}{\mathcal{N}} C_l^\nu(1) \sum_{m=1}^{\mathcal{N}} S_{lm}(\xi) S_{lm}(\omega), \tag{3.10}$$

where $C_l^\nu(t)$ is a Gegenbauer polynomial (see Appendix C), $\nu = (n-2)/2$, and $\mathfrak{a}_n$ is the surface area of an n-dimensional sphere. (See Remark 1 later in this section.) These manipulations applied to (3.9) yield

$$\sum_{m=1}^{\mathcal{N}} S_{lm}(\xi) S_{lm}(\xi) I_l\left(\frac{p}{r}\right) = \int_{|\omega|=1} \sum_{m=1}^{\mathcal{N}} S_{lm}(\xi) S_{lm}(\omega) \delta\left(\frac{p}{r} - \xi \cdot \omega\right) d\omega.$$

The left-hand side may be evaluated by setting $\omega = \xi$ in (3.10), and the right-hand side may be simplified by application of (3.10):

$$\frac{\mathcal{N}}{\mathfrak{a}_n} I_l\left(\frac{p}{r}\right) = \frac{\mathcal{N}}{\mathfrak{a}_n C_l^\nu(1)} \int_{|\omega|=1} C_l^\nu(\xi \cdot \omega) \delta\left(\frac{p}{r} - \xi \cdot \omega\right) d\omega.$$

To evaluate the integral on the right-hand side, we make use of the identity [John (1955); Courant and Hilbert (1962)]

$$\int_{|\omega|=1} F(\xi \cdot \omega)\, d\omega = \mathfrak{a}_{n-1} \int_{-1}^{+1} (1 - t^2)^{(n-3)/2} F(t)\, dt, \tag{3.11}$$

where $\xi \cdot \omega = t$. Thus

$$I_l\left(\frac{p}{r}\right) = \frac{\mathfrak{a}_{n-1}}{C_l^\nu(1)} \int_{-1}^{+1} C_l^\nu(t)(1 - t^2)^{(n-3)/2} \delta\left(\frac{p}{r} - t\right) dt.$$

Finally, since p/r lies between -1 and $+1$,

$$I_l\left(\frac{p}{r}\right) = \frac{\mathfrak{a}_{n-1}}{C_l^\nu(1)} C_l^\nu\left(\frac{p}{r}\right)\left(1 - \frac{p^2}{r^2}\right)^{\nu - 1/2}, \tag{3.12}$$

with $\nu = (n-2)/2$. Consequently, it follows that (3.8) becomes

$$\check{f}(p, \xi) = \frac{\mathfrak{a}_{n-1} S_{lm}(\xi)}{C_l^\nu(1)} \int_{|p|}^{\infty} r^{2\nu} g_l(r) C_l^\nu\left(\frac{p}{r}\right)\left(1 - \frac{p^2}{r^2}\right)^{\nu - 1/2} dr. \tag{3.13}$$

It will be convenient to write the preceding fundamental result as

$$\check{f}(p, \xi) = \mathscr{R}\{g_l(r) S_{lm}(\omega)\} = \check{g}_l(p) S_{lm}(\xi), \tag{3.14}$$

where $\check{g}_l(p)$ is given by a Gegenbauer transform:

$$\check{g}_l(p) = \frac{\mathfrak{a}_{n-1}}{C_l^\nu(1)} \int_{|p|}^{\infty} r^{2\nu} g_l(r) C_l^\nu\left(\frac{p}{r}\right)\left(1 - \frac{p^2}{r^2}\right)^{\nu - 1/2} dr. \tag{3.15}$$

The symmetry condition (3.3) along with the basic property $\check{f}(-p, \xi) = \check{f}(p, -\xi)$ yield another symmetry result:

$$\check{g}_l(-p) = (-1)^l \check{g}_l(p), \tag{3.16}$$

which will serve to define $\check{g}_l(-p)$. Note that while $\check{g}_l$ is defined for negative argument, thus far $g_l(r)$ is not, since we have only considered $r \geqslant 0$. In some of the following work, it will prove both useful and consistent to emphasize this by replacing $g_l(r)$ by $\mathfrak{u}(r)g_l(r)$

$$g_l(r) \to \mathfrak{u}(r)g_l(r), \tag{3.17}$$

where $\mathfrak{u}(r)$ is the unit step function. That is, $\mathfrak{u}$ is zero for negative argument and unity when the argument is positive,

$$\mathfrak{u}(t - t_0) = \begin{cases} 0 & t < t_0 \\ 1 & t \geqslant t_0. \end{cases} \tag{3.18}$$

Remark 1. An easy way to obtain an explicit expression for $\mathfrak{a}_n$ [Schwartz (1966)] is to observe that for $\mathbf{x} \in \mathbb{R}^n$, $t \in \mathbb{R}^1$,

$$\int e^{-\mathbf{x}\cdot\mathbf{x}}\, d\mathbf{x} = \int_0^\infty dr\, r^{n-1} e^{-r^2} \int_{|\omega|=1} d\omega$$

$$\left(\int_{-\infty}^{\infty} e^{-t^2}\, dt\right)^n = \mathfrak{a}_n \int_0^\infty r^{n-1} e^{-r^2}\, dr$$

$$(\sqrt{\pi})^n = \tfrac{1}{2}\mathfrak{a}_n \Gamma(\tfrac{n}{2}).$$

Thus

$$\mathfrak{a}_n = \frac{2\pi^{n/2}}{\Gamma(\frac{n}{2})}, \tag{3.19}$$

where Γ is the gamma function

$$\Gamma(z) = \int_0^\infty e^{-t} t^{z-1}\, dt.$$

Specifically, $\mathfrak{a}_1 = 2$, $\mathfrak{a}_2 = 2\pi$, $\mathfrak{a}_3 = 4\pi$, $\mathfrak{a}_4 = 2\pi^2$, $\mathfrak{a}_5 = 8\pi^2/3$, $\mathfrak{a}_6 = \pi^3$. □

Remark 2. The coefficient in front of (3.15) may be evaluated by use of (3.17), the explicit expression

$$C_l^\nu(1) = C_l^{(n-2)/2}(1) = \binom{l+n-3}{l},$$

and some gamma function identities [Hochstadt (1971)],

$$\frac{\mathfrak{a}_{n-1}}{C_l^\nu(1)} = \frac{(4\pi)^\nu l!\,\Gamma(\nu)}{(l+2\nu-1)!} (= \kappa_l^\nu \text{ for convenience}). \quad \square \tag{3.20}$$

The basic results (3.13)–(3.15) suggest, but of course do not prove, the

Hecke–Funk theorem [Hochstadt (1971); Erdélyi et al., (1953b)]

$$\int_{|\omega|=1} F(\xi\cdot\omega)S_{lm}(\omega)\,d\omega = \frac{\mathfrak{a}_{n-1}S_{lm}(\xi)}{C_l^\nu(1)}\int_{-1}^{+1}F(t)C_l^\nu(t)(1-t^2)^{\nu-1/2}\,dt. \tag{3.21}$$

Another Method

In fact, using (3.21), the major result (3.13) follows if a different order of integration is employed. Suppose the δ function in (3.6) is factored so the r integration may be done first, then

$$\begin{aligned}\check{f}(p,\xi) &= \int_{|\omega|=1} d\omega\, S_{lm}(\omega)\int_0^\infty dr\, r^{n-1}\mathfrak{u}(r)g_l(r)\frac{\delta\left(\frac{p}{\xi\cdot\omega}-r\right)}{|\xi\cdot\omega|}\\ &= \int_{|\omega|=1}\left(\frac{p}{\xi\cdot\omega}\right)^{n-1}\mathfrak{u}\left(\frac{p}{\xi\cdot\omega}\right)g_l\left(\frac{p}{\xi\cdot\omega}\right)\frac{S_{lm}(\omega)}{|\xi\cdot\omega|}d\omega. \end{aligned} \tag{3.22}$$

Note that the step function ensures the appropriate contribution after doing the integration over the δ function. At this point, it is appropriate to make use of (3.21), but there are two cases to consider: (i) $p>0$, and (ii) $p<0$.

If $p>0$ in (3.22), we have, from (3.21) [with $\kappa_l^\nu = \mathfrak{a}_{n-1}/C_l^\nu(1)$ for convenience],

$$\begin{aligned}\check{f}(p,\xi) &= \kappa_l^\nu S_{lm}(\xi)\int_{-1}^{+1}\left(\frac{p}{t}\right)^{n-1}\mathfrak{u}\left(\frac{p}{t}\right)g_l\left(\frac{p}{t}\right)C_l^\nu(t)(1-t^2)^{\nu-1/2}\frac{dt}{|t|}\\ &= \kappa_l^\nu S_{lm}(\xi)\int_0^1\left(\frac{p}{t}\right)^{n-1}g_l\left(\frac{p}{t}\right)C_l^\nu(t)(1-t^2)^{\nu-1/2}\frac{dt}{t}.\end{aligned}$$

By making the change of variables $t=p/r$, it follows immediately that

$$\check{f}(p,\xi) = \kappa_l^\nu S_{lm}(\xi)\int_p^\infty r^{n-2}g_l(r)C_l^\nu\left(\frac{p}{r}\right)\left(1-\frac{p^2}{r^2}\right)^{\nu-1/2}dr. \tag{3.23}$$

If $p<0$ in (3.22), then we may write $p=-|p|$, and (3.21) yields

$$\begin{aligned}\check{f}(p,\xi) &= \kappa_l^\nu S_{lm}(\xi)\int_{-1}^{+1}\left(\frac{-|p|}{t}\right)^{n-1}\mathfrak{u}\left(\frac{-|p|}{t}\right)g_l\left(\frac{-|p|}{t}\right)\\ &\quad\times C_l^\nu(t)(1-t^2)^{\nu-1/2}\frac{dt}{|t|}\\ &= \kappa_l^\nu S_{lm}(\xi)\int_{-1}^{0}\left(\frac{-|p|}{t}\right)^{n-1}g_l\left(\frac{-|p|}{t}\right)C_l^\nu(t)(1-t^2)^{\nu-1/2}\frac{dt}{|t|}.\end{aligned}$$

The change of variables $t \to -t$ gives

$$\check{f}(p, \xi) = \kappa_l^\nu S_{lm}(\xi) \int_0^1 \left(\frac{|p|}{t}\right)^{n-1} g_l\left(\frac{|p|}{t}\right) C_l^\nu(-t)(1 - t^2)^{\nu - 1/2} \frac{dt}{t},$$

and the change $t = |p|/r$ gives (for $p < 0$)

$$\check{f}(p, \xi) = (-1)^l \kappa_l^\nu S_{lm}(\xi) \int_{|p|}^{\infty} r^{n-2} g_l(r) C_l^\nu\left(\frac{|p|}{r}\right)\left(1 - \frac{p^2}{r^2}\right)^{\nu - 1/2} dr,$$

where the relation $C_l^\nu(-t) = (-1)^l C_l^\nu(t)$ has been used. By comparison with (3.23), it is clear that if the lower integration limit in (3.23) is replaced by $|p|$, then (3.23) serves for either positive or negative p, and we have reproduced (3.13) as desired.

[4.4] RELATION TO THE HOUGH TRANSFORM

Hough (1962) developed a transform for detecting straight lines in digital pictures. This transform has been discussed recently by several authors who are interested in image analysis [Pratt (1978); Duda and Hart (1972); Shapiro and Iannino (1979), and references therein]. It is clear now that this transform is a special case of the Radon transform [Deans (1981)]. The basic idea behind using the Hough transform or the Radon transform to identify lines in digital pictures is quite straightforward and follows directly from the definition.

Recall from [3.5] that the Radon transform of a function concentrated at a point $\delta(x - x_0)\delta(y - y_0)$ yields a sinusoidal curve

$$p = x_0 \cos\phi + y_0 \sin\phi$$

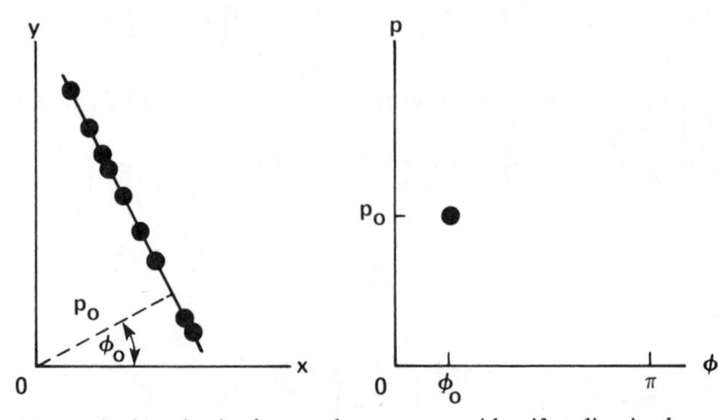

Figure 4.3 A single point in the ϕp plane serves to identify a line in the xy plane.

in the ϕp plane. Furthermore, all collinear points in the xy plane along the line determined by the fixed values ϕ_0 and p_0 also map to sinusoidal curves in the ϕp plane—and *they all intersect in the same point*: $(\phi, p) = (\phi_0, p_0)$. Thus, if we choose a suitable method for plotting $\check{f}$ as a function of ϕ and p, it follows that the Radon transform may be regarded as a line-to-point transformation, as indicated in Fig. 4.3. A single point in the ϕp plane (Radon space) contains information about the line in the xy plane (feature space).

An ongoing effort is being made to apply these ideas in other areas, including general curve and shape detection [Iannino and Shapiro (1978); Shapiro (1978), Sklansky (1978); Ballard (1981)], line, curve, and structure detection in noisy pictures [Cohen and Toussaint (1977); Shapiro (1975, 1978, 1979a); Zabele and Koplowitz (1979)], scene analysis [Dudani and Luk (1978); Duda, Nitzan, and Barrett (1979)], and image data compression [Shapiro (1979b)].

Chapter 5

Inversion

[5.1] INTRODUCTION

As mentioned in the preceding sections, in order to recover desired information about internal structure, it is necessary to invert the Radon transform, that is, to solve for f in terms of $\check{f}$. It will be convenient to consider the inversion formula in two parts, depending on whether $\mathbb{R}^n$ is of odd or even dimension. A unified formula valid for even or odd dimension can be given; it is most readily obtained by working from the two-part formula. Hence, in [5.2], the odd case is developed, and in [5.3], the more difficult even case is discussed.

The inversion method employed in [5.2] and [5.3] is patterned after that used by John (1955) with considerably more detail at certain stages. The formal unification in [5.4] and the Fourier methods of [5.5] are similar to the approach of Ludwig (1966), and the full unification in [5.4] is due to Helgason (1965, 1980).

Although stated earlier, it bears repeating that our present concern is with Euclidean space and the functions defined on $\mathbb{R}^n$ are from $\mathscr{S}$ or $\mathscr{D}$. This is especially important now to ensure that the various integrations in the following sections have meaning.

General References

Important general references for this chapter include John (1955), Ludwig (1966), Helgason (1965), and Courant and Hilbert (1962).

[5.2] ODD DIMENSION

The notation here is the same as that established in Chapter 2. In particular, $\mathbf{x}$, $\mathbf{y}$, and $\mathbf{z}$ are general vectors in $\mathbb{R}^n$, ξ is a unit vector, p is scalar, and $\Delta_{\mathbf{x}}$ is the Laplacian operator.

First, observe that for arbitrary $\mathbf{x} \in \mathbb{R}^n$

$$\int f(\mathbf{y})|\xi \cdot (\mathbf{y} - \mathbf{x})|\, d\mathbf{y} = \int d\mathbf{y}\, f(\mathbf{y}) \int_{-\infty}^{\infty} dp\, |p| \delta[p - \xi \cdot (\mathbf{y} - \mathbf{x})]$$

$$= \int_{-\infty}^{\infty} dp\, |p| \int d\mathbf{z}\, f(\mathbf{x} + \mathbf{z}) \delta(p - \xi \cdot \mathbf{z})$$

$$= \int_{-\infty}^{\infty} |p| \check{f}(p + \xi \cdot \mathbf{x}, \xi)\, dp, \qquad (2.1)$$

where the change $\mathbf{y} = \mathbf{x} + \mathbf{z}$ has been made in the second step and (5.2) of Chapter 3 has been used to obtain the final step. Next, integrate (2.1) over the unit* sphere S^{n-1}.

After integrating, (2.1) becomes

$$\int_{|\xi|=1} d\xi \int d\mathbf{y}\, f(\mathbf{y})|\xi \cdot (\mathbf{y} - \mathbf{x})| = \int_{|\xi|=1} d\xi \int_{-\infty}^{\infty} dp\, |p| \check{f}(p + \xi \cdot \mathbf{x}, \xi), \quad (2.2)$$

where $d\xi$ is the surface element on the unit sphere defined by $|\xi| = 1$. The integral on the left-hand side in (2.2) may be evaluated by use of the identity [Courant and Hilbert (1962)]

$$4(2\pi)^{n-1}(-1)^{(n-1)/2} f(\mathbf{x}) = \Delta_{\mathbf{x}}^{(n+1)/2} \int_{|\xi|=1} d\xi \int d\mathbf{y}\, f(\mathbf{y})|\xi \cdot (\mathbf{y} - \mathbf{x})|, \quad (2.3)$$

which is valid for odd $n \geqslant 3$.

After decomposing the Laplacian operator as

$$\Delta_{\mathbf{x}}^{(n+1)/2} = \Delta_{\mathbf{x}}^{(n-1)/2} \Delta_{\mathbf{x}},$$

(2.2) and (2.3) yield

$$4(2\pi)^{n-1}(-1)^{(n-1)/2} f(\mathbf{x})$$

$$= \Delta_{\mathbf{x}}^{(n-1)/2} \Delta_{\mathbf{x}} \int_{|\xi|=1} d\xi \int_{-\infty}^{\infty} dp\, |p| \check{f}(p + \xi \cdot \mathbf{x}, \xi). \qquad (2.4)$$

*The final result does not depend on the radius of the sphere; thus to carry the radius through the calculation would be unnecessarily cumbersome.

By a change of variables $p = t - \xi \cdot \mathbf{x}$,

$$\int_{-\infty}^{\infty} |p|\check{f}(p + \xi \cdot \mathbf{x}, \xi)\, dp = \int_{-\infty}^{\infty} |t - \xi \cdot \mathbf{x}|\check{f}(t, \xi)\, dt$$

$$= \int_{-\infty}^{\xi \cdot \mathbf{x}} |t - \xi \cdot \mathbf{x}|\check{f}(t, \xi)\, dt + \int_{\xi \cdot \mathbf{x}}^{\infty} (t - \xi \cdot \mathbf{x})\check{f}(t, \xi)\, dt$$

$$= \int_{\xi \cdot \mathbf{x}}^{\infty} (t - \xi \cdot \mathbf{x})\check{f}(t, \xi)\, dt - \int_{-\infty}^{\xi \cdot \mathbf{x}} (t - \xi \cdot \mathbf{x})\check{f}(t, \xi)\, dt.$$

Now, use of the Leibnitz rule for differentiating an integral gives

$$\Delta_{\mathbf{x}} \int_{-\infty}^{\infty} |p|\check{f}(p + \xi \cdot \mathbf{x}, \xi)\, dp = \Delta_{\mathbf{x}} \int_{\xi \cdot x}^{\infty} (t - \xi \cdot \mathbf{x})\check{f}(t, \xi)\, dt$$

$$- \Delta_{\mathbf{x}} \int_{-\infty}^{\xi \cdot \mathbf{x}} (t - \xi \cdot \mathbf{x})\check{f}(t, \xi)\, dt$$

$$= 2\xi \cdot \xi\, \check{f}(\xi \cdot \mathbf{x}, \xi).$$

Since $\xi \cdot \xi = 1$,

$$\Delta_{\mathbf{x}} \int_{-\infty}^{\infty} |p|\check{f}(p + \xi \cdot \mathbf{x}, \xi)\, dp = 2\check{f}(\xi \cdot \mathbf{x}, \xi). \tag{2.5}$$

Combining (2.5) with (2.4) gives the inversion formula for odd $n \geqslant 3$:

$$f(\mathbf{x}) = \mathcal{C}_n \Delta_{\mathbf{x}}^{(n-1)/2} \int_{|\xi|=1} \check{f}(\xi \cdot \mathbf{x}, \xi)\, d\xi. \tag{2.6}$$

Here

$$\mathcal{C}_n = \frac{(-1)^{(n-1)/2}}{2(2\pi)^{(n-1)}} = \frac{1}{2} \frac{1}{(2\pi i)^{n-1}} \tag{2.7}$$

and $i = \sqrt{-1}$.

An alternative way to write (2.6) is with the Laplacian operator taken inside the integral, then

$$f(\mathbf{x}) = \mathcal{C}_n \int_{|\xi|=1} \left(\frac{\partial}{\partial p}\right)^{n-1} \check{f}(\xi \cdot \mathbf{x}, \xi)\, d\xi. \tag{2.8}$$

Here it is understood that

$$\left(\frac{\partial}{\partial p}\right)^{n-1} \check{f}(p, \xi)\bigg|_{p=\xi \cdot \mathbf{x}} = \left(\frac{\partial}{\partial p}\right)^{n-1} \check{f}(\xi \cdot \mathbf{x}, \xi).$$

Remark. If the method used to go from (2.6) to (2.8) is not obvious, consider a less general case. For $n = 3$, $\mathbf{x} = (x_1, x_2, x_3) = (x, y, z)$, and

$$\Delta_{\mathbf{x}} \int_{|\xi|=1} \check{f}(\xi \cdot \mathbf{x}, \xi)\, d\xi = \int_{|\xi|=1} \left(\frac{\partial^2}{\partial x^2} + \frac{\partial^2}{\partial y^2} + \frac{\partial^2}{\partial z^2} \right) \check{f}(\xi \cdot \mathbf{x}, \xi)\, d\xi$$

$$= \int_{|\xi|=1} (\xi_1^2 + \xi_2^2 + \xi_3^2) \left. \frac{\partial^2 \check{f}}{\partial p^2}(p, \xi) \right|_{p=\xi \cdot \mathbf{x}} d\xi$$

$$= \int_{|\xi|=1} \frac{\partial^2 \check{f}}{\partial p^2}(\xi \cdot \mathbf{x}, \xi)\, d\xi.$$

The fact that ξ is a unit vector has been used to obtain the final result. The extension from 3 to n should be clear now. □

Example 1. Suppose $n = 3$, then $\mathcal{C}_n = \mathcal{C}_3 = -1/8\pi^2$, and the inversion formula is

$$f(\mathbf{x}) = \frac{-1}{8\pi^2} \Delta_{\mathbf{x}} \int_{|\xi|=1} \check{f}(\xi \cdot \mathbf{x}, \xi)\, d\xi$$

$$= \frac{-1}{8\pi^2} \int_{|\xi|=1} \check{f}_{pp}(\xi \cdot \mathbf{x}, \xi)\, d\xi,$$

where $\check{f}_{pp} = \partial^2 \check{f} / \partial p^2$. Even more explicitly,

$$f(x, y, z) = \frac{-1}{8\pi^2} \left(\frac{\partial^2}{\partial x^2} + \frac{\partial^2}{\partial y^2} + \frac{\partial^2}{\partial z^2} \right) \int_{|\xi|=1} \check{f}(\xi_1 x + \xi_2 y + \xi_3 z, \xi)\, d\xi.$$

Of course, ξ and $d\xi$ may be written in terms of usual $\mathbb{R}^3$ spherical coordinate polar angle θ and azimuthal angle ϕ:

$$\xi = (\sin\theta \cos\phi, \sin\theta \sin\phi, \cos\theta)$$

and

$$d\xi = \sin\theta \, d\theta \, d\phi.$$ □

Example 2. If $n > 3$, the *hyperspherical polar coordinates* defined in Chapter XI of Erdélyi et al. (1953b) may be used. These coordinates

$$(r, \theta_1, \theta_2, \ldots, \theta_{n-2}, \phi)$$

are defined by:

$$
\begin{aligned}
x_1 &= r\cos\theta_1\\
x_2 &= r\sin\theta_1\cos\theta_2\\
x_3 &= r\sin\theta_1\sin\theta_2\cos\theta_3\\
&\vdots\\
x_{n-2} &= r\sin\theta_1\sin\theta_2\ldots\sin\theta_{n-3}\cos\theta_{n-2}\\
x_{n-1} &= r\sin\theta_1\sin\theta_2\ldots\sin\theta_{n-2}\cos\phi\\
x_n &= r\sin\theta_1\sin\theta_2\ldots\sin\theta_{n-2}\sin\phi
\end{aligned}
$$

$$r \geqslant 0, \qquad 0 \leqslant 0_j \leqslant \pi, \qquad 0 \leqslant \phi < 2\pi.$$

Volume element:

$$d\mathbf{x} = r^{n-1}(\sin\theta_1)^{n-2}(\sin\theta_2)^{n-3}\ldots(\sin\theta_{n-2})\,dr\,d\theta_1\ldots d\theta_{n-2}\,d\phi.$$

Surface element on unit sphere: (times factor of r^{n-1} if $r \neq 1$)

$$d\omega = (\sin\theta_1)^{n-2}(\sin\theta_2)^{n-3}\ldots(\sin\theta_{n-2})\,d\theta_1\ldots d\theta_{n-2}\,d\phi. \qquad \square$$

[5.3] EVEN DIMENSION

The derivation for even n is not very different from that for odd n except that the identity (2.3) is different. Let us start with the observation

$$\int f(\mathbf{y})\log|\xi\cdot(\mathbf{y}-\mathbf{x})|\,d\mathbf{y} = \int d\mathbf{y} f(\mathbf{y})\int_{-\infty}^{\infty} dp\log|p|\,\delta[p-\xi\cdot(\mathbf{y}-\mathbf{x})]$$

$$= \int_{-\infty}^{\infty}\log|p|\,\check{f}(p+\xi\cdot\mathbf{x},\xi)\,dp. \tag{3.1}$$

Integrate (3.1) over a unit sphere S^{n-1} and invoke the even-n identity [Courant and Hilbert (1962)]

$$(2\pi)^n(-1)^{(n-2)/2}f(\mathbf{x}) = \Delta_{\mathbf{x}}^{n/2}\int_{|\xi|=1} d\xi\int d\mathbf{y}\,f(\mathbf{y})\log|\xi\cdot(\mathbf{y}-\mathbf{x})|. \tag{3.2}$$

From (3.1) and (3.2), it follows that

$$(2\pi)^n(-1)^{(n-2)/2}f(\mathbf{x}) = \Delta_{\mathbf{x}}^{(n-2)/2}\Delta_{\mathbf{x}}\int_{|\xi|=1} d\xi \int_{-\infty}^{\infty} dp \log|p|\, \check{f}(p + \xi\cdot\mathbf{x}, \xi). \tag{3.3}$$

Equation (3.3) is one form of the desired result, but it can be put in a different form by examining

$$\Delta_{\mathbf{x}} I(\mathbf{x}) = \Delta_{\mathbf{x}} \int_{-\infty}^{\infty} \log|p|\, \check{f}(p + \xi\cdot\mathbf{x}, \xi)\, dp. \tag{3.4}$$

This equation defines $I(\mathbf{x})$. There are two ways to proceed with simplification of (3.4). One may change variables $p = t - \xi\cdot\mathbf{x}$ and write

$$I(\mathbf{x}) = \int_{-\infty}^{\infty} \log|t - \xi\cdot\mathbf{x}|\, \check{f}(t, \xi)\, dt \tag{3.5}$$

before applying $\Delta_{\mathbf{x}}$, or one may apply $\Delta_{\mathbf{x}}$ first and then make a change of variables. The former approach leads to the inversion formula given on page 11 of Gel'fand et al. (1966). We shall employ the latter method and write

$$\Delta_{\mathbf{x}} I(\mathbf{x}) = \int_{-\infty}^{\infty} \log|p|\, \check{f}_{tt}(p + \xi\cdot\mathbf{x}, \xi)\, dp, \tag{3.6}$$

where we have used the relationship

$$\begin{aligned}
\Delta_{\mathbf{x}} \check{f}(p + \xi\cdot\mathbf{x}, \xi) &= \sum_{j=1}^{n} \frac{\partial}{\partial x_j}\frac{\partial}{\partial x_j}\check{f}(p + \xi\cdot\mathbf{x}, \xi) \\
&= |\xi|^2 \frac{\partial^2}{\partial t^2}\check{f}(t, \xi)\bigg|_{t=p+\xi\cdot\mathbf{x}} \\
&= \check{f}_{tt}(p + \xi\cdot\mathbf{x}, \xi).
\end{aligned}$$

The change of variables $p = t - \xi\cdot\mathbf{x}$ in (3.6) followed by an integration by parts yields

$$\begin{aligned}
\Delta_{\mathbf{x}} I(\mathbf{x}) &= \int_{-\infty}^{\infty} \log|t - \xi\cdot\mathbf{x}|\, \check{f}_{tt}(t, \xi)\, dt \\
&= -\int_{-\infty}^{\infty} \frac{\check{f}_p(p, \xi)}{p - \xi\cdot\mathbf{x}}\, dp.
\end{aligned} \tag{3.7}$$

Here, the Cauchy principal value is understood and $\check{f}_p(p, \xi) = \partial\check{f}(p, \xi)/\partial p$.

When the result obtained in (3.7) is substituted into (3.3), the inversion formula for even n is obtained,

$$f(\mathbf{x}) = \frac{\mathcal{C}_n}{i\pi} \Delta_{\mathbf{x}}^{(n-2)/2} \int_{|\xi|=1} d\xi \int_{-\infty}^{\infty} dp \, \frac{\check{f}_p(p,\xi)}{p - \xi \cdot \mathbf{x}}, \tag{3.8}$$

where $\mathcal{C}_n$ is still defined as in (2.7). If we now observe that $\Delta_{\mathbf{x}}$ operating on the integral over p followed by two integrations by parts yields

$$\Delta_{\mathbf{x}} \int_{-\infty}^{\infty} \frac{\check{f}_p(p,\xi)}{p - \xi \cdot \mathbf{x}} dp = \int_{-\infty}^{\infty} \frac{\check{f}_{ppp}(p,\xi)}{p - \xi \cdot \mathbf{x}} dp,$$

it follows that

$$f(\mathbf{x}) = \frac{\mathcal{C}_n}{i\pi} \int_{|\xi|=1} d\xi \int_{-\infty}^{\infty} dp \, \frac{\left(\dfrac{\partial}{\partial p}\right)^{n-1} \check{f}(p,\xi)}{p - \xi \cdot \mathbf{x}}. \tag{3.9}$$

Example 1. Suppose $n = 2$. Then

$$f(\mathbf{x}) = f(x, y) = \frac{-1}{4\pi^2} \int_0^{2\pi} d\phi \int_{-\infty}^{\infty} dp \, \frac{\check{f}_p(p,\xi)}{p - \xi \cdot \mathbf{x}}, \tag{3.10}$$

where $\xi = (\cos\phi, \sin\phi)$ and $\xi \cdot \mathbf{x} = x\cos\phi + y\sin\phi$. By making the appropriate changes of variables of integration, (3.10) may be written as

$$f(x, y) = \frac{-1}{2\pi^2} \int_0^{\pi} d\phi \int_{-\infty}^{\infty} dp \, \frac{\check{f}_p(p,\xi)}{p - \xi \cdot \mathbf{x}}, \tag{3.11}$$

or

$$f(x, y) = \frac{-1}{2\pi^2} \int_{-\pi/2}^{\pi/2} d\phi \int_{-\infty}^{\infty} dp \, \frac{\check{f}_p(p,\xi)}{p - \xi \cdot \mathbf{x}}. \tag{3.12}$$

If the replacements $x = r\cos\theta$, $y = r\sin\theta$ are made, then f in polar coordinates becomes

$$f(r, \theta) = \frac{-1}{2\pi^2} \int_0^{\pi} d\phi \int_{-\infty}^{\infty} dp \, \frac{\check{f}_p(p,\xi)}{p - r\cos(\phi - \theta)}. \qquad \square \tag{3.13}$$

These formulas for $n = 2$ constitute a solution to the problem of *reconstruction from projections.*

[5.4] UNIFICATION AND THE ADJOINT

Odd Dimension

It appears that the inversion formula (2.8) for odd n is different from the corresponding formula (3.9) for even n. It would be useful to have a single formula valid for even or odd dimension. Such unification is possible in both formal and fundamental ways [Ludwig (1966); Helgason (1965, 1980)].

First consider an arbitrary function of ξ and $p = \xi \cdot \mathbf{x}$ such that $\psi(p, \xi) = \psi(-p, -\xi)$. Define the integral operator $\mathscr{R}^\dagger$ and the new function of $\mathbf{x}$ by the equation

$$\hat{\psi}(\mathbf{x}) = \mathscr{R}^\dagger \psi = \int_{|\xi|=1} \psi(\xi \cdot \mathbf{x}, \xi)\, d\xi. \tag{4.1}$$

Later in this section, we will see that $\mathscr{R}^\dagger$ may be interpreted as the *adjoint* operator for the Radon transform.

Example 1. Let $n = 3$ and consider the Laplacian $\Delta = \Delta_{\mathbf{x}}$ operating on $\hat{\psi}(\mathbf{x})$,

$$\begin{aligned} \Delta\hat{\psi}(\mathbf{x}) &= \int_{|\xi|=1} |\xi|^2 \left.\frac{\partial^2 \psi}{\partial p^2}(p, \xi)\right|_{p=\xi\cdot\mathbf{x}} d\xi \\ &= \mathscr{R}^\dagger \Box \psi(p, \xi), \end{aligned}$$

where we have used $|\xi| = 1$ and reintroduced the $\Box$ operator defined at the end of [3.6]. Another way to write this result is

$$\Delta\hat{\psi} = (\Box\psi)\hat{}\,.$$

When this equation is compared with (6.12) from [3.6],

$$(\Delta f)\check{} = \Box \check{f}$$

one can observe how the $\mathscr{R}$ and $\mathscr{R}^\dagger$ operations *intertwine* Δ and $\Box$ [Helgason (1980)]. □

Example 2. The inversion formulas (2.6) and (2.8) may be expressed as

$$f(\mathbf{x}) = \mathcal{C}_n \Delta_{\mathbf{x}}^{(n-1)/2} \mathscr{R}^\dagger \check{f}(p, \xi) = \mathcal{C}_n \Delta_{\mathbf{x}}^{(n-1)/2} (\check{f})\hat{} = \mathcal{C}_n (\Box^{(n-1)/2} \check{f})\hat{}\,. \qquad \Box$$

Example 3. (See Appendix A.) Let $n = 3$ and consider $\mathscr{R}^\dagger$ operating on $\check{f}(p, \xi)$. Define

$$F(\mathbf{x}) = \frac{1}{4\pi} \mathscr{R}^\dagger \check{f} = \frac{1}{4\pi} \int_{|\xi|=1} \check{f}(\xi \cdot \mathbf{x}, \xi)\, d\xi.$$

However, $\check{f}(p, \xi) = \mathcal{R}f(\mathbf{x})$, so

$$F(\mathbf{x}) = \frac{1}{4\pi}\int_{|\xi|=1} d\xi \int d\mathbf{x}' f(\mathbf{x}')\delta(\xi \cdot \mathbf{x} - \xi \cdot \mathbf{x}')$$

$$= \frac{1}{4\pi}\int d\mathbf{x}' \frac{f(\mathbf{x}')}{|\mathbf{x} - \mathbf{x}'|}\int_{|\xi|=1} d\xi\, \delta(\xi \cdot \mathbf{n}),$$

where $\mathbf{n}$ is the unit vector $\mathbf{n} = (\mathbf{x} - \mathbf{x}')/|\mathbf{x} - \mathbf{x}'|$. The integral over the unit sphere yields 2π. Suppose $\mathbf{n}$ is along the z axis,

$$\int_{|\xi|=1} d\xi\, \delta(\xi \cdot \mathbf{n}) = \int_0^{2\pi}\int_0^{\pi} \sin\theta\, \delta(\cos\theta)\, d\theta\, d\phi$$

$$= 2\pi\int_{-1}^{+1}\sqrt{1 - x^2}\,\delta(x)\, dx$$

$$= 2\pi.$$

Due to the symmetry of the problem, the same result must follow regardless of the direction of $\mathbf{n}$. Thus

$$F(\mathbf{x}) = \frac{1}{2}\int \frac{f(\mathbf{x}')}{|\mathbf{x} - \mathbf{x}'|} d\mathbf{x}',$$

or, in terms of $\mathbf{x} = (x, y, z)$ and $\mathbf{x}' = (x', y', z')$,

$$F(x, y, z) = \frac{1}{2}\iiint_{-\infty}^{+\infty} \frac{f(x', y', z')\, dx'\, dy'\, dz'}{\left[(x - x')^2 + (y - y')^2 + (z - z')^2\right]^{1/2}}.$$

Hence F may be written as a three-dimensional convolution

$$F = f{*}{*}{*}\frac{1}{r}.$$

Furthermore, observe that $2F$ is precisely the analog of the electrostatic potential where the electric charge density is given by f. It follows from electrostatics that [Jackson (1975), p. 39]

$$f = \frac{-1}{4\pi}\Delta(2F),$$

where Δ is the Laplacian. Explicitly,

$$f(x, y, z) = \frac{-1}{2\pi}\left(\frac{\partial^2 F}{\partial x^2} + \frac{\partial^2 F}{\partial y^2} + \frac{\partial^2 F}{\partial z^2}\right).$$

The result here may be compared with Example 1 of [5.2]. □

Next, define the operator Υ_o such that Υ_o operating on any function of the real variable p yields a function $\bar{g}$ of the real variable t,

$$\bar{g}(t) = \Upsilon_o g(p) = \mathcal{C}_n \left(\frac{\partial}{\partial p} \right)^{n-1} g(p) \bigg|_{p=t} \tag{4.2a}$$

$$= \mathcal{C}_n \Box^{(n-1)/2} g(p)|_{p=t}. \tag{4.2b}$$

Clearly, t may be replaced by an arbitrary scalar, for example, p or $\xi \cdot \mathbf{x}$. In terms of the newly defined operators, the inversion formula (2.8) may be written as

$$f(\mathbf{x}) = \int_{|\xi|=1} \bar{\check{f}}(\xi \cdot \mathbf{x}, \xi)\, d\xi$$

$$= \mathcal{R}^{\dagger} \bar{\check{f}}(t, \xi)$$

$$= \mathcal{R}^{\dagger} \Upsilon_o \check{f}(p, \xi). \tag{4.3}$$

Here we use the notation

$$\Upsilon_o \check{f} = \bar{\check{f}}. \tag{4.4}$$

Moreover, since $\check{f} = \mathcal{R}f$, we may write

$$f = \mathcal{R}^{\dagger} \Upsilon_o \mathcal{R} f, \tag{4.5}$$

which suggests the operator identity

$$\mathcal{R}^{\dagger} \Upsilon_o \mathcal{R} = I \tag{4.6}$$

when applied to $f = f(\mathbf{x})$ defined on $\mathbb{R}^n$.

Remark. It is amusing to observe that the notation has been established in such a way that

$$f = \hat{\bar{\check{f}}}. \qquad \Box \tag{4.7}$$

Example 4. Consider the function of $\mathbf{x}$ obtained by applying $\mathcal{R}^{\dagger}$ to the differentiable function ψ (on $\mathbb{R}^1 \times \mathsf{S}^{n-1}$). Suppose ψ has the same properties as the Radon transform of some function, that is, $\psi(-p, -\xi) = \psi(p, \xi)$. Now apply (4.6) to $\hat{\psi}(\mathbf{x}) = \mathcal{R}^{\dagger}\psi$ [note that I of (4.6) *must* act on functions of $\mathbf{x}$]:

$$\mathcal{R}^{\dagger}\psi = I\mathcal{R}^{\dagger}\psi = \mathcal{R}^{\dagger}\Upsilon_o \mathcal{R}\mathcal{R}^{\dagger}\psi.$$

Thus

$$\psi = \Upsilon_o \mathcal{R} \mathcal{R}^\dagger \psi. \tag{4.8}$$

Hence another operator identity is

$$\Upsilon_o \mathcal{R} \mathcal{R}^\dagger = I. \tag{4.9}$$

This operator may be applied only to symmetric differentiable functions such as ψ defined on $\mathbb{R}^1 \times \mathsf{S}^{n-1}$. □

A final observation is in order. Note that (4.8) may be written in the form

$$\psi = C_n \square^{(n-1)/2} \mathcal{R} \hat{\psi} = \mathcal{C}_n \square^{(n-1)/2} (\hat{\psi})^{\vee}. \tag{4.10}$$

This formula along with the formula from Example 2,

$$f = \mathcal{C}_n \Delta^{(n-1)/2} (\check{f})^{\wedge}, \tag{4.11}$$

constitute major results of this chapter.

Even Dimension

The result corresponding to (4.3) for even n may be obtained by starting with (3.9) and the definition of the Hilbert transform [Bracewell (1978)]:

$$g_{\mathrm{H}}(t) = \mathcal{H} g = \frac{1}{\pi} \int_{-\infty}^{\infty} \frac{g(p)}{p-t}\, dp, \tag{4.12}$$

where the Cauchy principle value is understood. Note that the function $g(p)$ is converted to a function of t in this case. Of course if $t = \boldsymbol{\xi} \cdot \mathbf{x}$, then $g_{\mathrm{H}} = \mathcal{H} g$ is a function of $\boldsymbol{\xi} \cdot \mathbf{x}$. This may be denoted $g_{\mathrm{H}}(\boldsymbol{\xi} \cdot \mathbf{x})$ or $[\mathcal{H} g](\boldsymbol{\xi} \cdot \mathbf{x})$. Thus (3.9) may be written as

$$f(\mathbf{x}) = \frac{\mathcal{C}_n}{i} \int_{|\xi|=1} d\xi \left[\mathcal{H} \left\{ \left(\frac{\partial}{\partial p} \right)^{n-1} \check{f}(p, \xi) \right\} \right] (\xi \cdot \mathbf{x}, \xi). \tag{4.13}$$

If we define the operator Υ_e acting on some arbitrary function $g(p)$ by the prescription

$$\bar{g}(t) = \Upsilon_e g(p) = \frac{\mathcal{C}_n}{i} \left[\mathcal{H} \left\{ \left(\frac{\partial}{\partial p} \right)^{n-1} g(p) \right\} \right] (t), \tag{4.14}$$

then it is clear that for even n, $f(\mathbf{x})$ may be expressed as

$$\begin{aligned} f(\mathbf{x}) &= \int_{|\xi|=1} \bar{\check{f}}(\xi \cdot \mathbf{x}, \xi)\, d\xi \\ &= \mathscr{R}^{\dagger}\bar{\check{f}}(t, \xi) \\ &= \mathscr{R}^{\dagger}\Upsilon_e \check{f}(p, \xi), \end{aligned} \tag{4.15}$$

which is formally in agreement with (4.3).

Formal Unification

Suppose we define Υ by

$$\bar{g}(t) = \Upsilon g = \begin{cases} \Upsilon_o g = \mathcal{C}_n \left(\dfrac{\partial}{\partial p}\right)^{n-1} g(p)\Big|_{p=t} & (n \text{ odd}) \\ \Upsilon_e g = \dfrac{\mathcal{C}_n}{i}\left[\mathcal{H}\left\{\left(\dfrac{\partial}{\partial p}\right)^{n-1} g(p)\right\}\right](t) & (n \text{ even}) \end{cases} \tag{4.16}$$

with $\mathcal{C}_n = \frac{1}{2}(2\pi i)^{1-n}$ as in (2.7). Then symbolically,

$$f = \mathscr{R}^{\dagger}\Upsilon\mathscr{R}f = \mathscr{R}^{\dagger}\Upsilon\check{f} = \mathscr{R}^{\dagger}\bar{\check{f}} = \hat{\bar{\check{f}}} \tag{4.17}$$

for even or odd n. These results are equivalent to those developed by Ludwig (1966). There are some obvious notational differences and a sign difference in the definition of the Hilbert transform.

Full Unification

The unification has been carried even further by Helgason (1965), where in part (ii) of his theorem (4.5), he shows that the inversion formula

$$f(\mathbf{x}) = \mathcal{C}_n \Delta_{\mathbf{x}}^{(n-1)/2} \int_{|\xi|=1} \check{f}(\xi \cdot \mathbf{x}, \xi)\, d\xi$$

also holds for even n if one uses fractional powers of the Laplacian. Also, see page 123 of Helgason (1973) and page 20 of Helgason (1980). The fractional derivatives are defined in terms of Riesz potentials, which are discussed in detail in Helgason (1980).

For the special case where $f(\mathbf{x})$ can be factored into a product of a radial function multiplied by an angular function, Deans (1978, 1979) has given an inversion formula valid for even or odd dimension that does not involve fractional derivatives. If $f(\mathbf{x})$ is written as in (3.1) of [4.3]:

$$f(\mathbf{x}) = g_l(r)S_{lm}(\omega),$$

then $\check{f}(p, \xi)$ also admits to a similar decomposition, as can be seen from (3.14) of [4.3]:

$$\check{f}(p, \xi) = \check{g}_l(p)S_{lm}(\xi).$$

The functions $g_l(r)$ and $\check{g}_l(p)$ are a Gegenbauer transform pair [Deans (1978, 1979)]:

$$\check{g}_l(p) = \frac{(4\pi)^\nu \Gamma(l+1)\Gamma(\nu)}{\Gamma(l+2\nu)} \int_p^\infty r^{2\nu} g_l(r) C_l^\nu\left(\frac{p}{r}\right)\left(1 - \frac{p^2}{r^2}\right)^{\nu - 1/2} dr, \tag{4.18a}$$

$$g_l(r) = \frac{(-1)^{2\nu+1}\Gamma(l+1)\Gamma(\nu)}{2\pi^{\nu+1}\Gamma(l+2\nu)r} \int_r^\infty \check{g}_l^{(2\nu+1)}(t) C_l^\nu\left(\frac{t}{r}\right)\left(\frac{t^2}{r^2} - 1\right)^{\nu - 1/2} dt, \tag{4.18b}$$

with both r and $p \geqslant 0$ and $\check{g}_l^{(2\nu+1)}(t) = (d/dt)^{2\nu+1}\check{g}_l(t)$. This result is valid for even or odd values of n; as usual, $\nu = (n-2)/2$. A direct verification that (4.18b) satisfies (4.18a) can be accomplished using the following formula [Deans (1978); Montaldi (1979)] for $0 < s \leqslant t$:

$$\int_s^t x^{2\nu-1} C_l^\nu\left(\frac{s}{x}\right) C_l^\nu\left(\frac{t}{x}\right)\left[1 - \frac{s^2}{x^2}\right]^{\nu-1/2}\left[\frac{t^2}{x^2} - 1\right]^{\nu-1/2} dx$$

$$= \frac{\pi}{2^{2\nu-1}}\left[\frac{\Gamma(l+2\nu)}{\Gamma(l+1)\Gamma(\nu)}\right]^2 \frac{(t-s)^{2\nu}}{\Gamma(2\nu+1)}. \tag{4.19}$$

The Adjoint

Let $f(\mathbf{x})$ and $h(\mathbf{x})$ be C^∞ functions on $\mathbb{R}^n$, and let $\langle \cdot, \cdot \rangle$ designate an inner product on $\mathbb{R}^n$. In the transform domain, let $[\cdot, \cdot]$ designate an inner product on $\mathbb{R}^1 \times \mathsf{S}^{n-1}$, then $\mathscr{R}^\dagger$ is the adjoint operator in the sense that

$$\langle h, \mathscr{R}^\dagger \psi \rangle = [\mathscr{R}h, \psi], \tag{4.20}$$

where $\psi(p, \xi) = \mathscr{R}f(\mathbf{x})$. To see why this result is expected, consider the following integrations (neglecting possible weight functions)

$$\begin{aligned}
\langle h, \mathscr{R}^{\dagger}\psi\rangle &= \int d\mathbf{x}\, h(\mathbf{x})\hat{\psi}(\mathbf{x}) \\
&= \int d\mathbf{x}\, h(\mathbf{x}) \int_{|\xi|=1} d\xi\, \psi(\xi\cdot\mathbf{x}, \xi) \\
&= \int d\mathbf{x}\, h(\mathbf{x}) \int_{|\xi|=1} d\xi \int_{-\infty}^{\infty} dp\, \psi(p,\xi)\delta(p-\xi\cdot\mathbf{x}) \\
&= \int_{|\xi|=1} d\xi \int_{-\infty}^{\infty} dp \int d\mathbf{x}\, h(\mathbf{x})\delta(p-\xi\cdot\mathbf{x})\psi(p,\xi) \\
&= \int_{|\xi|=1} d\xi \int_{-\infty}^{\infty} dp\, \check{h}(p,\xi)\psi(p,\xi) \\
&= [\mathscr{R}h, \psi].
\end{aligned}$$

This same result can be expressed as $\langle h, \hat{\psi}\rangle = [\check{h}, \psi]$.

Note that this provides a convenient way to consider the Radon transform of distributions (Appendix B). Suppose, for example, that

$$h(\mathbf{x}) = \delta(\mathbf{x} - \mathbf{a}),$$

where $\mathbf{a}$ is a constant vector. Then

$$[\mathscr{R}\delta, \psi] = \langle \delta, \hat{\psi}\rangle = \hat{\psi}(\mathbf{a}).$$

For further discussion of transforms of distributions, see [8.2].

Finally, observe that if $\mathscr{R}^{\dagger}\psi = h$ in (4.20), we obtain a *Parseval relation* for the Radon transform:

$$\langle h, h\rangle = [\mathscr{R}h, \psi] = [\mathscr{R}h, \Upsilon\mathscr{R}h]. \tag{4.21}$$

Note: To solve for ψ in (4.21), use the identity

$$\psi = \Upsilon\mathscr{R}\mathscr{R}^{\dagger}\psi, \tag{4.22}$$

which follows immediately from (4.17) with $f = \mathscr{R}^{\dagger}\psi$.

[5.5] FOURIER METHODS

We have already seen in [4.2] that the Fourier transform of $\check{f}(p, \xi)$ along the radial direction yields the transformed function $\tilde{f}(k\xi)$

$$\tilde{f}(k\xi) = \int_{-\infty}^{\infty} \check{f}(p,\xi) e^{-i2\pi kp}\, dp. \tag{5.1}$$

In terms of $\mathbf{k} = k\boldsymbol{\xi}$, $\tilde{f}(k\boldsymbol{\xi}) = \tilde{f}(\mathbf{k})$. The inverse Fourier transform gives

$$f(\mathbf{x}) = \int \tilde{f}(\mathbf{k}) e^{i2\pi \mathbf{k}\cdot\mathbf{x}}\, d\mathbf{k}. \tag{5.2}$$

Thus, given $\check{f}$, it is possible to recover f by a radial Fourier transform followed by an n-dimensional inverse Fourier transform. Symbolically,

$$f = \mathscr{F}_n^{-1}\mathscr{F}_1 \check{f}. \tag{5.3}$$

Note that an interpolation in Fourier space is implied by (5.3) since one must go from $\tilde{f}(k\boldsymbol{\xi})$ to $\tilde{f}(\mathbf{k})$ in order to make use of (5.2).

On the other hand, by working in hyperspherical polar coordinates in Fourier space, it is possible to obtain the same inversion formulas as those already developed in the preceding sections. This approach [Ludwig (1966)] will be presented in the interest of completeness and to further emphasize both the similarities and differences between Radon and Fourier transforms.

As a starting point, observe that (5.2) may be written as

$$f(\mathbf{x}) = \int_{|\boldsymbol{\xi}|=1}\int_0^\infty \tilde{f}(k\boldsymbol{\xi}) e^{i2\pi k\boldsymbol{\xi}\cdot\mathbf{x}} k^{n-1}\, dk\, d\boldsymbol{\xi}.$$

If $\tilde{f}(k\boldsymbol{\xi})$ is replaced by the right-hand side of (5.1), it follows that

$$f(\mathbf{x}) = \int_{|\boldsymbol{\xi}|=1} d\boldsymbol{\xi}\left[\int_0^\infty dk\, k^{n-1} e^{i2\pi k\boldsymbol{\xi}\cdot\mathbf{x}} \int_{-\infty}^\infty dp\, e^{-i2\pi kp} \check{f}(p, \boldsymbol{\xi})\right].$$

Clearly, the integrals over k and p (in square brackets) leave some function of $\boldsymbol{\xi}\cdot\mathbf{x} = t$ and $\boldsymbol{\xi}$, which may be written as $\mathfrak{g}(t, \boldsymbol{\xi})$. Moreover, since this function is to be integrated over the unit sphere, it is sufficient to work with its even part in $\boldsymbol{\xi}$. Thus we replace $\mathfrak{g}(t, \boldsymbol{\xi})$ by

$$\mathfrak{g}_e(t, \boldsymbol{\xi}) = \tfrac{1}{2}[\mathfrak{g}(t, \boldsymbol{\xi}) + \mathfrak{g}(-t, -\boldsymbol{\xi})]$$

and

$$\mathfrak{g}_e(t, \boldsymbol{\xi}) = \frac{1}{2}\int_0^\infty dk\, k^{n-1}\int_{-\infty}^\infty dp\, \check{f}(p, \boldsymbol{\xi}) e^{i2\pi k(t-p)}$$

$$+ \frac{1}{2}\int_0^\infty dk\, k^{n-1}\int_{-\infty}^\infty dp\, \check{f}(p, -\boldsymbol{\xi}) e^{-i2\pi k(t+p)}.$$

The change of variables $k \to -k$, $p \to -p$ in the second term coupled with the symmetry condition $\check{f}(-p, -\boldsymbol{\xi}) = \check{f}(p, \boldsymbol{\xi})$ and the observation that k^{n-1} may be replaced by $|k|^{n-1}$ in the first integral leads to the formula

$$\mathfrak{g}_e(t, \boldsymbol{\xi}) = \frac{1}{2}\int_{-\infty}^\infty dk\, |k|^{n-1}\int_{-\infty}^\infty dp\, \check{f}(p, \boldsymbol{\xi}) e^{i2\pi k(t-p)}. \tag{5.4}$$

Until now, nothing has been mentioned about n being even or odd. If n is odd, $n - 1$ integrations by parts with respect to the p variable yield

$$\mathcal{g}_e(t, \xi) = \frac{1}{2} \frac{1}{(2\pi i)^{n-1}} \int_{-\infty}^{\infty} dk \int_{-\infty}^{\infty} dp\, e^{-i2\pi k(p-t)} \left(\frac{\partial}{\partial p} \right)^{n-1} \check{f}(p, \xi)$$

$$= \mathcal{C}_n \int_{-\infty}^{\infty} \left(\frac{\partial}{\partial p} \right)^{n-1} \check{f}(p, \xi) \delta(p - t)\, dp$$

$$= \mathcal{C}_n \left(\frac{\partial}{\partial t} \right)^{n-1} \check{f}(t, \xi),$$

where use has been made of Appendix B. By comparison with (4.16), it is clear that

$$\mathcal{g}_e(t, \xi) = \bar{\check{f}}(t, \xi),$$

and $f(\mathbf{x})$ follows from $\mathcal{R}^\dagger \bar{\check{f}}$ as in [5.4].

If n is even, the $n - 1$ integrations by parts applied to (5.4) lead to

$$\mathcal{g}_e(t, \xi) = \mathcal{C}_n \int_{-\infty}^{\infty} dk\, \mathrm{sgn}(k) \int_{-\infty}^{\infty} dp\, e^{-i2\pi k(p-t)} \left(\frac{\partial}{\partial p} \right)^{n-1} \check{f}(p, \xi)$$

$$= \frac{\mathcal{C}_n}{i\pi} \int_{-\infty}^{\infty} \frac{(\partial/\partial p)^{n-1} \check{f}(p, \xi)}{p - t}\, dp. \tag{5.5}$$

Appendix B has been used, and the Cauchy principle value is understood. Thus (4.16) still holds, and $f = \mathcal{R}^\dagger \bar{\check{f}}$.

Example 1. Suppose $n = 2$, then ξ is the unit vector $\xi = (\cos\phi, \sin\phi)$, and from (5.5),

$$\mathcal{g}_e(t, \xi) = \bar{\check{f}}(t, \xi) = \frac{-1}{4\pi^2} \int_{-\infty}^{\infty} \frac{\partial \check{f}(p, \xi)/\partial p}{p - t}\, dp$$

$$= \frac{-1}{4\pi} \mathcal{H} \left[\frac{\partial \check{f}}{\partial p}(p, \xi) \right],$$

where $\mathcal{H}$ designates the Hilbert transform. Thus

$$f(x, y) = \mathcal{R}^\dagger \bar{\check{f}}(p, \xi) = \int_0^{2\pi} \bar{\check{f}}(\xi \cdot \mathbf{x}, \xi)\, d\phi. \qquad \square$$

This result should be compared with the result obtained in Example 1 of [5.3].

Recall that the convolution of two functions $f_1(t)$ and $f_2(t)$ yields a new function of t:

$$f_1(t)*f_2(t) = \int_{-\infty}^{\infty} f_1(p) f_2(t-p)\, dp. \tag{5.6}$$

It follows that (5.5) may be expressed as the convolution

$$\bar{\check{f}}(t,\xi) = \frac{-\mathcal{C}_n}{i\pi} f_1(t)*f_2(t), \tag{5.7}$$

where

$$f_1(t) = \left(\frac{\partial}{\partial t}\right)^{n-1} \check{f}(t,\xi) \tag{5.8}$$

and

$$f_2(t) = \frac{1}{t}. \tag{5.9}$$

The Cauchy principle value is understood. Also recall that the derivative of a convolution may be expressed as [Bracewell (1978)]

$$\frac{d}{dt}[f_1(t)*f_2(t)] = f_1'(t)*f_2(t) = f_1(t)*f_2'(t). \tag{5.10}$$

An application of this observation is contained in Example 2.

Example 2. Let $n = 2$ and let f_1 and f_2 be defined as in (5.8) and (5.9). Here $C_2 = 1/4\pi i$ and

$$\begin{aligned}
\bar{\check{f}}(t,\xi) &= \frac{-1}{4\pi}\mathcal{H}\left(\frac{\partial \check{f}}{\partial p}\right) = \frac{1}{4\pi^2}\mathcal{P}\int_{-\infty}^{\infty} \frac{\dfrac{\partial \check{f}(p,\xi)}{\partial p}}{t-p}\, dp \\
&= \frac{1}{4\pi^2}\frac{\partial \check{f}}{\partial t} * \mathcal{P}\left(\frac{1}{t}\right) \\
&= \frac{1}{4\pi^2}\frac{\partial}{\partial t}\mathcal{P}\left[\check{f}(t,\xi) * \frac{1}{t}\right] \\
&= \frac{1}{4\pi^2}\check{f}(t,\xi) * \frac{\partial}{\partial t}\mathcal{P}\left(\frac{1}{t}\right).
\end{aligned}$$

The Cauchy principle value has been shown explicitly this time (see Appendix B). □

Chapter 6

Recent Development of Inversion Methods

[6.1] INTRODUCTION

Although the basic formulas for inversion of the Radon transform were worked out as early as 1917 [Appendix A and Radon (1917)], very little if any effort was devoted to implementing the inversion in a practical situation prior to the pioneer work in radio astronomy by Bracewell (1956a). Thus, "recent" in the chapter title means since the advent of applications.

The formula given by Radon (Theorem III in Appendix A),

$$f(P) = \frac{-1}{\pi} \int_0^\infty \frac{d\bar{F}_P(q)}{q},$$

is equivalent to the formulas of [5.3] and is a rigorous solution to the problem of *reconstruction from projections.* Shepp and Kruskal (1978) point out that some individuals have the mistaken notion that with such an inversion formula little more needs to be done. They further emphasize that an inversion formula is only a beginning for an applied problem.

The preceding formula is rigorously valid where f is continuous with compact support and the projections $\check{f}$ are given for all lines—an infinite set of projections rather than a discrete set. The compact support condition gives no trouble since any objects we might wish to image can be considered to be of finite extent and completely contained within some unit circle. Even the continuity condition might not be so serious for some situations, but the discrete nature of the projections in any realistic problem gives rise to subtle and difficult questions. In fact, there is a theorem by Smith, Solmon, and Wagner (1977) that states, after the rephrasing by Marr (1982): *A function f of compact support in $\mathbb{R}^2$ is uniquely determined by any infinite set, but by no finite set, of its projections.** Thus it is clear that we have to sacrifice uniqueness in applications. How serious is this? Well, that depends on just how close one can

*Gordon (1979) has suggested that this "Indeterminacy Theorem" may be the most important result of pure mathematics in this area since Radon's original work.

approximate the actual f with a nonunique approximation. (Note the parallel here with the nonuniqueness of δ-convergent sequences discussed in Appendix B.) One can try to "beat" the theorem by imposing appropriate a priori and/or optimality conditions on the solution: Shepp and Kruskal (1978); Artzy, Elfving, and Herman (1979); Hamaker, Smith, Solmon, and Wagner (1980). Also, encouragement may be provided by a theorem which in effect says that although the unknown function cannot be reconstructed exactly, arbitrarily good approximations can be found by utilizing an increasingly large number of projections [Hamaker, Smith, Solmon, and Wagner (1980)]. In summary, even though you can't win, you must not give up.

There are other things to consider: This particular inverse problem is technically *illposed* [Marr (1982)]. Thus the stability of the solution with respect to perturbations in the projection data may be such that extremely precise measurements (which are perhaps physically impossible) of the projections are required in order to obtain a satisfactory estimate of f.

Several authors have considered these technical matters of uniqueness, stability, use of a priori information, existence of solutions, and optimality from a rigorous point of view; see Marr (1982); Hamaker, Smith, Solmon, and Wagner (1980); Leahy, Smith, Solmon (1982); Natterer (1982); Petersen, Smith, and Solmon (1979); Shepp and Kruskal (1978); Hamaker and Solmon (1978); and other references contained in these articles. The bottom line seems to be that a certain amount of experimentation with various numerical approaches and different algorithms is essential. Clearly, this informed experimentation is paying off in usable reconstruction algorithms, as evidenced by the figures in [1.2].

Several algorithms will be discussed, and *all may be regarded as methods for approximating the inversion of the Radon transform*. However, they are by no means all equivalent due to various approximations and numerical methods for computer implementation.* In fact, even methods that have been shown to be theoretically equivalent can produce very different numerical results in practice [Herman and Rowland (1973)].

The lack of practical equivalence among reconstruction methods has encouraged many individuals to contribute to algorithm development. Also, in realistic situations, when working with discrete versions of formulas and actual observations, there are always questions concerning collection and preprocessing of data, numerical implementation of formulas, and postprocessing of images. In addition, statistical and experimental limitations vary widely with application. For an X-ray beam probe, for example, a few of these considerations include finite width of the probe, finite width of the detectors, various noise sources, relation of noise to different algorithms, resolution in reconstructed image, artifacts in reconstructed image, quantitative interpretation of image, and the relation between signal-to-noise ratio and radiation dose.

*A flowchart illustrating several methods is presented by Barrett and Swindell (1977).

It is understandable that the development, comparison, and analyses of algorithms is still an ongoing effort. Much of the recent work on algorithm development relates to optimal discrete versions, statistical limitations, resolution, artifacts, radiation dose versus signal-to-noise ratio, and of course computational speed. It is not our purpose here to get involved in the intricate details of these important technical matters. In this same spirit, we shall omit a discussion of the computational implementation of these algorithms. Such a discussion here is unnecessary since very comprehensive treatments of computer implementation of image reconstruction formulas have appeared recently [Rowland (1979); Budinger, Gullberg, and Huesman (1979)] and libraries of computer programs are available to utilize various algorithms [Herman and Rowland (1977); Huesman, Gullberg, Greenberg, and Budinger (1977)].

Over the past few years, contributions to algorithm development and associated technical considerations have been made by literally hundreds of individuals. An important subset of these contributions appears in categorized form in section [6.8]. These classifications are intended for general guidance only, and many of the articles overlap into more than one category. Most of the omitted papers are referenced in the review articles.

With the exception of [6.7], it will be observed that the following algorithms are for the two-dimensional problem of reconstruction from projections where the projection data constitute a sampling of the Radon transform on $\mathbb{R}^2$, $\check{f}(p, \phi) = \mathscr{R} f(x, y)$. The third dimension is usually obtained by appropriate stacking of several adjacent two-dimensional transverse sections rather than by directly attempting a genuine three-dimensional reconstruction. However, some work has been done on *direct three-dimensional reconstruction*. References are given in [6.8].

There are several ways to classify image reconstruction algorithms. These methods can be divided into four main categories:*

1. Direct Fourier methods;
2. Signal space convolution and frequency space filtering;
3. Iterative;
4. Series methods and orthogonal functions.

Summation or backprojection is often included as a fifth category; see, for example, Marr (1982). The idea here is to consider backprojection as an intermediate step rather than a separate approach. This is discussed further in [6.3].

The direct Fourier methods, which follow from the projection-slice theorem, are discussed in [6.2]. Category 2 actually contains four possible approaches;

*In keeping with the terminology of signal processing, we equate signal space with Radon space and frequency space with Fourier space.

these are discussed in [6.4] and [6.5]. Briefly, the approaches can be categorized according to whether the backprojection is done first or last:

(i) signal space convolution then backproject
(ii) frequency space filtering then backproject } [6.4]
(iii) backproject then signal space convolution
(iv) backproject then frequency space filtering } [6.5]

A brief discussion of matrix and iterative approaches appears in [6.6], and Category 4 is covered in Chapter 7. The iterative approaches are still of interest due to their applicability to a wide variety of problems of the inverse type. Although these methods are no longer used in computerized tomography, they were used by the earliest scanners developed for EMI Ltd. by Hounsfield (1972, 1973). A recent application in geophysics is discussed by Mason (1981).

[6.2] PROJECTION-SLICE THEOREM

Two-dimensional Version

The two-dimensional version of the *projection-slice theorem* [Mersereau and Oppenheim (1974)] or, equivalently, the *central slice theorem* [Marr (1982)] provides a connection between the two-dimensional Fourier transform of $f(x, y)$ and the one-dimensional Fourier transform of the Radon transform of $f(x, y)$. This result was derived by Bracewell (1956a) with a view toward applications in radio astronomy; however, the result was already known in the areas of probability theory [Cramér and Wold (1936); Rényi (1952)] and differential equations [John (1934)]. This version is a special case of the situation discussed in [4.2]. In operator form,

$$\mathscr{F}_2 f = \mathscr{F}_1 \mathscr{R} f = \mathscr{F}_1 \check{f}. \tag{2.1}$$

Note that there is an implied interpolation between rectangular and polar coordinates in Fourier space since

$$\mathscr{F}_2 f(x, y) = \tilde{f}(k_x, k_y) \tag{2.2}$$

and

$$\mathscr{F}_1 \check{f}(p, \phi) = \tilde{\check{f}}(k, \phi). \tag{2.3}$$

This situation is illustrated in Fig. 6.1. It is interesting to compare this figure with Fig. 5 in the original paper by Bracewell (1956a).

It follows that in terms of a single projection of f at a fixed angle $\phi = \Phi$, designated by $\check{f}_\Phi(p)$, the same information is contained in the projection $\check{f}_\Phi(p)$

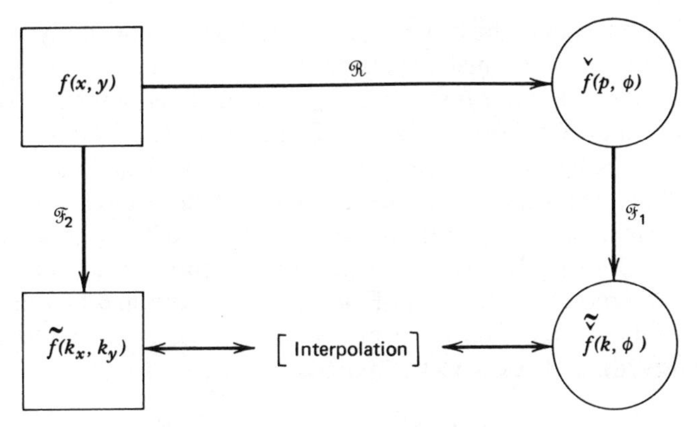

Figure 6.1 Projection-slice theorem in two dimensions. After interpolation $\tilde{f}(k_x, k_y) = \tilde{\check{f}}(k, \phi)$ where $k = (k_x^2 + k_y^2)^{1/2}$ and $\phi = \tan^{-1}(k_y/k_x)$.

as in a slice at the same angle Φ of the two-dimensional Fourier transform of f. That is, $\tilde{f}(k_x, k_y)$ evaluated along a line passing through the origin at angle Φ in Fourier space (coordinate axes k_x, k_y) is identical to the one-dimensional Fourier transform of the projection $\check{f}_\Phi(p)$, where the projection angle is the same angle Φ (see Fig. 6.2). After performing the interpolation, $\tilde{f}(k_x, k_y)$ may be written as $\tilde{f}(k, \phi)$, where $k = (k_x^2 + k_y^2)^{1/2}$ and $\phi = \tan^{-1}(k_y/k_x)$. At $\phi = \Phi$,

$$\tilde{f}(k, \Phi) = \tilde{\check{f}}_\Phi(k).$$

When a reconstruction is obtained by using the projection-slice theorem, the method is often referred to as a *direct Fourier method*. Although Bracewell (1956a) developed the required equations in connection with celestial bright-

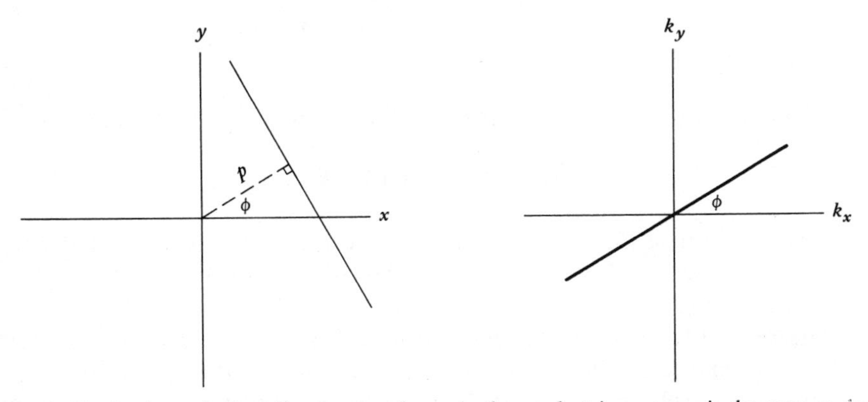

Figure 6.2 In the projection-slice theorem the projection angle ϕ in xy space is the same as the angle that defines the slice in Fourier space.

ness distribution studies in radio astronomy, he did not apply the method because of the computational unfeasibility (at that time) of doing the two-dimensional inverse Fourier transform. A few years later, Klug and co-workers [Klug, Crick, and Wyckoff (1958); DeRosier and Klug (1968); Crowther, DeRosier, and Klug (1970); Crowther, Amos, Finch, DeRosier, and Klug (1970)] rediscovered Bracewell's method and adapted it for use in electron microscopy. Since that early work, several algorithms of the direct Fourier type have been developed. The main distinguishing feature among these relates to the way the interpolation is done in Fourier space (see Fig. 6.1). For a detailed discussion of these direct Fourier techniques, see Stark, Paul, and Sarna (1979), Mersereau (1976), and references contained therein.

Three-dimensional Version

The three-dimensional version of the projection-slice theorem equates the two-dimensional Fourier transform of a plane projection of $f(x, y, z)$ to the central section of the three-dimensional Fourier transform of $f(x, y, z)$ perpendicular to the direction of view. To see this, let $\tilde{f}(k_x, k_y, k_z)$ be the three-dimensional Fourier transform of $f(x, y, z)$, and let $\sigma(y, z)$ represent the projection of $f(x, y, z)$ along the x direction,

$$\sigma(y, z) = \int_{-\infty}^{\infty} f(x, y, z)\, dx. \tag{2.4}$$

This integral is assumed known for all y and z. Thus, if $f(x, y, z)$ is a three-dimensional density distribution, then σ is a projection or two-dimensional density distribution. Now, evaluate $\tilde{f}(k_x, k_y, k_z)$ at $k_x = 0$,

$$\begin{aligned} \tilde{f}(0, k_y, k_z) &= \iiint_{-\infty}^{\infty} f(x, y, z) e^{-i2\pi(yk_y + zk_z)}\, dx\, dy\, dz \\ &= \iint_{-\infty}^{\infty} \sigma(y, z) e^{-i2\pi(yk_y + zk_z)}\, dy\, dz. \end{aligned}$$

It follows from the definition of the two-dimensional Fourier transform that

$$\tilde{f}(0, k_y, k_z) = \tilde{\sigma}(k_y, k_z). \tag{2.5}$$

This is the form of the projection-slice theorem used by DeRosier and Klug (1968); also, see DeRosier (1971) for three-dimensional reconstruction in molecular biology utilizing electron micrographs.

The result (2.5) may be extended to n dimensions in an obvious fashion [Mersereau and Oppenheim (1974)]. Observe, however, that this extension is not the same result obtained in [4.2] except for the special case $n = 2$. On the other hand, it is straightforward to obtain (2.5) from (2.1), as will be shown.

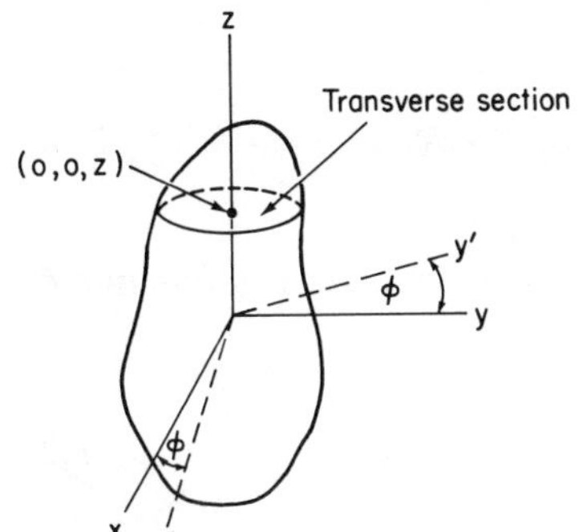

Figure 6.3 Geometry for obtaining the three dimensional projection theorem by stacking the transverse sections.

Suppose the projections are taken perpendicular to the $y'z$ plane along the x' direction as indicated in Fig. 6.3. The Radon transform of an arbitrary transverse section that intersects the z axis at some point z may be written

$$\check{f}(p, \phi; z) = \iint_{-\infty}^{\infty} f(x, y, z)\, \delta(p - x\cos\phi - y\sin\phi)\, dx\, dy.$$

Note that z is being carried along as a parameter to indicate which transverse section is being transformed. From (2.1) or by Fourier transforming the preceding equation,

$$\int_{-\infty}^{\infty} e^{-i2\pi kp}\check{f}(p, \phi; z)\, dp = \iint_{-\infty}^{\infty} f(x, y, z) e^{-i2\pi(xk_x + yk_y)}\, dx\, dy. \tag{2.6}$$

These results are assumed known for all z. For $\phi = \pi$, the variable p is just y, $k = k_y$, and $k_x = 0$. Thus $\check{f}(y, \pi; z)$ is just $\sigma(y, z)$ from (2.4). Hence (2.6) becomes

$$\int_{-\infty}^{\infty} \sigma(y, z) e^{-i2\pi yk_y}\, dy = \iint_{-\infty}^{\infty} f(x, y, z) e^{-i2\pi yk_y}\, dx\, dy. \tag{2.7}$$

Finally, the Fourier transform of both sides of (2.7) with respect to the z variable yields the desired result

$$\tilde{\sigma}(k_y, k_z) = \tilde{f}(0, k_y, k_z).$$

[6.3] BACKPROJECTION

As usual, let $\xi = (\cos\phi, \sin\phi)$ be a unit vector. Consider an arbitrary function $\psi(t, \xi)$ where $t = \xi \cdot \mathbf{x} = x\cos\phi + y\sin\phi$. The backprojection operator $\mathscr{B}$ is defined by [Rowland (1979)]

$$\mathscr{B}\psi = \int_0^{\pi} \psi(x \cos \phi + y \sin \phi, \xi)\, d\phi. \tag{3.1}$$

Since ξ is completely determined by specifying ϕ and since $\mathscr{B}\psi$ is a function of x and y, it may be useful to write

$$[\mathscr{B}\psi](x, y) = \int_0^{\pi} \psi(x \cos \phi + y \sin \phi, \xi)\, d\phi, \tag{3.2}$$

or, in terms of polar coordinates (r, θ), where $x = r \cos \theta$ and $y = r \sin \theta$,

$$[\mathscr{B}\psi](r, \theta) = \int_0^{\pi} \psi[r \cos(\theta - \phi), \phi]\, d\phi. \tag{3.3}$$

The backprojection operation for only two projections is illustrated in Fig. 6.4, and the operation for fixed ϕ is illustrated in Fig. 6.5. Observe that if

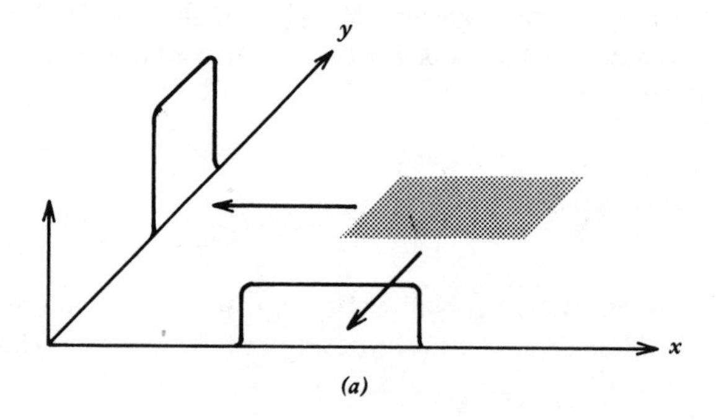

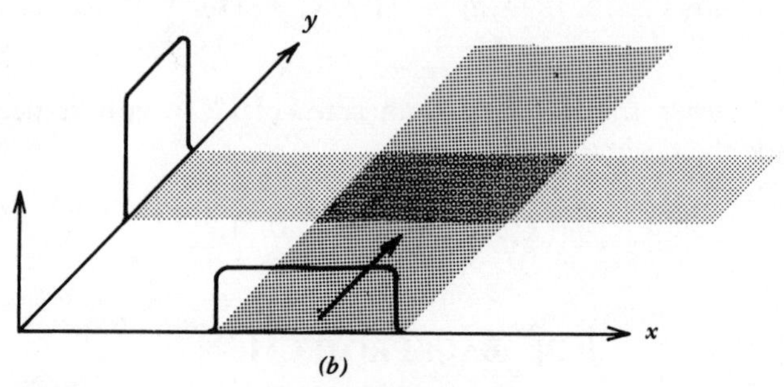

Figure 6.4 (*a*) Two profiles of a rectangular object. (*b*) Backprojection of profiles and superposition to form an approximation to original object. Reprinted with permission, Brooks and Di Chiro (1976a). Courtesy of R. A. Brooks.

$\psi(p, \phi)$ is identified with the projection function $\check{f}(p, \phi)$ obtained from $f(x, y)$ by the Radon transform, then for fixed angle ϕ, the incremental contribution to $\mathscr{B}\psi$ at the point (x, y) is just the value of $\check{f}(p, \phi)$ multiplied by $d\phi$ when p is computed from $p = x\cos\phi + y\sin\phi$. Of course, that value may be found by integrating f along the line that passes through (x, y) and is at a distance $p = x\cos\phi + y\sin\phi$ from the origin. If this ϕ-backprojection, also defined in (3.12), is known for all angles ϕ, then the complete backprojection is obtained by integration over ϕ as indicated by (3.2).

The term *backprojection* was introduced by Crowther, DeRosier, and Klug (1970) and expanded on by Gilbert (1972a). The same operation was used by

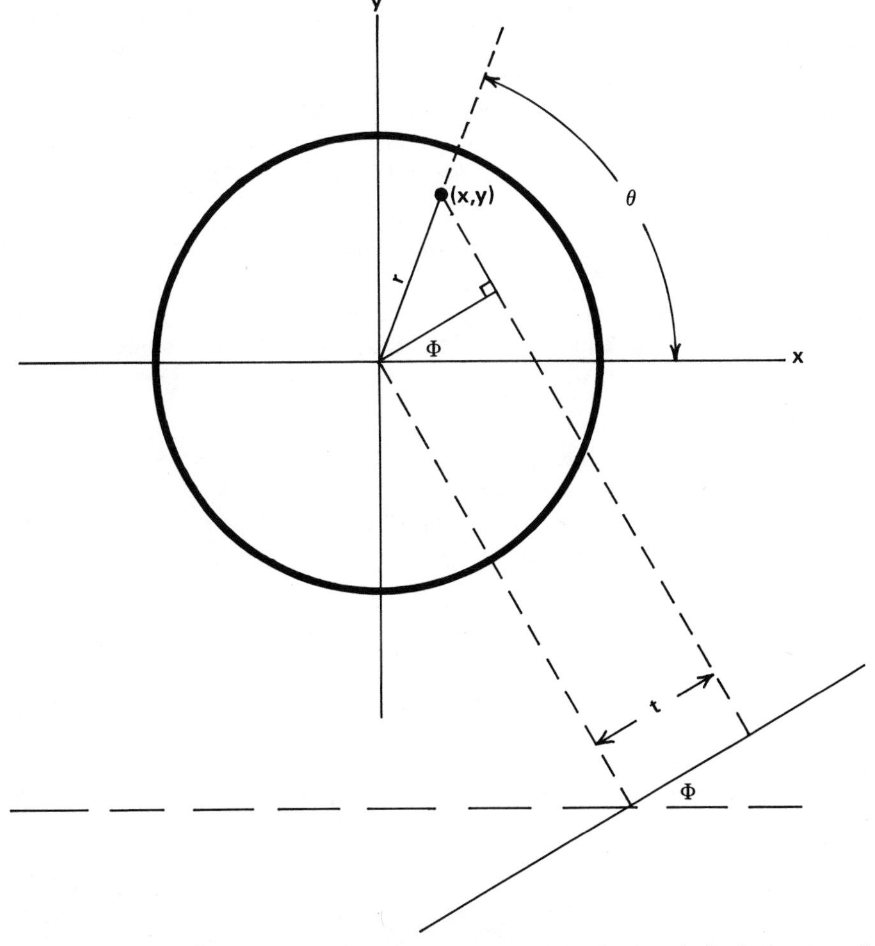

Figure 6.5 Geometry for obtaining the ϕ-backprojection. For a fixed angle Φ, the incremental contribution $d(\mathscr{B}\psi)$ to $\mathscr{B}\psi$ at the point (x, y) or, equivalently, (r, θ) is given by $\psi(t, \Phi)\, d\phi$. The full contribution to $\mathscr{B}\psi$ at (x, y) is found by integrating over ϕ as indicated in (3.2). Note that $t = x\cos\Phi + y\sin\Phi = r\cos(\theta - \Phi)$.

Bates and Peters (1971) and Smith, Peters, and Bates (1973). These authors suggested that the term *layergram* would be more appropriate than backprojection. Other terms used for this operation, especially the discrete version, include *summation* and *linear superposition* [Budinger and Gullberg (1974); Gordon and Herman (1974)]. Oldendorf (1961) used this approach in X-ray imaging to approximate a reconstruction. It was also used in transverse section scanning in nuclear medicine [Kuhl and Edwards (1963, 1968a, b)] and in the area of electron microscopy [Hart (1968); Vainshtein (1970)]. More detail on backprojection as an approximate method for reconstruction may be found in the review by Brooks and Di Chiro (1976a).

We shall see that in some algorithms backprojection is an important intermediate step in the reconstruction process. Also, as emphasized by Marr (1982), it is important for historical reasons, since, as can be seen from (3.6), the backprojected image represents a blurred representation of the true image. Efforts to eliminate or correct the blurring led many individuals to algorithms for reconstructing an image by using its projections. For a review of these algorithms and a general discussion on reconstruction from projections, see the articles by Gordon and Herman (1974) and Budinger and Gullberg (1974).

Relation to Adjoint

It is easy to relate the backprojection operator to the adjoint $\mathscr{R}^\dagger$ for the two-dimensional case. Suppose ψ is the Radon transform of some function; that is, $\psi = \check{f}$. Then ψ satisfies the symmetry property $\psi(-t, -\xi) = \psi(t, \xi)$ and

$$\begin{aligned}\mathscr{B}\psi &= \frac{1}{2}\int_0^{2\pi} \psi(\xi \cdot \mathbf{x}, \xi)\, d\phi \\ &= \tfrac{1}{2}\mathscr{R}^\dagger \psi.\end{aligned}$$

Therefore, for the two-dimensional case,

$$\mathscr{R}^\dagger = 2\mathscr{B}. \tag{3.4}$$

Example 1. In terms of $\mathscr{B}$ and operators discussed in [5.4], the inversion formula has the form

$$f = \frac{-1}{2\pi}\mathscr{B}\,\mathcal{H}\,\frac{\partial}{\partial p}\mathscr{R} f. \qquad \square \tag{3.5}$$

This is just Radon's formula expressed in operator form.

Relation to True Image

A useful result related to the backprojected image, defined by

$$b(x, y) = \mathscr{B}\check{f}(p, \phi)$$

$$= \int_0^{\pi} \check{f}(x \cos \phi + y \sin \phi, \phi)\, d\phi, \tag{3.6}$$

is that the true image $f(x, y)$ convolved with $1/r$ yields the backprojected image

$$b(x, y) = f(x, y) ** \frac{1}{r}$$

$$= \iint_{-\infty}^{\infty} \frac{f(x', y')\, dx'\, dy'}{\left[(x - x')^2 + (y - y')^2\right]^{1/2}}, \tag{3.7}$$

where $r = (x^2 + y^2)^{1/2}$ and $**$ means two-dimensional convolution [Bracewell (1978)]. This may be seen by considering the $\mathscr{B}$ operator acting on $\check{f}$, where (2.1) in the form $\check{f} = \mathscr{F}_1^{-1}\mathscr{F}_2 f$ is used for the projections [Budinger and Gullberg (1974)]:

$$\mathscr{B}\check{f} = \mathscr{B}\mathscr{F}_1^{-1}\mathscr{F}_2 f$$

or

$$b(r, \theta) = \mathscr{B}\mathscr{F}_1^{-1}\tilde{f}(k, \phi). \tag{3.8}$$

Here $\phi = \tan^{-1}(k_y/k_x)$, $k = (k_x^2 + k_y^2)^{1/2}$, and $\mathscr{F}_1^{-1}$ transforms the k variable. Explicitly,

$$b(r, \theta) = \mathscr{B}\int_{-\infty}^{\infty} \tilde{f}(k, \phi) e^{i2\pi tk}\, dk$$

$$= \int_0^{\pi} d\phi \int_{-\infty}^{\infty} dk\, \tilde{f}(k, \phi) e^{i2\pi kr\cos(\theta - \phi)}$$

$$= \int_0^{2\pi} \int_0^{\infty} k^{-1}\tilde{f}(k, \phi) e^{i2\pi kr\cos(\theta - \phi)} k\, dk\, d\phi$$

$$= \mathscr{F}_2^{-1}\{k^{-1}\tilde{f}\}. \tag{3.9}$$

Then, from the convolution theorem,

$$b(r,\theta) = \mathscr{F}_2^{-1}\{\tilde{f}\} ** \mathscr{F}_2^{-1}\{k^{-1}\}$$

$$= f(r,\theta) ** \frac{1}{r}, \tag{3.10}$$

which is equivalent to (3.7). The result $\mathscr{F}_2^{-1}\{k^{-1}\} = r^{-1}$ follows if we reduce $\mathscr{F}_2^{-1}\{k^{-1}\}$ to a zero-order Hankel transform [Bracewell (1978)]

$$\mathscr{F}_2^{-1}\{k^{-1}\} = 2\pi \int_0^\infty J_0(2\pi k r)\, dk = \frac{1}{r}, \tag{3.11}$$

where J_0 is the Bessel function of order zero.

Natural Transpose

It is often useful to define the *ϕ-backprojection* or *natural transpose* of some function $h(p)$ [Marr (1982)] by

$$\mathscr{B}_\phi h(p) = h(x\cos\phi + y\sin\phi), \tag{3.12}$$

for fixed angle ϕ. Then

$$\int f(\mathbf{x})\mathscr{B}_\phi h\, d\mathbf{x} = \int d\mathbf{x} \int_{-\infty}^{\infty} dp\, f(\mathbf{x}) h(p)\, \delta(p - \xi\cdot\mathbf{x})$$

$$= \int_{-\infty}^{\infty} \check{f}_\phi(p) h(p)\, dp, \tag{3.13}$$

which relates the ϕ-backprojection to the projection $\check{f}_\phi(p)$ at fixed ϕ.

Example 2. Suppose $h(p) = e^{-i2\pi k p}$ for real k. Then (3.13) becomes

$$\iint_{-\infty}^{\infty} f(x,y) e^{-i2\pi k(x\cos\phi + y\sin\phi)}\, dx\, dy = \int_{-\infty}^{\infty} \check{f}_\phi(p) e^{-i2\pi k p}\, dp,$$

which is just the projection-slice theorem discussed in [6.2] with $k_x = k\cos\phi$ and $k_y = k\sin\phi$. □

[6.4] BACKPROJECTION OF FILTERED PROJECTIONS

Basic Formulas

Perhaps the most direct method for obtaining the desired formula for the backprojection of filtered projections (also called filtered backprojection to add to the confusion) is to start with the inversion formula in the form

$$f = 2\mathscr{B}\bar{\check{f}} = 2\mathscr{B}\mathscr{F}_1^{-1}\mathscr{F}_1\bar{\check{f}} \tag{4.1}$$

and make use of Example 2 in [5.5]. (Note that the identity has been inserted between $\mathscr{B}$ and $\bar{\check{f}}$.) Thus

$$f = \frac{2\mathscr{B}}{4\pi^2}\mathscr{F}_1^{-1}\mathscr{F}_1\frac{\partial}{\partial t}\left[\frac{1}{t} * \check{f}(t,\phi)\right]$$

$$= \frac{\mathscr{B}}{2\pi^2}\mathscr{F}_1^{-1}(i2\pi k)\left[\mathscr{F}_1\left(\frac{1}{t}\right)\right]\left[\mathscr{F}_1\check{f}(t,\phi)\right]$$

$$= \frac{\mathscr{B}}{2\pi^2}\mathscr{F}_1^{-1}(i2\pi k)(-i\pi \operatorname{sgn} k)\tilde{\check{f}}(k,\phi)$$

$$= \mathscr{B}\mathscr{F}_1^{-1}\left[k \operatorname{sgn} k\tilde{\check{f}}(k,\phi)\right]$$

$$= \mathscr{B}\mathscr{F}_1^{-1}\left[|k|\tilde{\check{f}}(k,\phi)\right]. \tag{4.2}$$

Here, use was made of the derivative theorem, the convolution theorem, and $\mathscr{F}(1/t) = -i\pi \operatorname{sgn} k$ from Bracewell (1978).

It is helpful to define the modified projection function

$$f^*(s,\phi) = \mathscr{F}_1^{-1}\left[|k|\tilde{\check{f}}(k,\phi)\right]. \tag{4.3}$$

The function f is then recovered by backprojection,

$$f(x,y) = \mathscr{B}f^* = \int_0^{\pi} f^*(x\cos\phi + y\sin\phi, \phi)\, d\phi. \tag{4.4}$$

The entire procedure is illustrated in Fig. 6.6.

Equations (4.3) and (4.4) serve as the basis for the most widely used and computationally efficient algorithms for reconstruction from projections [Rowland (1979); Marr (1982)]. Early work on filtered backprojection algorithms for parallel projection data was done by Bracewell and Riddle (1967), Ramachandran and Lakshminarayanan (1971), and Shepp and Logan (1974). An example of a computer program utilizing filtered backprojection for image reconstruction is given in the article by Brooks and Di Chiro (1975). It is also possible to develop similar algorithms that apply to fan-beam (divergent beam) projection data. Almost all commercial scanners in X-ray computed tomography perform the reconstructions directly from divergent-beam projections. Much of the early work on these algorithms was done by Lakshminarayanan (1975) Herman, Lakshminarayanan, and Naparstek (1976, 1977), Herman,

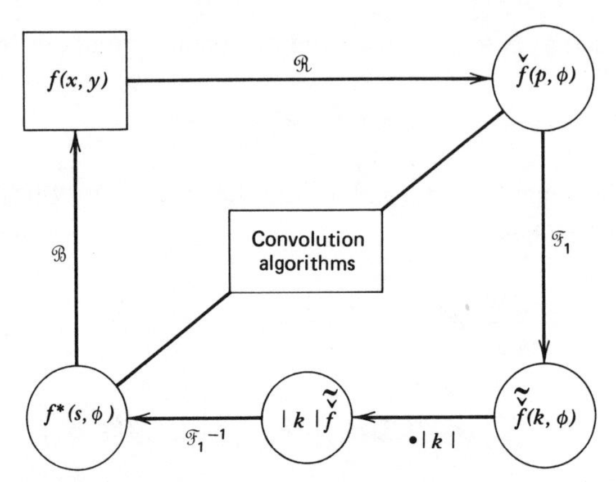

Figure 6.6 Backprojection of filtered projections: the implementation of (4.2). The alternative path to f^* by use of convolution methods is discussed near the end of [6.4]; see (4.9), for example.

Lakshminarayanan, Ritman, Robb, and Wood (1976), and Herman and Naparstek (1977). Work on both parallel and divergent beam algorithms continues. The recent review by Kak (1979) emphasizes the use of algorithms of the filtered backprojection type and provides numerous references to the literature on this subject. Also, see the references under "Fan-Beam Methods" in [6.8]. The optimality of filtered backprojection algorithms is discussed by Hanson (1980) and Tretiak (1978).

Convolution Methods

It is worthwhile investigating some of the implications of (4.3) and (4.4). Equation (4.3) was obtained by using generalized functions (distribution theory) in steps leading to (4.2). This presents no theoretical difficulty, however, there is a serious problem with numerical implementation due to the presence of the $|k|$ factor. What we would like is some *well-behaved* function $g(t)$ such that $\mathscr{F}g = |k|$ [Shepp and Kruskal (1978)]. Then, by the convolution theorem, it would follow that

$$\begin{aligned} f^*(s,\phi) &= \mathscr{F}_1^{-1}\left[(\mathscr{F}_1 g)(\mathscr{F}_1 \check{f})\right] \\ &= g * \check{f} = \check{f} * g \\ &= \int_{-\infty}^{\infty} \check{f}(t,\phi) g(s-t)\, dt. \end{aligned} \tag{4.5}$$

Then f^* as it appears in the integrand of (4.4) could be computed:

$$f^*(x\cos\phi + y\sin\phi, \phi) = \int_{-\infty}^{\infty} \check{f}(t,\phi) g(x\cos\phi + y\sin\phi - t)\, dt.$$

Finally, f could be recovered by use of backprojection (4.4).

It turns out that a function g exists so that $\tilde{g} = |k|$, but rather than being well behaved, it is a singular distribution (Appendix B). This can be seen by comparison of (4.5) and Example 2 of [5.5],

$$\mathscr{B}[\check{f} * g] = f = 2\mathscr{B}\bar{\check{f}} = \mathscr{B}\left[\check{f}(s, \phi) * \frac{1}{2\pi^2}\frac{\partial}{\partial s}\mathscr{P}\left(\frac{1}{s}\right)\right]. \tag{4.6}$$

Thus g may be identified with the singular distribution [Barrett and Swindell (1977)]

$$g(s) = \frac{1}{2\pi^2}\frac{\partial}{\partial s}\mathscr{P}\left(\frac{1}{s}\right),$$

which serves to point out the illposedness of the inverse problem [Marr (1982)].

Remark 1. To verify that $\tilde{g} = |k|$, consider the following:

$$\begin{aligned}\mathscr{F}\left[\frac{1}{2\pi^2}\frac{\partial}{\partial s}\mathscr{P}\left(\frac{1}{s}\right)\right] &= \frac{2\pi ik}{2\pi^2}\mathscr{F}\left[\mathscr{P}\left(\frac{1}{s}\right)\right]\\ &= \left(\frac{ik}{\pi}\right)(-i\pi\,\mathrm{sgn}\,k)\\ &= |k|. \qquad \square\end{aligned}$$

The way around the difficulty just described is to try to find a well-behaved function $\mathrm{c}(s)$ whose Fourier transform approximates $|k|$ rather than a function $g(s)$, whose Fourier transform equals $|k|$. Moreover, in physically interesting situations, it is reasonable to assume that $\tilde{\check{f}}(k, \phi)$ is small for sufficiently large values of $|k|$, and since it is the product $|k|\tilde{\check{f}}$ which appears in (4.3), an even less stringent condition on c is adequate:

$$\tilde{\mathrm{c}} \simeq |k| \qquad \text{for} \quad |k| < \kappa,$$

where κ is left unspecified until the nature of $\tilde{\check{f}}$ is known from the physical situation,

$$\tilde{\check{f}}(k, \phi) \text{ is small for } |k| > \kappa.$$

In the language of signal analysis, $\check{f}$ has no important high-frequency components.

The usual approach is to define a *filter* function

$$\tilde{\mathrm{c}}(k) = \mathscr{F}_1[\mathrm{c}(s)] = |k|\mathrm{w}(k) \tag{4.7}$$

in terms of a *window* function $\mathrm{w}(k)$. [This is a standard procedure in signal processing. For a discussion of filters and windows, see Harris (1978), Hamming (1977), Rabiner and Gold (1975), or Oppenheim and Schafer (1975).] Now there are still two basic ways to proceed. One can filter in frequency space, that is, use (4.3) as it stands:

$$f^*(s,\phi) = \mathscr{F}_1^{-1}\left[|k|\mathrm{w}(k)\tilde{\check{f}}(k,\phi)\right]. \tag{4.8}$$

Alternatively, one can use the convolution theorem and work in signal space

$$f^*(s,\phi) = \mathrm{c}(s) * \check{f}(s,\phi), \tag{4.9}$$

before backprojecting. Primarily because of (4.9), the filtered backprojection algorithms are often referred to as *convolution algorithms* (see Fig. 6.6).

The obvious advantage in (4.9) is that one need not Fourier transform the projection data $\check{f}$; however, the convolver must be computed from

$$\mathrm{c}(s) = \mathscr{F}_1^{-1}[|k|\mathrm{w}(k)] \tag{4.10}$$

prior to implementation of (4.9). This signal space convolution approach is widely used and discussed in much detail by Rowland (1979).

On the other hand, it is easier to implement the filter if one works in Fourier space, that is, directly with (4.8). This approach has been emphasized by Budinger, Gullberg, and Huesman (1979), especially for situations where noise is an important consideration.

Clearly, the quality of reconstruction depends on the choice of the window function w, and there are many possible choices. The Hann, Hamming, Parzen, and rectangular windows are discussed and illustrated by Budinger, Gullberg, and Huesman (1979); effects of band limiting, low-pass sinc, low-pass cosine, and generalized Hamming windows are treated by Rowland (1979). An alternative to selecting the convolver by studying the performance characteristics of various windows is to attempt to find convolvers or classes of convolvers that are in some sense optimal. Work has been done in this area by Cho, Ahn, Bohm, and Huth (1974), Marr (1974), Logan and Shepp (1975), Tanaka and Iinuma (1975, 1976), Miller (1978), Davison and Grünbaum (1979), Lewitt (1979), and Kenue and Greenleaf (1979a, b). Further discussion of the filtered backprojection algorithm aimed especially toward implementation is contained in Rosenfeld and Kak (1982).

[6.5] FILTER OF BACKPROJECTIONS

Basic Formulas

This method, developed by Bates and Peters (1971) and Smith, Peters, and Bates (1973), may be derived by first taking the Fourier transform of (3.7) and using the two-dimensional convolution theorem

$$\tilde{b}(k_x, k_y) = \mathscr{F}_2 b(x, y)$$

$$= \mathscr{F}_2\left[f(x, y) ** \frac{1}{r}\right]$$

$$= |k|^{-1}\tilde{f}(k_x, k_y), \tag{5.1}$$

where $|k| = (k_x^2 + k_y^2)^{1/2}$. Thus, from (5.1) and (3.6),

$$\tilde{f}(k_x, k_y) = |k|\tilde{b}(k_x, k_y)$$

$$= |k|\mathscr{F}_2 b(x, y)$$

$$= |k|\mathscr{F}_2\mathscr{B}\check{f}. \tag{5.2}$$

The inverse Fourier transform of both sides yields

$$f(x, y) = \mathscr{F}_2^{-1}\left[|k|\mathscr{F}_2\mathscr{B}\check{f}\right]. \tag{5.3}$$

Thus, in this case the backprojection comes first and the filtering follows; in [6.4], the filtering came first and backprojection last. Figure 6.7 illustrates the situation.

Although (5.3) may be somewhat more cumbersome than (4.4) when it comes to computer implementation, Bates and Peters (1971) and Peters (1974)

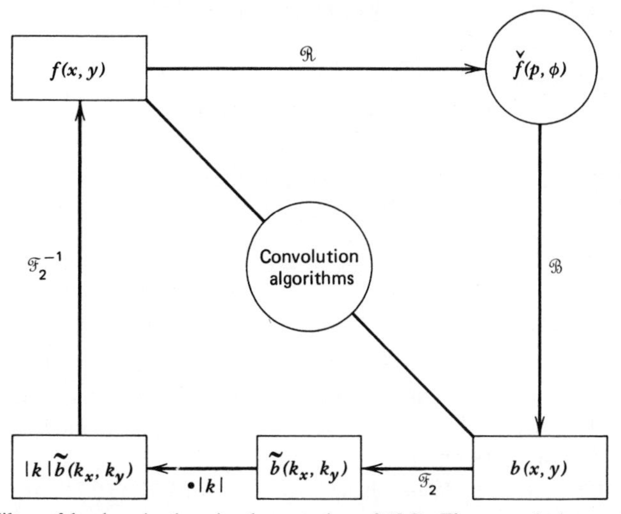

Figure 6.7 Filter of backprojection: implementation of (5.3). The convolution path corresponds to (5.5).

have demonstrated that (5.3) serves as the basis for reconstruction utilizing an optical system rather than a computer, the lenses in the optical system serving as Fourier transform devices [Goodman (1968)].

Convolution Methods

As in the preceding section, define the filter function $\tilde{c}$ in terms of a window function w, but this time $\tilde{c}$ and w must be treated as functions of two variables rather than one as in (4.7):

$$\tilde{c}(k_x, k_y) = |k| w(k_x, k_y). \tag{5.4}$$

If $|k|$ is replaced by $\tilde{c}$ in (5.3), then one obtains a convolution algorithm

$$\begin{aligned} f(x, y) &= \mathscr{F}_2^{-1}\left[\tilde{c}\mathscr{F}_2(\mathscr{B}\check{f})\right] \\ &= c(x, y) ** b(x, y). \end{aligned} \tag{5.5}$$

Here, the two stars $**$ indicate two-dimensional convolution, $b = \mathscr{B}\check{f}$ by definition (3.6), and

$$c(x, y) = \mathscr{F}_2^{-1}\tilde{c}(k_x, k_y). \tag{5.6}$$

The filter-of-the-backprojection algorithm has also been adapted to fan-beam geometry for data collection. [Budinger and Gullberg (1975); Gullberg (1979a)]. For a digital implementation (parallel-beam or fan-beam), see Huesman, Gullberg, Greenberg, and Budinger (1977), and for further discussion of this method, see Budinger, Gullberg, and Huesman (1979).

[6.6] ITERATIVE METHODS

Fundamentally, iterative algorithms for reconstruction from projections are schemes for solving a system of linear equations of the form [Marr (1982)]

$$\mathsf{AX} = \mathsf{B}, \tag{6.1}$$

where A is a rectangular $M \times M$ matrix and X and B are $M \times 1$ column matrices. Actually, A may not be naturally square, but by an appropriate choice of data collection, it can be made square. The dimension of A may be quite large, with M of order 10,000; however, many of the elements of A are zero. That is, A is a sparse matrix.

Budinger and Gullberg (1974) show how equations like (6.1) can arise. Let F represent a digitized version of the image $f(x, y)$. Assume that the object fits inside some square region and this large square is subdivided into m times m

($= m^2$) little squares. In each little square or *pixel* (for picture element), f, which in general varies over the pixel, is replaced by a constant over the entire pixel. This approximates the (variable) value of f over that small region. Thus $f(x, y)$ is approximated by an $m \times m$ array of numbers. These numbers are the elements of the $m \times m$ matrix F.

Now, consider a single projection at angle Φ_j, say the ith projection out of m projections for the angle Φ_j. (It is not essential to select m projections when the matrix F is m by m, but it can be done and our purposes are served here by that selection.) Figure 6.8 illustrates the arrangement. The ith projection, temporarily designated by P_{i,Φ_j}, may be written as

$$\mathsf{P}_{i,\Phi_j} = \sum_{I=1}^{m} \sum_{J=1}^{m} W_{IJ}(i, \Phi_j)\mathsf{F}_{IJ}. \tag{6.2}$$

The $W_{IJ}(i, \Phi_j)$ are weight factors that depend on both the projection number i and the angle Φ_j. Since I and J both range from 1 to m, there are m^2 weight factors for a fixed i and Φ_j. Furthermore, note that the vast majority of the weight factors are zero. Only those weights associated with pixels through which the line passes are nonzero. These nonzero weights can be selected in

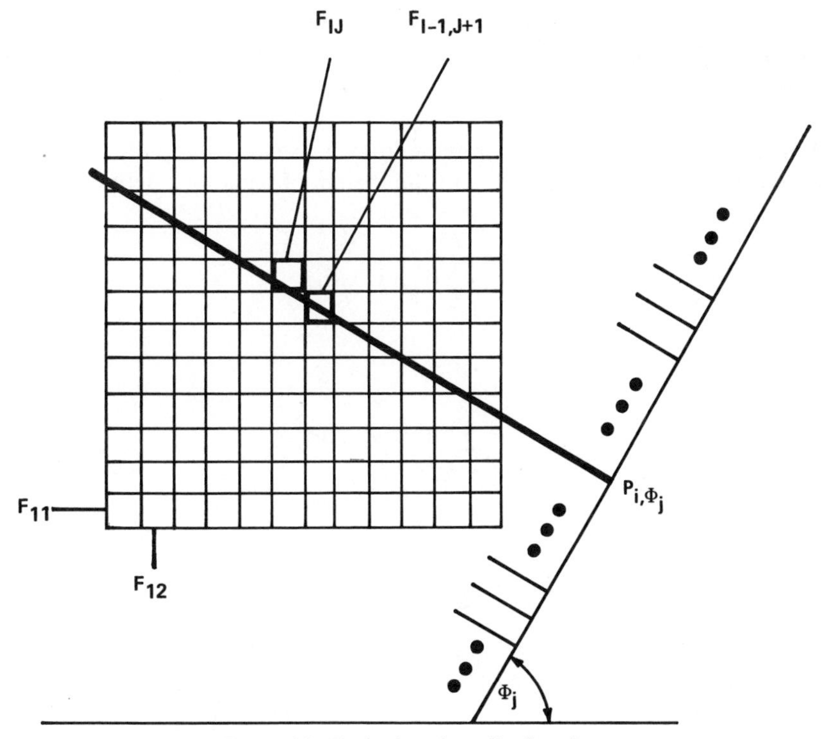

Figure 6.8 Projections for a fixed angle.

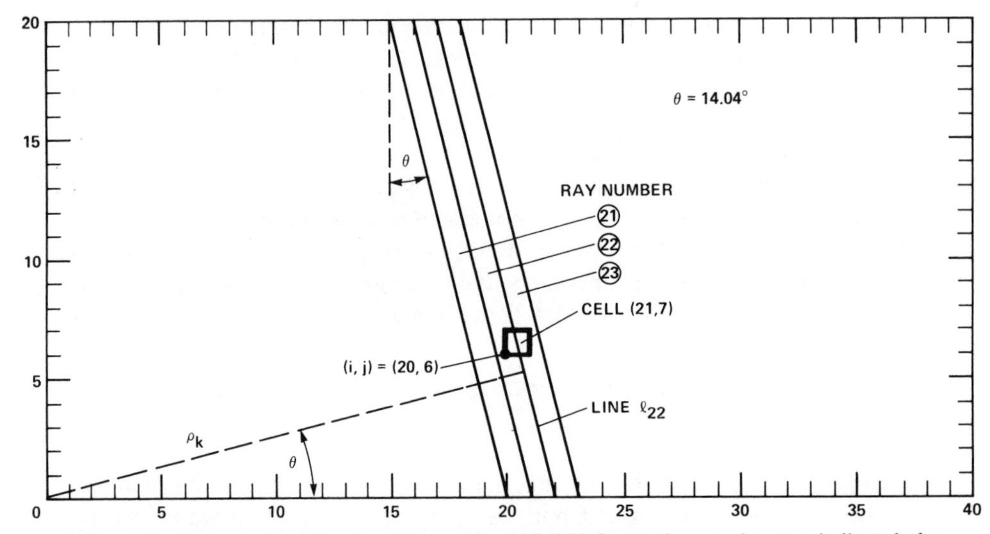

Figure 6.9 Rays can be defined as "lines with width." If the numbers work out as indicated, the weight factor contribution for cell (21, 7) to ray 22 might be chosen proportional to the area of the cell that overlaps ray 22. This is illustrated in more detail in Fig. 6.10.

various ways. One obvious way is to make them proportional to the length of the line through the corresponding pixel.* Thus, in Fig. 6.8, $W_{11}(i, \Phi_j) = 0$, $W_{12}(i, \Phi_j) = 0$, and $0 < W_{IJ}(i, \Phi_j) < W_{I-1, J+1}(i, \Phi_j)$, since the line does not go through pixels (1, 1) and (1, 2) and the line is longer in pixel $(I - 1, J + 1)$ than in pixel (I, J).

For a given Φ_j, we assume i ranges from 1 to m. If there are k different angles, then j ranges from 1 to k and the system of equations defined by (6.2) may be arranged as

$$\mathsf{P}_{11} = \sum_{I, J} W_{IJ}(1, 1)\mathsf{F}_{IJ}$$

$$\vdots$$

$$\mathsf{P}_{m1} = \sum_{I, J} W_{IJ}(m, 1)\mathsf{F}_{IJ}$$

$$\vdots$$

$$\mathsf{P}_{mk} = \sum_{I, J} W_{IJ}(m, k)\mathsf{F}_{IJ},$$

where for brevity Φ_j is replaced by j. This system may be written in the form

*An alternative method is described in Figs. 6.9 and 6.10.

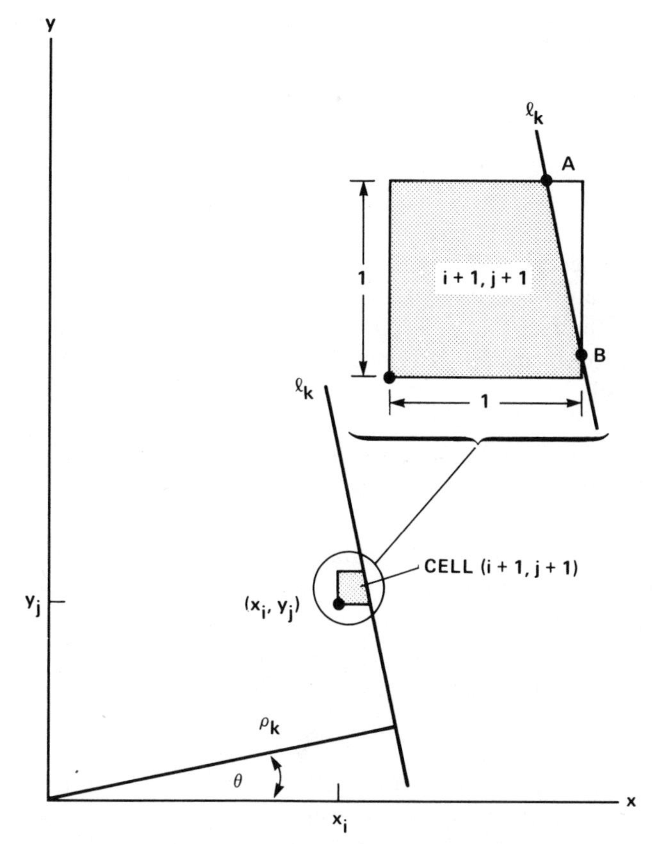

Figure 6.10 Relative to the kth ray the weight factor for cell $(i + 1, j + 1)$ is proportional to the shaded area. It is assumed that line l_{k-1} does not intersect cell $(i + 1, j + 1)$.

$\mathsf{AX} = \mathsf{B}$, where B is the column matrix with mk elements

$$\mathsf{B} = \begin{bmatrix} \mathsf{P}_{11} \\ \vdots \\ \mathsf{P}_{m1} \\ \vdots \\ \mathsf{P}_{mk} \end{bmatrix},$$

A is the $mk \times m^2$ rectangular matrix

$$\mathsf{A} = \begin{bmatrix} W_{11}(11) & W_{12}(11) & \cdots & W_{mm}(11) \\ \cdots & \cdots & \cdots & \cdots \\ W_{11}(m1) & W_{12}(m1) & \cdots & W_{mm}(m1) \\ \cdots & \cdots & \cdots & \cdots \\ W_{11}(mk) & W_{12}(mk) & \cdots & W_{mm}(mk) \end{bmatrix},$$

and X is the column matrix with m^2 elements

$$\mathsf{X} = \begin{bmatrix} \mathsf{F}_{11} \\ \mathsf{F}_{12} \\ \vdots \\ \mathsf{F}_{mm} \end{bmatrix}.$$

If angles are selected such that $k = m$, then A is square (m^2 by m^2), and $M = m^2$. Typically m is of order 10^2 so M is of order 10^4.

Other than the fact that A is rather large, this looks like a straightforward linear algebra problem, but it is not for the following reasons [Gordon and Herman (1974)]:

1. The number of unknowns, elements of B, and the number of equations (m^4) may be so large that standard techniques for solving the equations become unfeasible.
2. The number of equations available from the projection data may be fewer than the number of unknowns. Thus one may be faced with selecting one solution from an infinite number of possible solutions.
3. Since the elements of B are obtained experimentally and contain errors, it is possible that the system of equations is inconsistent and no exact solution even exists.

Efforts to get around these difficulties, all of which may be present at the same time, have led to various iterative schemes to approximate a solution. These include algebraic reconstruction techniques (ART), Bayesian methods, maximum entropy concepts [Gullberg (1975); Wernecke and D'Addario (1977); Minerbo (1979b)] and generalized inverse matrix methods [Tewarson (1972); Tewarson and Narain (1974a, b); Nashed (1976)]. Several of the algebraic and matrix methods have been reviewed by Gordon (1974), Gordon and Herman (1974), Budinger and Gullberg (1974), and Herman and Lent (1976a). These reviews refer to the earlier work in this area. For more recent work on algebraic reconstruction techniques and Bayesian approaches, see Herman (1979), Herman, Hurwitz, Lent, and Lung (1979), and Herman, Hurwitz, and Lent (1977).

Finally, it should be pointed out that although the iterative schemes have the drawback of high computational cost, there are some advantages too [Marr (1982)]:

(i) General applicability to a wide range of problems;
(ii) Applicability to standard numerical analysis techniques;
(iii) Applicability to arbitrary configurations of rays in two or three dimensions;
(iv) Ease of incorporation of a priori information.

As mentioned earlier, the matrix A in (6.1) need not be square; suppose it is M

by N, with $M > N$, both M and N large and of the same order of magnitude. A very attractive iterative method for solving such a system was developed by Kaczmarz (1937). This method has been discussed further by Tanabe (1971), and it is nicely illustrated by Rosenfeld and Kak (1982). Many other iterative optimization procedures have been developed [6.10], several of which have been shown to be equivalent to that of Kaczmarz.

[6.7] THREE-DIMENSIONAL METHODS

Shepp (1980) has developed two fully three-dimensional algorithms in connection with applications to nuclear magnetic resonance zeugmatography. The most efficient of the two is based on the inversion formula of [5.2] written in the form

$$f(x, y, z) = \frac{-1}{8\pi^2} \int_0^{2\pi} d\phi \int_0^{\pi} d\theta \sin\theta \, \check{f}_{pp}(p, \theta, \phi). \tag{7.1}$$

Here, we have written the integral of Example 1 in [5.2] explicitly, and it is understood that following the differentiation p is replaced by

$$p = \xi \cdot \mathbf{x} = x \sin\theta \cos\phi + y \sin\theta \sin\phi + z \cos\theta \tag{7.2}$$

prior to the integration. Note that the reconstruction (7.1) is in the form of a filtered backprojection where the filter accomplishes the operation of second-order differentiation. Shepp (1980) has given computer programs for implementation of this algorithm and has done simulation experiments to demonstrate its potential.

Shepp's second algorithm performed less satisfactorily. It is based on the result of Example 2 of [5.4]. An explicit discrete version of this algorithm also appears in Shepp (1980).

For reference to other three-dimensional algorithms, see [6.8].

[6.8] CATEGORIZED REFERENCES

The references are divided into several categories; in many cases these divisions are not rigid and often there is overlap. To form a more extensive list see, in particular, those followed by "and references therein." Also, the review articles contain additional references, as does the list by major areas in [1.10].

Fourier Methods, Including Frequency Space Techniques Utilizing Filters

Bracewell (1956a); Bracewell and Riddle (1967); DeRosier and Klug (1968); Hoppe, Langer Knesch, and Poppe (1968); Tretiak, Eden, and Simon (1969); Rowley (1969); Crowther, DeRosier, and Klug (1970); DeRosier and Moore (1970); Crowther, Amos, Finch, DeRosier, and Klug

(1970); DeRosier (1971); Crowther (1971); Klug (1971); Crowther and Amos (1971); Bates and Peters (1971); Tretiak, Ozonoff, Klopping, and Eden (1971); Lake (1972); Sweeney and Vest (1973); Smith, Peters, and Bates (1973); Mersereau (1973); Peters, Smith, and Gibson (1973). Mersereau and Oppenheim (1974); Kay, Keyes, and Simon (1974); Mersereau (1976), and references therein; Wee and Prakash (1979); Stark, Paul, and Sarna (1979); Tsui and Budinger (1979); Budinger, Gullberg, and Huesman (1979), and references therein; Stark, Woods, Paul, and Hingorani (1980), and references therein.

Signal Space and Convolution Methods

Bracewell (1956a); Bracewell and Riddel (1967); Ramachandran and Lakshminrayanan (1971); Mikhailov and Vainshtein (1971); Vainshtein (1970, 1971); Vainshtein and Orlov (1972); Vainshtein and Mikhailov (1972); Gilbert (1972a); Herman and Rowland (1973); Friedman, Beattie, and Laughlin (1974); Cho, Ahn, Bohm, and Huth (1974); Shepp and Logan (1974); Chesler and Riederer (1975); Wagner (1976); Drieke and Boyd (1976); Herman, Lakshminarayanan, and Naparstek (1976); Baba and Murata (1977); Horn (1978); Natterer (1978); Levitan, Degani, and Zak (1979); Minerbo (1979a); Tanaka (1979); Davidson and Grünbaum (1979); Lewitt (1979); Kenue and Greenleaf (1979a, b); Herman (1980b); Naparstek (1980); Buonocore, Brody and Macovski (1981); Nahamoo, Crawford, and Kak (1981); Gilbert, Kenue, Robb, Chu, Lent, and Swartzlander (1981), and references therein; Rosenfeld and Kak (1982).

Three-dimensional Methods

Orlov (1975a, b); Schlindwein (1978); Altschuler, Herman, and Lent (1978); Nalcioglu and Cho (1978); Denton, Friedlander, and Rockmore (1979); Minerbo (1979a); Tanaka (1979); Altschuler (1979); Kowalski (1979); Levitan (1979); Colsher (1980); Shepp (1980); Chiu, Barrett, and Simpson (1980); Lai and Lauterbur (1980); Lauterbur (1980b); Ra and Cho (1981); Schorr and Townsend (1981).

Fan-Beam Methods

Lakshminarayanan (1975); Beattie (1975); Dreike and Boyd (1976); Peters and Lewitt (1977); Herman and Naparstek (1977); Edelheit, Herman, and Lakshminarayanan (1977); Herman, Lakshminarayanan, and Naparstek (1977); Wang (1977); Reed, Truong, Chang, and Kwoh (1978); Kowalski (1978); Smith, Solmon, Wagner, and Hamaker (1978); Weinstein (1978); Kak (1979), and references therein; Gullberg (1979a); Horn (1979); Smith (1979); Glover and Eisner (1979b); Weinstein (1980); Naparstek (1980); Joseph and Schulz (1980); Hamaker, Smith, Solmon and Wagner (1980); Nassi, Brody, Cipriano, and Macovski (1981); Gilbert, Kenue, Robb, Chu, Lent, and Swartzlander (1981), and references therein; Rosenfeld and Kak (1982).

Iterative and Matrix Inversion Methods, Including Algebraic Reconstruction Techniques, Generalized Inverse Matrix Methods, and Maximum Entropy Approaches

Gordon, Bender, and Herman (1970); Bender, Bellman, and Gordon (1970); Herman and Rowland (1971); Tanabe (1971); Hounsfield (1972); Tewarson (1972); Goitein (1972); Gilbert (1972b); Krishnan, Prabhu, and Krishnamurthy (1973); Gordon (1974), and references therein; Oppenheim (1974); Guenther, Kerber, Killian, Smith, and Wagner (1974); Kashyap and Mittal (1975); Hurwitz (1975); Gullberg (1975, 1976); Herman and Lent (1976a, b, c), and references therein; Colsher (1977); Wernecke and D'Addario (1977); Schlindwein (1978); Herman, Lent and Lutz (1978); Natterer (1978); Herman, Hurwitz, Lent, and Lung (1979); Wee and Prakash (1979);

Minerbo (1979b); Herman (1979); Smith (1979); Duck and Hill (1979); Artzy, Elfving, and Herman (1979), and references therein; Herman (1980b); Mason (1981); Walters, Simon, Chesler, and Correia (1981); Wood and Morf (1981), and references therein; Buonocore, Brody and Macovski (1981), and references therein; Rosenfeld and Kak (1982).

Improvements, Optimization, Physical Considerations, Artifacts, Interpolation, Corrections, Limited Views, Evaluations and Comparisons, Error analysis and Statistical studies

Herman and Rowland (1971, 1973); Herman (1973); Cho, Ahn, Bohm, and Huth (1974); McCullough, Baker, Houser, and Reese (1974); Friedman, Beattie, and Laughlin (1974); Logan and Shepp (1975); Cho, Chan, Hall, Kruger, and McCaughey (1975); Chesler and Riederer (1975); McDavid, Waggener, Payne, and Dennis (1975); McCullough (1975); Hurwitz (1975); Tanaka and Iinuma (1975, 1976); Barrett, Gordon, and Hershell (1976); Herman and Lent (1976b, c); Brooks and Weiss (1976); Wagner (1976); Gordon (1976), and references therein; Huang and Wu (1976); Rowland (1976); Alvarez and Macovski (1976); Brooks and Di Chiro (1976b, c); Herman, Lakshminarayanan, Naparstek, Ritman, Robb, and Wood (1976); Tasto (1976, 1977); Huesman (1977); Reed, Kwoh, Truong, and Hall (1977); Shepp and Stein (1977); Bracewell (1977); Kwoh, Reed, and Truong (1977a, b); Brooks (1977); Chesler, Riederer, and Pelc (1977); Schulz, Olson, and Han (1977); Kowalski (1977a, b); Hanson (1977); Kowalski (1978); Gore and Tofts (1978); Kijewski and Bjarngard (1978); Cormack (1978); Brooks, Weiss, and Talbert (1978); Frei (1978); Tsui and Budinger (1978); Horn (1978); Joseph and Spital (1978); Tretiak (1978); Holden and Ip (1978); Riederer, Pelc, and Chesler (1978); Lewitt and Bates (1978a, b, c); Lewitt, Bates, and Peters (1978); Duerinckx and Macovski (1978); Brooks, Glover, Talbert, Eisner, and DiBianca (1979); Wee and Prakash (1979); Inouye (1979); Lewitt (1979); Crawford and Kak (1979); Stark (1979a, b); Brooks, Sank, Talbert, and Di Chiro (1979); Bottomley (1979); Brunner and Ernst (1979); Glover and Eisner (1979a); Hanson (1979a); Tsui and Budinger (1979); Vezzetti and Aks (1979); Shepp, Hilal, and Schulz (1979); Stockham (1979); Artzy, Elfving, and Herman (1979); Davison and Grünbaum (1979); Wagner (1979); Herman, Rowland, and Yau (1979); Duerinckx and Macovski (1979a, b), and references therein; Kenue and Greenleaf (1979a, b); Robb, Ritman, Gilbert, Kinsey, Harris, and Wood (1979); Wood, Kinsey, Robb, Gilbert, Harris, and Ritman (1979); Tam, Perez-Mendez, and Macdonald (1979), and references therein; Chiu, Barrett, Simpson, Chou, Arendt, and Gindi (1979); Chang (1979); Weaver and Goodenough (1979); Levitan, Degani, and Zak (1979); Strohbehn, Yates, Curran, and Sternick (1979); Tanaka (1979); Verly and Bracewell (1979); Wood, Kinsey, Robb, Gilbert, Harris, and Ritman (1979), and references therein; Hanson (1980); Morgenthaler, Brooks, and Talbert (1980); Duerinckz and Macovski (1980); Talbert, Brooks, and Morgenthaler (1980); Joseph, Hilal, Schulz, and Kelcz (1980), and references therein; Baker and Sullivan (1980); Brooks, Keller, O'Connor, and Sheridan (1980); Durrani and Goutis (1980); Herman (1980a); Sheridan, Keller, O'Connor, Brooks, and Hanson (1980); Joseph, Spital, and Stockham (1980); McCullough (1980); Rutt and Fenster (1980); Reed, Glenn, Kwoh, and Truong (1980); Swartzlander and Gilbert (1980); Louis (1980, 1981); Nassi, Brody, Cipriano, and Macovski (1981); Judy, Swensson, and Szulc (1981); Southon (1981); Davison and Grünbaum (1981); Riederer (1981); Gilbert, Kenue, Robb, Chu, Lent, and Swartzlander (1981); Schorr and Townsend (1981); Wood and Morf (1981), and references therein; Buonocore, Brody, and Macovski (1981), and references therein; McKinnon and Bates (1981); Stonestron, Alvarez, and Macovski (1981), and references therein; Gullberg and Budinger (1981), and references therein.

Reviews

Vainshtein (1973); Gordon (1974); Gordon and Herman (1974); Budinger and Gullberg (1974); Cho (1974); Mersereau and Oppenheim (1974); Brouw (1975); Crowther and Klug (1975); Gordon, Herman, and Johnson (1975); Brooks and Di Chiro (1976a); Barrett and Swindell

(1977); Swindell and Barrett (1977); Phelps (1977); Shepp and Kruskal (1978); Vainshtein (1978); Scudder (1978); Brownell, Correia, and Zamenhof (1978); Hawkes (1978); Hoppe and Typke (1978); Wade, Mueller, and Kaveh (1979); Gullberg (1979b); Mueller, Kaveh, and Wade (1979); Bracewell (1979); Budinger (1979a); Greenleaf, Johnson, Bahn, Rajagopalan, and Kenue (1979); Budinger, Gullberg, and Huesman (1979); Dines and Lytle (1979); Kak (1979); Vest (1979); Waggener and McDavid (1979); Andrew (1980b); Ter-Pogossian, Raichle, and Sobel (1980); Ellingson and Berger (1980); Herman (1980b); Jaszczak, Coleman, and Lim (1980); Hoppe and Hegerl (1980); Gilbert, Kenue, Robb, Chu, Lent, and Swartzlander (1981); Brooks, Saik, Friauf, Leighton, Cascio, and Di Chiro (1981); Greenleaf and Bahn (1981); Klepper, Brandenburger, Mimbs, Sobel, and Miller (1981); Norton and Linzer (1981); Redington and Berninger (1981); Gullberg and Budinger (1981); Lindgren and Rattey (1981).

Chapter 7

Series Methods

[7.1] INTRODUCTION

Many series approaches yield an approximation to a function $f(\mathbf{x})$ if sufficient information about the Radon transform $\check{f}(p, \xi)$ is available. Most of these approaches depend on the particular physical situation and the quality of the data. It is not our purpose to review these technical matters; rather, we will concentrate on some general results that form the basis for most of the series approaches in use. Hopefully, this will serve to help place the special uses in better perspective.

The general framework is outlined in [7.2] and various special cases for $n = 2$ and $n = 3$ are discussed in Sections [7.3] through [7.7]. Finally, in [7.8] a few other approaches are mentioned, some of which are currently being developed.

[7.2] GEGENBAUER TRANSFORM PAIR

Background material for this section appears in Chapters 4 and 5. Suppose $f(\mathbf{x})$ may be written as a sum

$$f(\mathbf{x}) = \sum_{l,m} A_{lm} g_l(r) S_{lm}(\omega), \tag{2.1}$$

where the A_{lm} are constants. In the general case $\Sigma_{l,m}$ means $\Sigma_{m=1}^{\mathcal{N}} \Sigma_{l=0}^{\infty}$ with S_{lm} and $\mathcal{N} = \mathcal{N}(n, l)$ defined in [4.3]. In practice, the sum would be finite and represent an approximation to $f(\mathbf{x})$.

The Radon transform of (2.1) is given by

$$\check{f}(p, \xi) = \sum_{l,m} A_{lm} \check{g}_l(p) S_{lm}(\xi). \tag{2.2}$$

It is assumed that $\check{f}(p, \xi)$ is known. The constants may be determined by doing the angular integrations and making use of the orthogonality of the

$S_{lm}(\xi)$,

$$\int_{|\xi|=1} \check{f}(p,\xi) S_{l'm'}(\xi)\, d\xi = \int_{|\xi|=1} \sum_{l,m} A_{lm} \check{g}_l(p) S_{lm}(\xi) S_{l'm'}(\xi)\, d\xi$$

$$= \sum_{l,m} A_{lm} \check{g}_l(p) \delta_{ll'} \delta_{mm'}$$

$$= A_{l'm'} \check{g}_{l'}(p).$$

Equivalently,

$$A_{lm} \check{g}_l(p) = \int_{|\xi|=1} \check{f}(p,\xi) S_{lm}(\xi)\, d\xi. \tag{2.3}$$

The radial functions $g_l(r)$ in (2.1) and $\check{g}_l(p)$ in (2.3) are a Gegenbauer transform pair. From (4.18) in [5.4]

$$g_l(r) = \frac{(-1)^{2\nu+1}\Gamma(l+1)\Gamma(\nu)}{2\pi^{\nu+1}\Gamma(l+2\nu)r} \int_r^{\infty} \check{g}_l^{(2\nu+1)}(p) C_l^{\nu}\left(\frac{p}{r}\right)\left[\frac{p^2}{r^2}-1\right]^{\nu-1/2} dp. \tag{2.4}$$

The preceding approach looks straightforward enough. Determine $A_{lm}\check{g}_l(p)$ from (2.3) and use (2.4) to get $A_{lm}g_l(r)$, then compute $f(\mathbf{x})$ from (2.1). The actual numerical implementation of this procedure is quite difficult however, since there are rather large oscillatory cancellations which occur for certain ratios of p/r in (2.4). There is a need for further investigation of this difficulty.

Special cases of the general development here are considered in the following sections.

[7.3] CIRCULAR HARMONIC EXPANSION ($n = 2$)

Reduction of General Equation

Observe, from (3.4) of [4.3] that $\mathcal{N}(2, l) = 2$. As usual for the two-dimensional case, we let $\omega = (\cos\theta, \sin\theta)$. The circular functions $S_{lm}(\omega)$ may be selected as

$$S_{l1}(\omega) = \frac{1}{\sqrt{\pi}} \cos l\theta, \qquad S_{l2}(\omega) = \frac{1}{\sqrt{\pi}} \sin l\theta. \tag{3.1}$$

To emphasize the consistency of this choice, it is useful to show an example using (3.10) of [4.3] for this special case of $n = 2$.

Example 1. If $n = 2$, then $\nu = 0$, and

$$C_l^0(\xi \cdot \omega) = \frac{2\pi}{2} C_l^0(1) \sum_{m=1}^{2} S_{lm}(\xi) S_{lm}(\omega)$$

$$= C_l^0(1) \cos l(\theta - \phi).$$

Note that

$$\lim_{\alpha \to 0} \frac{C_l^\alpha(t)}{\alpha} = C_l^0(t) = \frac{2}{l} T_l(t),$$

where $T_l(t)$ is a Tchebycheff polynomial of the first kind. Now, if $t = \xi \cdot \omega = \cos(\theta - \phi)$, then

$$\frac{2}{l} T_l(t) = \frac{2}{l} T_l(1) \cos[l \arccos t]$$

or since $T_l(1) = 1$,

$$T_l(t) = \cos[l \arccos t].$$

This result is often taken as the definition of $T_l(t)$. □

When working in two dimensions it is often more convenient to use certain linear combinations of the S_{lm}; for example,

$$\sqrt{\pi}\,(S_{l1} + iS_{l2}) = e^{il\theta}$$

and

$$\sqrt{\pi}\,(S_{l1} - iS_{l2}) = e^{-il\theta}.$$

Also, observe that both the object $f(r, \theta)$ and the transform $\check{f}(p, \phi)$ may be assumed periodic with period 2π. Consequently, it is reasonable to expand $f(r, \theta)$ in the form

$$f(r, \theta) = \sum_{l=-\infty}^{\infty} g_l(r) e^{il\theta}. \tag{3.2}$$

Observe that this is of the same form as (2.1) with an appropriate linear combination of the S_{lm} to yield the $e^{il\theta}$ term. The constant has been absorbed in the function $g_l(r)$. Now, the Radon transform of (3.2) yields

$$\check{f}(p, \phi) = \sum_{l=-\infty}^{\infty} \check{g}_l(p) e^{il\phi}. \tag{3.3}$$

The function $\check{g}_l(p)$ can be computed by the usual procedure:

$$\int_0^{2\pi} \check{f}(p,\phi) e^{-il'\phi}\,d\phi = \sum_{l=-\infty}^{\infty} \check{g}_l(p) \int_0^{2\pi} e^{il\phi} e^{-il'\phi}\,d\phi$$

$$= \sum_{l=-\infty}^{\infty} \check{g}_l(p) 2\pi\delta_{ll'}$$

$$= 2\pi \check{g}_{l'}(p).$$

Equivalently,

$$\check{g}_l(p) = \frac{1}{2\pi}\int_0^{2\pi} \check{f}(p,\phi) e^{-il\phi}\,d\phi. \tag{3.4}$$

To find the equation relating $\check{g}_l(p)$ to $g_l(r)$, we evaluate (2.4) using $n = 2$. The procedure is straightforward if one first multiplies by ν/ν, takes the limit as $\nu \to 0$, and uses

$$\lim_{\nu\to 0} C_l^\nu(z) = \frac{2}{l} T_l(z).$$

The result is in the form of a Tchebycheff transform

$$g_l(r) = \frac{-1}{\pi r}\int_r^\infty \check{g}_l'(p) T_l\left(\frac{p}{r}\right)\left(\frac{p^2}{r^2} - 1\right)^{-1/2} dp, \tag{3.5}$$

where $\check{g}_l'(p) = \frac{d}{dp}\check{g}_l(p)$ and $r > 0$. The inverse equation that relates $g_l(r)$ and $\check{g}_l(p)$ follows from (4.18a) in [5.4]:

$$\check{g}_l(p) = 2\int_p^\infty g_l(r) T_l\left(\frac{p}{r}\right)\left(1 - \frac{p^2}{r^2}\right)^{-1/2} dr, \tag{3.6}$$

for $p \geqslant 0$. Note that in (3.5) $r \neq 0$. It turns out that for $r = 0$ the only contribution to $f(0, \theta)$ comes from the $l = 0$ part of the expansion (3.2). This is probably most readily verified by performing the steps outlined. Substitute (3.3) into the inversion formula (3.13) of [5.3], set $r = 0$, do the ϕ integration first for $l = 2, 4, 6, \ldots,$ perform the p integration first for $l = 1, 3, 5, \ldots,$ and make use of the symmetry condition (3.16) of [4.3].

The inversion formula (3.5) was discovered by Cormack (1963, 1964) and has been rederived by various methods including use of Mellin transforms; see Ein-Gal (1974), Deans (1977), and Hansen (1981).

Abel Transform

Consider equations (3.5) and (3.6) with $l = 0$. (This is the circularly symmetric case.) The Abel transform pair [Bracewell (1978)] emerges immediately:

$$g_0(r) = \frac{-1}{\pi} \int_r^\infty \check{g}_0'(p)\left[p^2 - r^2\right]^{-1/2} dp, \tag{3.7a}$$

$$\check{g}_0(p) = 2\int_p^\infty r g_0(r)\left[r^2 - p^2\right]^{-1/2} dr. \tag{3.7b}$$

Note that the $r = 0$ case may give a nonzero contribution to $f(0, \theta)$. For example, if $\check{g}_0(p) = \sqrt{\pi}\, e^{-p^2}$, then $g_0(0) = 1$ and $f(0, \theta) = 1$.

Direct Use of Radon's Formula

For reasons that will become clear as the discussion progresses, it is convenient to make direct use of Radon's inversion formula in finding a relationship between $g_l(r)$ and $\check{g}_l(p)$. We begin by observing that (3.13) in [5.3] can be written in the form

$$f(r, \theta) = \frac{-1}{2\pi^2} \int_0^\pi d\phi \int_{-1}^1 dp \frac{\partial \check{f}/\partial p}{p - r\cos(\phi - \theta)}, \qquad 0 < r \leqslant 1, \tag{3.8}$$

if it is assumed that f vanishes outside a disk of radius 1. Actually, it is not essential to work in a unit disk since the region could be scaled up or down as desired; however, it is much less cumbersome to use the unit disk throughout and assume that $f \in \mathscr{D}$.

It will be convenient to rewrite (3.8) in the form

$$f(r, \theta) = \frac{-1}{2\pi^2} \int_0^\pi d\phi \int_{-1}^1 dp \int_{-r}^r \frac{dt}{p - t} \frac{\partial \check{g}_l(p)}{\partial p} e^{il\phi} \delta\left[t - r\cos(\phi - \theta)\right], \tag{3.9}$$

where $|t/r| \leqslant 1$.

We replaced $\check{f}(p, \phi)$ by $\check{g}_l(p)e^{il\phi}$. [Keep in mind that $\check{g}_l(-p) = (-1)^l \check{g}_l(p)$.] The sum over l need not be carried along for the current arguments. First, consider the ϕ integral, designated by J

$$J = \int_0^\pi e^{il\phi} \delta\left[t - r\cos(\phi - \theta)\right] d\phi. \tag{3.10}$$

Make the change of variables $x = \cos(\phi - \theta)$, then J becomes

$$J = \int_{-\cos\theta}^{\cos\theta} \frac{\exp[il\theta + il\cos^{-1}x]\delta(t - rx)}{(1 - x^2)^{1/2}} dx. \tag{3.11}$$

Rewrite $\delta(t - rx)$ as $(1/r)\delta(t/r - x)$ with $0 < r \leqslant 1$ and $|t/r| \leqslant 1$. The integration over the δ function immediately yields

$$J = \frac{1}{r} \frac{\exp[il\theta + il\cos^{-1}(t/r)]}{(1 - t^2/r^2)^{1/2}}. \tag{3.12}$$

The exponential can be written in the form

$$e^{il\theta}\left\{\cos\left[l\cos^{-1}\left(\frac{t}{r}\right)\right] + i\sin\left[l\cos^{-1}\left(\frac{t}{r}\right)\right]\right\},$$

which leads to the introduction of Tchebycheff polynomials of the first and second kinds (see Appendix C):

$$J = e^{il\theta}\left\{\frac{T_l\left(\frac{t}{r}\right)}{(r^2 - t^2)^{1/2}} + \frac{i}{r}U_{l-1}\left(\frac{t}{r}\right)\right\}. \tag{3.13}$$

Now that the ϕ integration has been completed, (3.9) becomes

$$f(r, \theta) = g_l(r)e^{il\theta} = \frac{-1}{2\pi^2}\int_{-1}^{+1} dp\, \check{g}_l'(p)\int_{-r}^{r}\frac{dt}{p - t}J.$$

Another change of variables is in order. Let $t = rx$, $-1 \leqslant x \leqslant 1$, then

$$g_l(r)e^{il\theta} = \frac{e^{il\theta}}{2\pi^2 r}\int_{-1}^{1} dp\, \check{g}_l'(p)\int_{-1}^{1}\frac{dx}{\left(x - \frac{p}{r}\right)}\left\{\frac{T_l(x)}{(1 - x^2)^{1/2}} + iU_{l-1}(x)\right\}. \tag{3.14}$$

The term involving $U_{l-1}(x)$ may be dropped. We expect that since $g_l(r)$ is real. Also, it is easy to show that

$$K = \int_{-1}^{1} dp\, \check{g}_l'(p)\int_{-1}^{1} dx \frac{U_{l-1}(x)}{\left(x - \frac{p}{r}\right)}$$

vanishes. (Let $x \to -x$, $p \to -p$, and show that $K = -K$.) After dividing by the common factor of $e^{il\theta}$ and dropping the term involving $U_{l-1}(x)$, we have

$$g_l(r) = \frac{1}{2\pi^2 r}\int_{-1}^{1} dp\, \breve{g}_l'(p) \int_{-1}^{1} \frac{dx}{\left(x - \frac{p}{r}\right)} \frac{T_l(x)}{(1-x^2)^{1/2}}. \tag{3.15}$$

Select the x integration from (3.15) and define

$$I_l\left(\frac{p}{r}\right) = \int_{-1}^{1}\left(x - \frac{p}{r}\right)^{-1}(1-x^2)^{-1/2} T_l(x)\, dx. \tag{3.16}$$

Then,

$$g_l(r) = \frac{-1}{2\pi^2 r}\int_{-1}^{1} I\left(\frac{p}{r}\right)\breve{g}_l'(p)\, dp. \tag{3.17}$$

It is possible to simplify (3.17), but care must be taken since the evaluation of $I_l(p/r)$ depends on whether $|p/r| < 1$ or $|p/r| > 1$. To properly take this into consideration, we write

$$g_l(r) = \frac{1}{2\pi^2 r}\left\{\int_{-1}^{-r} I_l\left(\frac{p}{r}\right)\breve{g}_l'(p)\, dp + \int_{-r}^{0} I_l\left(\frac{p}{r}\right)\breve{g}_l'(p)\, dp + \int_{0}^{r} I_l\left(\frac{p}{r}\right)\breve{g}_l'(p)\, dp + \int_{r}^{1} I_l\left(\frac{p}{r}\right)\breve{g}_l'(p)\, dp\right\}.$$

By making the simultaneous change of variables $x \to -x$, $p \to -p$ in the first two integrals on the right-hand side, it follows that

$$g_l(r) = \frac{1}{\pi^2 r}\left\{\int_{0}^{r} I_l\left(\frac{p}{r}\right)\breve{g}_l'(p)\, dp + \int_{r}^{1} I_l\left(\frac{p}{r}\right)\breve{g}_l'(p)\, dp\right\}. \tag{3.18}$$

Note that $0 < r < 1$ and from $f(r, \theta) = f(-r, \theta + \pi)$, it follows that $g_l(-r) = (-1)^l g_l(r)$. It is now possible to evaluate $I_l(p/r)$ by use of tabulated results [Luke (1969)], since in the first integral of (3.18), $p/r \leqslant 1$, and in the second integral, $p/r \geqslant 1$. The result is

$$g_l(r) = \frac{1}{\pi r}\int_{0}^{r} \breve{g}_l'(p) U_{l-1}\left(\frac{p}{r}\right) dp - \frac{1}{\pi r}\int_{r}^{1}\left(\frac{p^2}{r^2} - 1\right)^{-1/2}\left(\frac{p}{r} + \sqrt{\frac{p^2}{r^2} - 1}\right)^{-l} \breve{g}_l'(p)\, dp. \tag{3.19}$$

This formula has been obtained by several authors [Perry (1975); Deans

(1977); Minerbo and Sanderson (1977)] by various methods. In particular, see Hansen (1981), where solution (3.19) is classified as the noncausal-stable form as opposed to (3.5), which is classified by Hansen (1981) as the causal-unstable form. Hansen's arguments are based upon inversion of the Mellin transform and will not be repeated here; however, it may be useful to indicate how (3.19) is equivalent to (3.5), given that (3.6) holds.

Conversion of (3.19) to (3.5)

Using the Tchebycheff polynomials in [C.2], it follows immediately that (3.19) may be modified to read

$$g_l(r) = \frac{1}{\pi r}\int_0^1 \check{g}_l'(p)U_{l-1}\left(\frac{p}{r}\right)dp - \frac{1}{\pi r}\int_r^1 \check{g}_l'(p)\frac{T_l\left(\frac{p}{r}\right)}{\left(\frac{p^2}{r^2}-1\right)^{1/2}}dp.$$

It remains to show that the first integral on the right-hand side,

$$K = \int_0^1 \check{g}_l'(p)U_{l-1}\left(\frac{p}{r}\right)dp,$$

vanishes. The details have been given for $n = 2$ [Deans (1977)] and for the general case [Deans (1978)]. The basic idea is to integrate by parts, which shifts the derivative from $\check{g}_l$ to U_{l-1}, and then insert the expression in (3.6) for $\check{g}_l(p)$. Finally, after some algebraic manipulations, the fact that $T_l(x)$ is orthogonal to any polynomial of degree less than l over the interval $[-1, 1]$ can be used to show that K vanishes.

Thus (3.5) with upper limit unity and (3.19) are equivalent. The important problem of comparing the numerical implementation remains. Hansen (1981) has promised that such a study is forthcoming.

An Alternative Form for $g_l(r)$

Following the derivation by Cormack (1963), an alternative form for $g_l(r)$ can be obtained. Start with

$$\check{g}_l(p) = 2\int_p^1 \frac{rg_l(r)T_l\left(\frac{p}{r}\right)dr}{(r^2-p^2)^{1/2}},$$

multiply both sides by $(t/p)(p^2 - t^2)^{-1/2}$ and integrate from $p = t$ to $p = 1$.

$$\int_t^1 \frac{t\check{g}_l(p)T_l\left(\frac{p}{t}\right)}{p(p^2-t^2)^{1/2}}dp = 2\int_t^1 dp\int_p^1 dr\frac{g_l(r)trT_l\left(\frac{p}{t}\right)T_l\left(\frac{p}{r}\right)}{p(r^2-p^2)^{1/2}(p^2-t^2)^{1/2}}.$$

Now change the order of integration in the double integral (see Fig. 7.1):

$$\int_t^1 \frac{t\check{g}_l(p)T_l\left(\frac{p}{t}\right)dp}{p(p^2-t^2)^{1/2}} = 2\int_t^1 g_l(r)\,dr \int_t^r \frac{rtT_l\left(\frac{p}{t}\right)T_l\left(\frac{p}{r}\right)dp}{p(r^2-p^2)^{1/2}(p^2-t^2)^{1/2}}.$$

The p integral on the right-hand side is a special case of more general results involving Gegenbauer polynomials [Deans (1978)] and Jacobi polynomials [Koornwinder (1982)]. Reduce (4.19) of [5.4] to the $\nu = 0$ case, make the variable change $x \to 1/x$, and finally redefine the constants s and t to obtain the desired form. The result is $\pi/2$, which immediately yields

$$\pi\int_t^1 g_l(r)\,dr = \int_t^1 \frac{t\check{g}_l(p)T_l\left(\frac{p}{t}\right)dp}{p(p^2-t^2)^{1/2}}.$$

The final step is to differentiate with respect to t. The left-hand side gives $-\pi g_l(t)$, and upon solving for $g_l(t)$,

$$g_l(t) = \frac{-1}{\pi}\frac{d}{dt}\int_t^1 \frac{t\check{g}_l(p)T_l\left(\frac{p}{t}\right)dp}{p(p^2-t^2)^{1/2}}.$$

Of course, the independent variable t can be replaced by r to give the inversion

$$g_l(r) = \frac{-1}{\pi}\frac{d}{dr}\int_r^1 \frac{\check{g}_l(p)T_l\left(\frac{p}{r}\right)dp}{p\left(\frac{p^2}{r^2}-1\right)^{1/2}}, \tag{3.20}$$

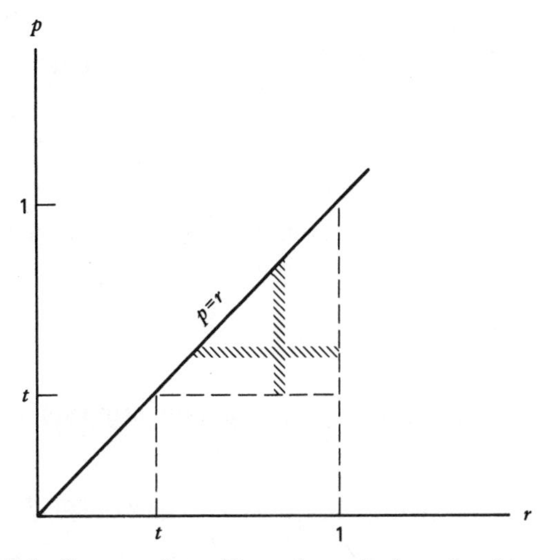

Figure 7.1 Geometry for making a change in the order of integration.

which is to be compared with

$$g_l(r) = \frac{-1}{\pi r} \int_r^1 \frac{\check{g}_l'(p) T_l\left(\frac{p}{r}\right) dp}{\left(\frac{p^2}{r^2} - 1\right)^{1/2}}.$$

[7.4] SPHERICAL HARMONIC EXPANSION ($n = 3$)

The Y_{lm} and the S_{lm}

If $n = 3$, the real orthonormal functions S_{lm} are related to the more commonly used spherical harmonics Y_{lm}. Before establishing the relationship, we give the basic relations that define the Y_{lm}. These definitions and phases coincide with standard usage [Condon and Shortley (1970); Jackson (1975); Arfken (1970)].

The Legendre polynomials $P_l(t)$ are defined by

$$P_l(t) = \frac{1}{2^l l!} \left(\frac{d}{dt}\right)^l (t^2 - 1)^l, \tag{4.1}$$

and the associated Legendre functions are defined by

$$P_l^m(t) = \frac{1}{2^l l!} (1 - t^2)^{m/2} \left(\frac{d}{dt}\right)^{l+m} (t^2 - 1)^l, \qquad 0 \leqslant m \leqslant l. \tag{4.2}$$

The spherical harmonics in terms of the associated Legendre functions are

$$Y_{lm}(\theta, \phi) = (-1)^m \left[\frac{(2l+1)}{4\pi} \frac{(l-m)!}{(l+m)!}\right]^{1/2} P_l^m(\cos\theta) e^{im\phi}, \qquad 0 \leqslant m \leqslant l, \tag{4.3}$$

where θ and ϕ are the polar and azimuthal angles, respectively. The Y_{lm} satisfy the symmetry condition

$$Y_{l,-m}(\theta, \phi) = (-1)^m Y_{lm}^*(\theta, \phi), \tag{4.4}$$

where the * means complex conjugation, and the orthogonality condition is

$$\langle Y_{lm}, Y_{l'm'} \rangle = \int_0^{2\pi} d\phi \int_0^{\pi} d\theta \, Y_{lm}^*(\theta, \phi) Y_{l'm'}(\theta, \phi) \sin\theta = \delta_{ll'} \delta_{mm'}. \tag{4.5}$$

One way to connect the S_{lm} and the Y_{lm} is to define the $\mathcal{N}(n, l) = \mathcal{N}(3, l) = 2l + 1$ functions S_{lm} by

$$S^e_{l,m} = \frac{Y_{lm} + Y^*_{lm}}{\sqrt{2}} = (-1)^m \left[\frac{(2l+1)}{2\pi} \frac{(l-m)!}{(l+m)!} \right]^{1/2} P_l^m(\cos\theta)\cos m\phi,$$

for $m = 1, 2, \ldots, l$

$$S^o_{l,m} = \frac{Y_{lm} - Y^*_{lm}}{i\sqrt{2}} = (-1)^m \left[\frac{(2l+1)}{2\pi} \frac{(l+m)!}{(l-m)!} \right]^{1/2} P_l^{-m}(\cos\theta)\sin m\phi,$$

for $m = -1, -2, \ldots, -l$

$$S_{l0} = Y_{l0} = \sqrt{\frac{2l+1}{4\pi}}\, P_l(\cos\theta), \qquad \text{for } m = 0.$$

The superscripts e (for even) and o (for odd) have been inserted to help identify the types of real spherical harmonics. The preceding equations may be solved for Y_{lm} and Y^*_{lm} if we first note that

$$S^o_{l,-m} = \frac{Y_{l,-m} - Y^*_{l,-m}}{i\sqrt{2}} = (-1)^m \frac{(Y^*_{lm} - Y_{lm})}{i\sqrt{2}}, \qquad \text{for } m = 1, 2, \ldots, l.$$

As an example we use the addition theorem (3.10) in [4.3].

Example 1. First observe that for $n = 3$, $\nu = \frac{1}{2}$ and $C_l^{1/2}(t) = P_l(t)$. Suppose ω and ξ are unit vectors characterized by (θ, ϕ) and (θ', ϕ'), respectively. Let $t = \omega \cdot \xi = \cos\gamma$, where γ is the angle between the two vectors. The addition theorem yields

$$C_l^{1/2}(t) = P_l(\cos\gamma) = \frac{4\pi}{2l+1} \sum_m S_{lm}(\omega) S_{lm}(\xi).$$

The sum over m can be expressed in three separate parts,

$$Y_{l0}(\theta, \phi) Y_{l0}(\theta', \phi')$$

$$+ \sum_{m=1}^{l} \left[\frac{Y_{lm}(\theta, \phi) + Y^*_{lm}(\theta, \phi)}{\sqrt{2}} \right] \left[\frac{Y_{lm}(\theta', \phi') + Y^*_{lm}(\theta', \phi')}{\sqrt{2}} \right]$$

$$+ \sum_{m=-1}^{-l} \left[\frac{Y_{lm}(\theta, \phi) - Y^*_{lm}(\theta, \phi)}{i\sqrt{2}} \right] \left[\frac{Y_{lm}(\theta', \phi') - Y^*_{lm}(\theta', \phi')}{i\sqrt{2}} \right].$$

After performing the indicated multiplication inside the summation and appropriate use of (4.4), the usual spherical harmonic addition theorem emerges:

$$P_l(\cos\gamma) = \frac{4\pi}{2l+1} \sum_{m=-l}^{l} Y_{lm}^*(\theta', \phi') Y_{lm}(\theta, \phi). \qquad \square$$

It should be clear now that we can use either the S_{lm} or the Y_{lm} in series expansions. The advantage in using the S_{lm} is that they are real and orthonormal, but the Y_{lm} are more commonly used and thus familiar to a wider audience.

A Legendre Transform Pair

Given that $f(\mathbf{x}) = f(r, \theta, \phi)$ can be expanded in the form

$$f(r, \omega) = \sum_{m=-l}^{l} \sum_{l=0}^{\infty} A_{lm} g_l(r) Y_{lm}(\theta, \phi), \tag{4.6}$$

it follows that

$$\check{f}(p, \xi) = \sum_{m=-l}^{l} \sum_{l=0}^{\infty} A_{lm} \check{g}_l(p) Y_{lm}(\theta', \phi'). \tag{4.7}$$

Here we have used unit vectors

$$\omega = (\sin\theta\cos\phi, \sin\theta\sin\phi, \cos\theta)$$

and

$$\xi = (\sin\theta'\cos\phi', \sin\theta'\sin\phi', \cos\theta').$$

As in [7.2], the constants and $\check{g}_l(p)$ are determined by integrating over angles,

$$A_{lm} \check{g}_l(p) = \int_0^{2\pi} \int_0^{\pi} \check{f}(p, \xi) Y_{lm}(\theta', \phi') \sin\theta' \, d\theta' \, d\phi'. \tag{4.8}$$

Since the Y_{lm} are being used rather than the S_{lm}, the constants A_{lm} will be complex in general. In this case ($n = 3$, $\nu = \frac{1}{2}$), the radial functions are Legendre transform pairs. From (4.18) in [5.4],

$$\check{g}_l(p) = 2\pi \int_p^{\infty} r g_l(r) P_l\left(\frac{p}{r}\right) dr \tag{4.9}$$

and

$$g_l(r) = \frac{1}{2\pi r} \int_r^{\infty} \check{g}_l''(p) P_l\left(\frac{p}{r}\right) dp. \tag{4.10}$$

As usual, $\check{g}_l''(p)$ means $(d/dp)^2 \check{g}_l(p)$.

[7.5] A TCHEBYCHEFF TRANSFORM PAIR OF THE SECOND KIND

Four-dimensional problems arise in connection with a study of diatomic molecules [Judd (1975)]. The S_{lm} can be related to the four-dimensional spherical harmonics used by Judd (1975). Series expansions follow the same program discussed in the preceding sections. The major difference is that the radial functions are Tchebycheff transform pairs of the second kind. If $n = 4$, then $\nu = 1$ and

$$C_l^1(t) = U_l(t).$$

The transform pair g_l and $\check{g}_l$ satisfy

$$\check{g}_l(p) = \frac{4\pi}{l+1}\int_p^\infty r^2 g_l(r) U_l\left(\frac{p}{r}\right)\left(1 - \frac{p^2}{r^2}\right)^{1/2} dr,$$

$$g_l(r) = \frac{-1}{2\pi^2(l+1)r}\int_r^\infty \check{g}_l'''(p) U_l\left(\frac{p}{r}\right)\left(\frac{p^2}{r^2} - 1\right)^{1/2} dp.$$

[7.6] ORTHOGONAL FUNCTION EXPANSIONS ON THE UNIT DISK

Desirable Properties of the Basis Sets

From the preceding sections, it might appear that the inversion problem is solved very nicely by series methods. One simply determines $\check{g}_l(p)$ from the data, computes $g_l(r)$ on a computer, and finally performs the appropriate sum (2.1) to recover f. The difficulty is that when $\check{g}_l(p)$ is experimentally determined, the operation leading to $g_l(r)$ is not a simple numerical operation. Consequently, it is advantageous to attempt to approximate $\check{g}_l(p)$ by some appropriate analytic functions. An infinite number of choices are available unless restrictions are placed on the properties of those ultimately selected. Two properties that immediately come to mind are orthogonality and completeness.

Let us focus on the two-dimensional case, $n = 2$, with $f(x, y)$ confined to the unit disk. There are two obvious possibilities: (i) assume $g_l(r)$ can be expanded in a series of orthogonal functions and use (3.6) to compute the corresponding expansion for $\check{g}_l(p)$; (ii) assume $\check{g}_l(p)$ can be expanded in a set of orthogonal functions and use (3.5) to find the corresponding expansion for $g_l(r)$. A variation on these approaches is to avoid using either (3.5) or (3.6) directly; this is what we shall do eventually. Note that selecting an orthogonal set in which to expand one of the functions $g_l(r)$ or $\check{g}_l(p)$ is no guarantee that the transformed set will also be orthogonal. Thus another desirable property is that both sets be orthogonal. For each fixed l, there should be an orthogonal expansion for $g_l(r)$ and $\check{g}_l(p)$.

Convenient sets of functions that have precisely the desired properties do exist. These functions are Zernike polynomials, which have a long history of use by workers in optics, and Tchebycheff polynomials of the second kind, which emerge by transforming the Zernike polynomials. Actually, the way the Zernike polynomials have been used in optics along with their invariance properties under a rotation of the disk adequately serve to motivate their consideration [Zernike (1934); Bhatia and Wolf (1954); Herlitz (1963); Born and Wolf (1975)]. Further motivation that also leads to the Zernike polynomials will be presented later. Also, see Marr (1974), Zeitler (1974), Lerche and Zeitler (1976), and Eggermont (1975).

Recall that it was assumed

$$f(x, y) = \sum_{l=-\infty}^{\infty} g_l(r) e^{il\theta}. \tag{6.1}$$

We now assume that $f(x, y)$ can be approximated by a sum of monomials of the form $x^k y^m$. Suppose $f(x, y) = x^k y^m$, then $f(r\cos\theta, r\sin\theta) = r^{k+m} f(\cos\theta, \sin\theta)$. This makes it plausible to further expand $g_l(r)$ in terms of the complete set [Born and Wolf (1975)] of Zernike polynomials to obtain the expansion

$$f(x, y) = \sum_{l=-\infty}^{\infty} g_l(r) e^{il\theta} = \sum_{s=0}^{\infty} \sum_{l=-\infty}^{\infty} A_{ls} Z_{l+2s}^{l}(r) e^{il\theta}, \tag{6.2}$$

where the A_{ls} are constants. The notation for the Zernike polynomials has been selected to conform to the conventional usage except for the use of Z rather than R, as in Born and Wolf (1975).

Some Properties of the Zernike Polynomials

The functions $B_s^l = Z_{|l|+2s}^{|l|} e^{il\theta}$ form a complete set of orthogonal functions over the unit disk.* The orthogonality relation for the Zernike polynomials is

$$\int_0^1 Z_{l+2s}^{l}(r) Z_{l+2t}^{l}(r) r\, dr = \frac{1}{2(|l| + 2s + 1)} \delta_{st}. \tag{6.3}$$

Thus the constants A_{ls} in (6.2) are given by

$$A_{ls} = \frac{2(|l| + 2s + 1)}{2\pi} \int_0^{2\pi} \int_0^1 f(r\cos\theta, r\sin\theta) Z_{l+2s}^{l}(r) e^{-il\theta} r\, dr\, d\theta, \tag{6.4}$$

with $A_{-l,s} = A_{ls}^*$.

*See Tables 7.1 and 7.2.

For a given value of $l \geqslant 0$, the members of the set of $Z_{l+2s}^{l}(r)$ include

$$\{Z_l^l, Z_{l+2}^l, Z_{l+4}^l, \ldots\}.$$

These polynomials may be obtained by orthogonalizing the powers $r^l, r^{l+2}, r^{l+4}, \ldots$ with weight factor r over the interval $0 \leqslant r \leqslant 1$. It follows that $Z_{l+2s}^l(r)$ is a polynomial in r of degree $l + 2s$ and contains no powers of r less than l. Since $Z_{l+2s}^l(r)$ is even or odd depending on whether l is even or odd, it follows that

$$Z_{l+2s}^l(-r) = (-1)^l Z_{l+2s}^l(r). \tag{6.5}$$

Note that this property is essential since we also must require that

$$g_l(-r) = (-1)^l g_l(r).$$

For $l < 0$, Z_{l+2s}^l is computed from

$$Z_{l+2s}^l(r) = Z_{|l|+2s}^{|l|}(r). \tag{6.6}$$

(If we keep this in mind, we can write l rather than $|l|$ in most of the following development.) In Table 7.1, we list the first few polynomials, and in Table 7.2, we indicate those basis functions of degree $\leqslant |l| + 2s$; compare these tables with Tables 3.1 and 3.2. A more complete list, similar to Table 7.1, is given by Born and Wolf (1975) along with many additional properties such as generating function, recursion formulas, integral relations, and relation to other special functions. Additional properties can be deduced from the fact that the $Z_{l+2s}^l(r)$ are related to the more general Jacobi polynomials $P_n^{(\alpha,\beta)}(z)$

$$Z_{l+2s}^l(r) = r^{|l|} P_s^{(0,|l|)}(2r^2 - 1), \tag{6.7}$$

which are extensively discussed in many sources [Szegö (1939); Erdélyi, Magnus, Oberhettinger, and Tricomi (1953b); Abramowitz and Stegun (1964)].

Table 7.1 The First Few Zernike Polynomials $Z_{l+2s}^l(r)$

	$l + 2s$					
l	0	1	2	3	4	5
0	1		$2r^2 - 1$		$6r^4 - 6r^2 + 1$	
1		r		$3r^3 - 2r$		$10r^5 - 12r^3 + 3r$
2			r^2		$4r^4 - 3r^2$	
3				r^3		$5r^5 - 4r^3$
4					r^4	
5						r^5

Table 7.2 **Linearly Independent Functions** $B_s^l = Z_{|l|+2s}^{|l|}(r)e^{il\theta}$

M	Degree $\lvert l\rvert + 2s \leqslant M$	Total Number
0	B_0^0	1
1	B_0^0, B_0^1, B_0^{-1}	3
2	$B_0^0, B_0^1, B_0^{-1}, B_1^0, B_0^2, B_0^{-2}$	6
3	$B_0^0, B_0^1, B_0^{-1}, B_1^0, B_0^2, B_0^{-2}, B_1^1, B_1^{-1}, B_0^3, B_0^{-3}$	10
⋮		⋮
m	$B_0^0, \ldots, B_0^{-m}$	$\frac{1}{2}(m+1)(m+2)$

Example 1. Suppose $f(x, y) = x$, then

$$f(r\cos\theta, r\sin\theta) = r\cos\theta.$$

Since the degree is 1, $|l| + 2s \leqslant 1$ and the series expansion

$$f(x, y) = \sum_{s=0}^{\infty} \sum_{l=-\infty}^{\infty} A_{ls} Z_{l+2s}^{l}(r) e^{il\theta}$$

by inspection of Table 7.1 reduces to

$$f(x, y) = A_{00} Z_0^0 + A_{10} Z_1^1 e^{i\theta} + A_{-10} Z_1^1 e^{-i\theta}.$$

Clearly $A_{00} = 0$, $A_{10} = A_{-10} = \frac{1}{2}$ leads to the desired expansion

$$f(x, y) = r\frac{e^{i\theta} + e^{-i\theta}}{2} = Z_1^1(r)\cos\theta. \qquad \square$$

Example 2. Suppose $f(x, y) = (x^2 + y^2)x$. This time the sum in terms of exponentials is slightly more difficult to find by inspection, but it can be found with a little trial and error:

$$f(x, y) = \left(A_{11}e^{i\theta} + A_{-11}e^{-i\theta}\right)Z_3^1 + \left(A_{10}e^{i\theta} + A_{-10}e^{-i\theta}\right)Z_1^1$$

with $A_{11} = A_{-11} = \frac{1}{6}$, $A_{10} = A_{-10} = \frac{1}{3}$. Thus

$$\begin{aligned} f(x, y) &= \tfrac{1}{3}Z_3^1(r)\cos\theta + \tfrac{2}{3}Z_1^1(r)\cos\theta \\ &= \tfrac{1}{3}(3r^3 - 2r)\cos\theta + \tfrac{2}{3}r\cos\theta \\ &= r^3\cos\theta = \left(x^2 + y^2\right)x. \end{aligned}$$

Of course, the coefficients could have been found by direct integration of (6.4); for example,

$$A_{11} = \frac{4}{\pi}\int_0^1 r^3 Z_3^1(r) r\,dr \int_0^{2\pi} \cos\theta\, e^{-i\theta}\,d\theta$$

$$= 4\int_0^1 (3r^7 - 2r^5)\,dr = \frac{1}{6}. \qquad \square$$

These simple examples illustrate the use of Table 7.1 and the formulas used to solve for the coefficients. We shall return to this briefly after we learn how to transform the Zernike polynomials.

Transform of the Basis Set

There are several ways one might go about finding the Radon transform of a function of the form

$$f(x, y) = Z_{l+2s}^l(r) e^{il\theta}. \tag{6.8}$$

(Here we consider $l \geqslant 0$. The $l < 0$ case can always be found by complex conjugation.) We already know that the angular part just transforms to $e^{il\phi}$ and that the radial part must satisfy (3.6), with upper limit unity,

$$\check{g}_l(p) = 2\int_p^1 Z_{l+2s}^l(r) T_l\left(\frac{p}{r}\right)\left(1 - \frac{p^2}{r^2}\right)^{-1/2} dr. \tag{6.9}$$

Inspection of standard references on integral formulas is not very helpful in evaluating this integral. Of the various approaches to this problem, those of Marr (1974) and Eggermont (1975) are highly recommended; also, an elegant approach pioneered by Cormack (1964) and further discussed by Zeitler (1974) will be expanded here. [Another elegant method based on the use of generating functions is presented by Lerche and Zeitler (1976).]

The idea is to make use of the approach illustrated in Fig. 4.2. First we transform f in "real space" to $\tilde{f}$ in "Fourier space" and then transform $\tilde{f}$ to $\check{f}$ in "Radon space."* For convenience, in working with the Hankel transforms we shall temporarily abandon our symmetric definition of the Fourier transform and revert to the old-fashioned definition. If $f(x)$ is a function of the real variable x, then the Fourier transform of f is

$$\tilde{f}(k) = \int_{-\infty}^{+\infty} f(x) e^{-ikx}\,dx,$$

*Other descriptive terminology is feature space, signal space, and frequency space for real space, Radon space, and Fourier space, respectively.

and the inverse transform is

$$f(x) = \frac{1}{2\pi}\int_{-\infty}^{+\infty} \tilde{f}(k)e^{ikx}\,dk.$$

Given that f has the form (6.8), we want to find $\tilde{f} = \mathscr{F}_2 f$. Since f has radial symmetry, we use the polar coordinate form,

$$\tilde{f} = \iint_{-\infty}^{\infty} f(x, y)e^{-i(xu+yv)}\,dx\,dy$$

$$= \int_0^{2\pi}\int_0^{\infty} Z_{l+2s}^{l}(r)e^{il\theta}e^{-iqr\cos(\theta-\phi)}r\,dr\,d\theta,$$

where $x + iy = re^{i\theta}$ and $u + iv = qe^{i\phi}$. Next, make the change of variables $\beta = \theta - \phi$, then

$$\tilde{f}(q, \phi) = e^{il\phi}\int_0^{\infty} dr\,r\,Z_{l+2s}^{l}(r)\int_0^{2\pi} d\beta\,e^{il\beta - iqr\cos\beta}.$$

The β integral is a standard integral representation for the Bessel functions [page 20 of Watson (1966)]:

$$J_l(x) = \frac{1}{2\pi}\int_{\alpha}^{\alpha+2\pi} e^{i(n\theta - x\sin\theta)}\,d\theta$$

$$= \frac{e^{il\pi/2}}{2\pi}\int_0^{2\pi} e^{i(l\beta - x\cos\beta)}\,d\beta,$$

where $\theta = \beta + \pi/2$ and α is selected equal to $\pi/2$. Thus $\tilde{f}(q, \phi)$ becomes

$$\tilde{f}(q, \phi) = 2\pi e^{il\phi}e^{-il\pi/2}\int_0^{\infty} Z_{l+2s}^{l}(r)J_l(qr)r\,dr.$$

The integral over r is just a Hankel transform. From page 47 formula (5) of Erdélyi, Magnus, Oberhettinger, and Tricomi (1954)

$$\int_0^{\infty} \frac{J_{l+2s+1}(q)}{q}J_l(qr)q\,dq = \begin{cases} 0 & \text{if } 1 < r < \infty \\ r^l P_s^{(l,0)}(1 - 2r^2) & \text{if } 0 < r < 1 \end{cases}.$$

However,

$$r^l P_s^{(l,0)}(1 - 2r^2) = (-1)^s r^l P_s^{(0,l)}(2r^2 - 1)$$

$$= (-1)^s Z_{l+2s}^{l}(r).$$

In view of the self-reciprocal property of Hankel transforms,

$$\int_0^{\infty} Z_{l+2s}^l(r) J_l(qr) r\, dr = (-1)^s \frac{J_{l+2s+1}(q)}{q}.$$

This yields the desired result,

$$\tilde{f}(q,\phi) = 2\pi e^{il\phi} e^{-il\pi/2} (-1)^s \frac{J_{l+2s+1}(q)}{q}. \tag{6.10}$$

The next step is to inverse Fourier transform (6.10) on the q variable; that is,

$$\begin{aligned}\check{f}(p,\phi) &= \mathscr{F}_1^{-1} \tilde{f}(q,\phi)\\ &= 2\pi e^{il\phi} e^{-il\pi/2} (-1)^s \frac{1}{2\pi} \int_{-\infty}^{+\infty} \frac{J_{l+2s+1}(q)}{q} e^{ipq}\, dq.\end{aligned}$$

The integral can be evaluated by use of tabulated results; from page 743 of Gradshteyn and Ryzhik (1965), along with the identity arc cos x + arc sin $x = \pi/2$,

$$\check{f}(p,\phi) = e^{il\phi} e^{-il\pi/2} (-1)^s \begin{cases} \dfrac{2(-1)^s}{l+2s+1} \cos\dfrac{l\pi}{2} \sqrt{1-p^2}\, U_{l+2s}(p), & l \text{ even} \\[2ex] \dfrac{2i(-1)^s}{l+2s+1} \sin\dfrac{l\pi}{2} \sqrt{1-p^2}\, U_{l+2s}(p), & l \text{ odd} \end{cases}$$

Since this yields the same result for even or odd l, we finally have (in a notation that holds for negative or positive l)

$$\begin{aligned}\check{f}(p,\phi) &= \mathscr{R}\{Z_{|l|+2s}^{|l|}(r) e^{\pm il\theta}\}\\ &= \frac{2}{|l|+2s+1} \sqrt{1-p^2}\, U_{|l|+2s}(p) e^{\pm il\phi}.\end{aligned} \tag{6.11}$$

The orthogonality of the U_n is discussed in Appendix C. Thus the Zernike polynomials transform to Tchebycheff polynomials of the second kind:

$$Z_{|l|+2s}^{|l|}(r) \xrightarrow{\mathscr{R}} \frac{2}{|l|+2s+1} \sqrt{1-p^2}\, U_{|l|+2s}(p). \tag{6.12}$$

It turns out that we have more than we bargained for here. The functions $J_{|l|+2s+1}(q)$ are also orthogonal and have been studied by Wilkins (1948),

$$\int_0^{\infty} (J_{|l|+2s+1}(q))(J_{|l|+2t+1}(q)) q^{-1}\, dq = \frac{1}{2(|l|+2s+1)} \delta_{st}. \tag{6.13}$$

The functions

$$F_s^l(q,\phi) = J_{|l|+2s+1}(q)e^{il\phi} \tag{6.14}$$

form a convenient basis set in Fourier space.

Thus three sets of orthogonal functions exist. The major results are illustrated in Fig. 7.2 and summarized in Table 7.3.

The transformation (6.12) leads to integral formulas deduced from (3.5) and (3.6) with upper limit unity, $p > 0$, $l \geqslant 0$,

$$\frac{2\sqrt{1-p^2}}{l+2s+1}U_{l+2s}(p) = 2\int_p^1 \frac{Z_{l+2s}^l(r)T_l(p/r)}{\left(1-\frac{p^2}{r^2}\right)^{1/2}}\,dr, \tag{6.15}$$

$$\frac{(l+2s+1)}{2}Z_{l+2s}^l(r) = \frac{-1}{\pi r}\int_r^1 \frac{\frac{d}{dp}\left\{\sqrt{1-p^2}\,U_{l+2s}(p)\right\}T_l\left(\frac{p}{r}\right)}{\left(\frac{p^2}{r^2}-1\right)^{1/2}}\,dp. \tag{6.16}$$

Table 7.3 Summary of Basic Results

Space	Basis Function[a]	Interval	Weight Function
Real	$B_s^l = Z_{\lvert l\rvert+2s}^{\lvert l\rvert}(r)e^{il\theta}$	$0 \leqslant r \leqslant 1$ $0 \leqslant \theta < 2\pi$	r
Fourier	$F_s^l = J_{\lvert l\rvert+2s+1}(q)e^{il\phi}$	$0 \leqslant q < \infty$ $0 \leqslant \phi < 2\pi$	q^{-1}
Radon	$R_s^l = U_{\lvert l\rvert+2s}(p)e^{il\phi}$	$-1 \leqslant p \leqslant 1$ $0 \leqslant \phi < 2\pi$	$(1-p^2)^{1/2}$

Orthogonality Integrals

Real

$$\int_0^{2\pi}\int_0^1 (B_s^l)^* B_{s'}^{l'}\, r\,dr\,d\theta = \frac{\pi}{|l|+2s+1}\delta_{ll'}\delta_{ss'}$$

Fourier

$$\int_0^{2\pi}\int_0^{\infty} (F_s^l)^* F_{s'}^{l'}\, q^{-1}\,dq\,d\phi = \frac{\pi}{|l|+2s+1}\delta_{ll'}\delta_{ss'}$$

Radon

$$\int_0^{2\pi}\int_{-1}^1 (R_s^l)^* R_{s'}^{l'}(1-p^2)^{1/2}\,dp\,d\phi = \pi^2\delta_{ll'}\delta_{ss'}$$

[a]Sometimes it is convenient to use F_s^l/q with weight function q, and $(1-p^2)^{1/2}R_s^l$ with weight function $(1-p^2)^{-1/2}$.

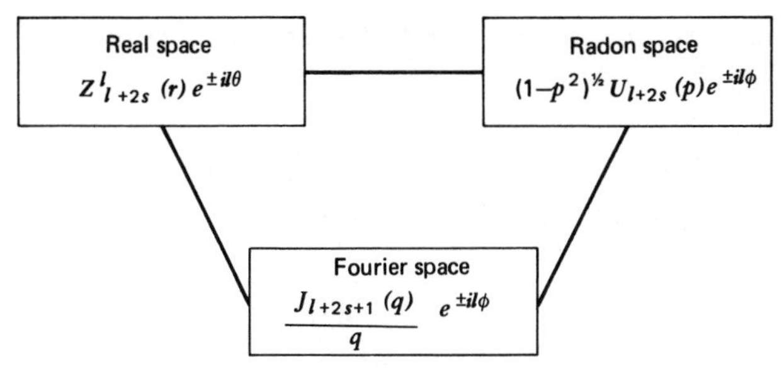

Figure 7.2 Transformation properties of radial functions among the three spaces. Normalization factors are not included.

Example 3. Suppose $f(x, y) = 1$ on the unit disk and zero elsewhere. In this case, $f(x, y) = Z_0^0$, and from (6.11), $\check{f}(p) = 2\sqrt{1 - p^2}$ for $|p| < 1$ and zero otherwise. This result was obtained by a different method in Example 5 of [3.4]. □

Example 4. Suppose $f(r, \theta) = 1 - r^2$ on the unit disk and zero elsewhere. In this case

$$f = Z_0^0 - \tfrac{1}{2}\left(Z_0^0 + Z_2^0\right)$$

and

$$\check{f} = \tfrac{1}{2}2\sqrt{1 - p^2}\,U_0 - \tfrac{1}{2}\tfrac{2}{3}\sqrt{1 - p^2}\,U_2$$

$$= \tfrac{4}{3}(1 - p^2)\sqrt{1 - p^2}, \qquad |p| \leqslant 1$$ □

Example 5. Suppose $f(r, \theta) = (x^2 + y^2)x$ on the unit disk and zero elsewhere. Using Example 2, we have

$$f = \tfrac{1}{3}\left(Z_3^1 + 2Z_1^1\right)\cos\theta,$$

and

$$\check{f}(p, \phi) = \frac{\sqrt{1 - p^2}}{3}\left[\tfrac{1}{2}U_3(p) + 2U_1(p)\right]\cos\phi$$

$$= \frac{2p}{3}(2p^2 + 1)\sqrt{1 - p^2}\cos\phi.$$

Note that this is *not* the same as if we had carelessly assumed that f could be written as $Z_3^3 \cos\theta$. □

Example 6. Find the result of the preceding example by directly doing the Radon transform. Start with the definition

$$\check{f}(p,\phi) = \iint_D f(x,y)\delta(p - x\cos\phi - y\sin\phi)\,dx\,dy,$$

where D represents the unit disk. Perform a rotation of coordinates through the angle ϕ such that

$$x = s\cos\phi - t\sin\phi$$

$$y = s\sin\phi + t\cos\phi.$$

Then, see (2.4) in [2.2],

$$\check{f}(p,\phi) = \iint_D f(s\cos\phi - t\sin\phi, s\sin\phi + t\cos\phi)\delta(p - s)\,ds\,dt,$$

or after the s integration

$$\check{f}(p,\phi) = \int_{-\sqrt{1-p^2}}^{+\sqrt{1-p^2}} f(p\cos\phi - t\sin\phi, p\sin\phi + t\cos\phi)\,dt. \quad (6.17)$$

Now let $f(x, y) = (x^2 + y^2)x$ as in the preceding example. Then (6.17) becomes

$$\begin{aligned} f(p,\phi) &= \int_{-\sqrt{1-p^2}}^{+\sqrt{1-p^2}} (p^2 + t^2)(p\cos\phi - t\sin\phi)\,dt \\ &= 2p\cos\phi\int_0^{\sqrt{1-p^2}} (p^2 + t^2)\,dt \\ &= \tfrac{2}{3}p(2p^2 + 1)\sqrt{1 - p^2}\cos\phi, \end{aligned}$$

which is in agreement with Example 5. □

Example 7. Find the Radon transform of $f(x, y) = x\sqrt{x^2 + y^2}$, where, as usual, the domain of f is the unit disk. Although this has the form $f(r,\theta) = rx = r^2\cos\theta$, the transform is not trivial in this case since f is not a simple sum over monomials of the form $x^k y^m$. Although we might be tempted to assume that an appropriate expansion for f is $Z_2^2(r)\cos\theta$, it is not difficult to demonstrate that such an assumption is not justified. The transform of $Z_2^2(r)\cos\theta$ is $\frac{2}{3}\sqrt{1-p^2}\,U_2(p)\cos\phi$, whereas the transform of $x\sqrt{x^2+y^2}$ is given by evaluation of (6.17):

$$\check{f}(p,\phi) = \int_{-\sqrt{1-p^2}}^{+\sqrt{1-p^2}} \sqrt{p^2+t^2}\,(p\cos\phi - t\sin\phi)\,dt$$

$$= 2p\cos\phi \int_0^{\sqrt{1-p^2}} \sqrt{p^2+t^2}\,dt$$

$$= 2p\cos\phi\left[\tfrac{1}{2}\sqrt{1-p^2} + \frac{p^2}{2}\log\left(\frac{1+\sqrt{1-p^2}}{p}\right)\right]. \qquad \square$$

Example 8. Find the transform of $f(x, y) = x^2$ by the direct integration method and by Zernike decomposition. By direct integration of (6.17),

$$\check{f}(p,\phi) = \int_{-\sqrt{1-p^2}}^{+\sqrt{1-p^2}} (p\cos\phi - t\sin\phi)^2\,dt$$

$$= 2p^2\cos^2\phi \int_0^{\sqrt{1-p^2}} dt + 2\sin^2\phi \int_0^{\sqrt{1-p^2}} t^2\,dt$$

$$= \sqrt{1-p^2}\left[2p^2\cos^2\phi + \tfrac{2}{3}(1-p^2)\sin^2\phi\right].$$

By orthogonal function expansion,

$$f(r,\theta) = r^2\left(\tfrac{1}{2} + \tfrac{1}{2}\cos 2\theta\right)$$

$$= \tfrac{1}{4}\left(Z_0^0 + Z_2^0\right) + \tfrac{1}{2}Z_2^2\cos 2\theta.$$

The transform is

$$\check{f}(p,\phi) = \sqrt{1-p^2}\left[\tfrac{1}{4}\left(2U_0 + \tfrac{2}{3}U_2\right) + \tfrac{1}{3}U_2\cos 2\phi,\right.$$

which reduces to the result obtained by direct integration after inserting the explicit expressions for the Tchebycheff polynomials. □

Example 9. Suppose $f(x, y) = xy^2$ on the unit disk and zero elsewhere. First observe that

$$f(r,\theta) = r^3\cos\theta\sin^2\theta = \frac{r^3}{4}(\cos\theta - \cos 3\theta).$$

The A_{ls} coefficients may be computed directly from (6.4):

$$A_{11} = A_{-11} = \tfrac{1}{24}, \qquad A_{10} = A_{-10} = \tfrac{1}{12}, \qquad A_{30} = A_{-30} = -\tfrac{1}{8}.$$

Hence

$$f = \tfrac{1}{8}\left[\tfrac{1}{3}\left(Z_3^1 + 2Z_1^1\right)\left(e^{i\theta} + e^{-i\theta}\right) - Z_3^3\left(e^{3i\theta} + e^{-3i\theta}\right)\right].$$

This transforms to

$$\begin{aligned}\check{f}(p,\phi) &= \frac{\sqrt{1-p^2}}{8}\left[\tfrac{1}{6}U_3(p) + \tfrac{2}{3}U_1(p)\right]\left[e^{i\phi} + e^{-i\phi}\right]\\ &\quad - \frac{\sqrt{1-p^2}}{8}\left[\tfrac{1}{2}U_3(p)\right]\left[e^{3i\phi} + e^{-3i\phi}\right]\\ &= \frac{\sqrt{1-p^2}}{4}\left\{\left[\tfrac{1}{6}U_3(p) + \tfrac{2}{3}U_1(p)\right]\cos\phi\right.\\ &\quad \left. - \left[\tfrac{1}{2}U_3(p)\right]\cos 3\phi\right\}.\end{aligned}$$ □

Caution. In the preceding example, it would have been incorrect to attempt to transform

$$Z_3^3(r)\cos\theta\sin^2\theta,$$

since the angle portion is not expressed in the appropriate form.

Although the preceding examples may have been somewhat tedious they serve to point out some delicate problems associated with Radon transforms of functions which are approximated by polynomial expansions.

Series Expansion in Radon Space

Given that

$$f(r,\theta) = \sum_{s=0}^{\infty}\sum_{l=-\infty}^{\infty} A_{ls} Z_{l+2s}^{l}(r) e^{il\theta}, \tag{6.18}$$

with $A_{-ls} = A_{ls}^*$, it follows that

$$\check{f}(p,\phi) = \sum_{s=0}^{\infty}\sum_{l=-\infty}^{\infty} \frac{2A_{ls}}{|l| + 2s + 1}\sqrt{1-p^2}\, U_{|l|+2s}(p) e^{il\phi}. \tag{6.19}$$

It is straightforward to solve for A_{ls} by multiplying (6.18) by $e^{-il'\phi}U_{l'+2s'}$ and integrating over p and ϕ:

$$A_{ls} = \frac{(|l| + 2s + 1)}{2\pi^2} \int_0^{2\pi} \int_{-1}^{1} \check{f}(p, \phi) e^{-il\phi} U_{|l|+2s}(p)\, dp\, d\phi.$$

In a more suggestive form,

$$A_{ls} = \frac{(|l| + 2s + 1)}{2\pi^2} \int_0^{2\pi} \int_{-1}^{1} \frac{\check{f}(p, \phi)}{\sqrt{1 - p^2}} e^{-il\phi} U_{|l|+2s}(p) \sqrt{1 - p^2}\, dp\, d\phi. \tag{6.20}$$

Suppose we can express f and $\check{f}$ in the forms

$$f = \sum_{l=-\infty}^{\infty} g_l(r) e^{il\theta} \tag{6.21a}$$

$$\check{f} = \sum_{l=-\infty}^{\infty} \check{g}_l(p) e^{il\phi}. \tag{6.21b}$$

From (6.18) and (6.19), it follows that for fixed l the polynomial expansions are

$$g_l(r) = \sum_{s=0}^{\infty} A_{ls} Z_{l+2s}^{l}(r) \tag{6.22}$$

and

$$\frac{\check{g}_l(p)}{2\sqrt{1 - p^2}} = \sum_{s=0}^{\infty} \frac{A_{ls}}{l + 2s + 1} U_{l+2s}(p). \tag{6.23}$$

Note that $\check{g}_l(p)$ has been divided by the projection of a uniform disk. From (6.23) and the orthogonality of the Tchebycheff polynomials,

$$A_{ls} = \frac{2(|l| + 2s + 1)}{\pi} \int_{-1}^{1} \frac{\check{g}_l(p)}{2\sqrt{1 - p^2}} U_{|l|+2s}(p) \sqrt{1 - p^2}\, dp. \tag{6.24}$$

Example 10. If f can be expressed as in (6.21a), then from (6.22) and (6.3)

$$A_{ls} = 2(|l| + 2s + 1) \int_0^1 g_l(r) Z_{l+2s}^{l}(r) r\, dr,$$

and by comparison with (6.24), we have

$$\pi \int_0^1 g_l(r) Z_{l+2s}^l(r)\, dr = \int_{-1}^{+1} \frac{\check{g}_l(p)}{2\sqrt{1-p^2}} U_{l+2s}(p)\sqrt{1-p^2}\, dp. \quad \square \quad (6.25)$$

In summary, if $\check{f}$ can be expressed as in (6.21b) and A_{ls} can be determined from (6.24) then $g_l(r)$ can be determined from (6.22) and f can be recovered from (6.21a). If it is not feasible to write $\check{f}$ in the form (6.21b) initially, then the constants A_{ls} must be determined from (6.20) and f recovered from (6.18).

Inner Products

Let $\langle f_1, f_2 \rangle$ designate the inner product

$$\langle f_1, f_2 \rangle = \iint_D f_1^*(x, y) f_2(x, y)\, dx\, dy$$

$$= \int_0^{2\pi} \int_0^1 f_1^*(r, \theta) f_2(r, \theta) r\, dr\, d\theta \quad (6.26)$$

in real space. Given that the Radon transforms of f_1 and f_2 are designated by $\check{f}_1$ and $\check{f}_2$, respectively, the inner product $[\check{f}_1, \check{f}_2]$ in Radon space is defined as

$$[\check{f}_1, \check{f}_2] = \int_0^{2\pi} \int_{-1}^1 \check{f}_1^*(p, \phi) \check{f}_2(p, \phi) \frac{dp}{\sqrt{1-p^2}}\, d\phi. \quad (6.27)$$

The choice for the form of the weight function $1/\sqrt{1-p^2}$ becomes clear if f_1 and f_2 are expanded as in (6.2). Then $\check{f}_1$ and $\check{f}_2$ have expansions of the form (6.19), and the following integral emerges:

$$[\check{f}_1, \check{f}_2] = \sum_{l,s} \sum_{l',s'} \frac{4A_{ls}^* A_{l's'}}{(|l| + 2s + 1)(|l'| + 2s + 1)} 2\pi\delta_{ll'}$$

$$\times \int_{-1}^1 U_{|l|+2s}(p) U_{|l'|+2s'}(p) \sqrt{1-p^2}\, dp.$$

Since the integral on the right-hand side is just $(\pi/2)\delta_{ss'}$, the sums over l' and s' yield

$$[\check{f}_1, \check{f}_2] = \sum_{l,s} \frac{4\pi^2 |A_{ls}|^2}{(|l| + 2s + 1)^2}. \quad (6.28)$$

In a similar fashion,

$$\langle f_1, f_2 \rangle = \sum_{l,s} \sum_{l',s'} A_{ls}^* A_{l's'} 2\pi\delta_{ll'} \int_0^1 Z_{l+2s}^l(r) Z_{l'+2s'}^{l'}(r) r\, dr$$

$$= \sum_{l,s} \frac{\pi}{(|l| + 2s + 1)} |A_{ls}|^2. \tag{6.29}$$

If we observe that

$$4\pi \sum_{l,s} \frac{\pi |A_{ls}|^2}{(|l| + 2s + 1)^2} \leqslant 4\pi \sum_{l,s} \frac{\pi |A_{ls}|^2}{(|l| + 2s + 1)},$$

it follows that

$$[\check{f}_1, \check{f}_2] \leqslant 4\pi \langle f_1, f_2 \rangle. \tag{6.30}$$

Motivation for the Basis

It follows from the Weierstrass approximation theorem [Courant and Hilbert (1953)] that constants A_{jk} exist such that any square integrable function defined on the unit disk D can be uniquely approximated by a sum of monomials of the form $x^j y^k$ with $j, k \geqslant 0$:

$$f(x, y) = \sum_{j,k} A_{jk} x^j y^k. \tag{6.31}$$

We present an argument similar to that of Eggermont (1975) [also see Marr (1974)] which points to the use of Tchebycheff polynomials of the second kind in Radon space. Of course, once this is accepted, the selection of Zernike polynomials in real space follows directly by transforming from $\check{f}$ to $\tilde{f}$ to f. These transforms can be done since the relevant Hankel transforms are tabulated [Erdélyi, Magnus, Oberhettinger, and Tricomi (1954)].

Let $P(D)$ represent the space of all polynomials on D and let $P_0(D), P_1(D), \ldots, P_M(D)$ be linear subspaces of $P(D)$ such that

1. $P_M(D)$ contains polynomials of exact degree M and the zero polynomial;
2. $P_M(D)$ is orthogonal to $P_K(D)$ for every $K \leqslant M - 1$;
3. $P(D)$ is the direct sum of the $P_M(D)$, $M = 0, 1, 2, \ldots$.

If $f_M(x, y) \in P_M(D)$ and $f_k(x, y) \in P_K(D)$ and $M \neq K$, then

$$\langle f_K, f_M \rangle = \iint_D f_K^*(x, y) f_M(x, y)\, dx\, dy = 0. \tag{6.32}$$

(For real polynomials $f_K^* = f_K$.) Given a general square integrable function $f \in L^2(D)$, there exist polynomials $f_M(x, y) \in P_M(D)$ such that f has a unique expansion

$$f(x, y) = \sum_{M=0}^{\infty} f_M(x, y). \tag{6.33}$$

A special case of (6.32) is

$$\iint_D (x \cos \phi + y \sin \phi)^l f_M(x, y)\, dx\, dy = 0, \qquad l \leqslant M - 1. \tag{6.34}$$

Remark. Observe that at this point we are taking on faith that polynomials $f_M(x, y)$ can be found such that (6.34) holds. From previous work, we recognize that the term $(x \cos \phi + y \sin \phi)^n$ with $x = r \cos \theta$ and $y = r \sin \theta$ can be expanded in terms of the basis functions $Z_{l+2s}^l(r) e^{il\theta}$ with total degree $l + 2s \leqslant n$ and the $f_M(x, y)$ consist of linear combination of basis functions $Z_{l+2s}^l(r) e^{il\theta}$ such that $l + 2s = M > n$. With this hindsight we know that (6.34) is possible, but there are severe restrictions on the possible form of the members of P_M. For our purpose here, we may ignore this, assume (6.34) holds, and see where it leads. □

Now make the change of variables

$$x = p \cos \phi - t \sin \phi$$

$$y = p \sin \phi + t \cos \phi.$$

This corresponds to a rotation of the disk through angle ϕ. After this transformation, (6.34) becomes

$$\iint_D p^l f_M(p \cos \phi - t \sin \phi, p \sin \phi + t \cos \phi)\, dp\, dt = 0,$$

or, more explicitly,

$$\int_{-1}^{1} p^l\, dp \int_{-\sqrt{1-p^2}}^{\sqrt{1-p^2}} f_M(p \cos \phi - t \sin \phi, p \sin \phi + t \cos \phi)\, dt = 0. \tag{6.35}$$

We recognize the integral over t as the Radon transform of f_M [see (6.17)]

$$\int_{-1}^{1} p^l \check{f}_M(p, \phi)\, dp = 0. \tag{6.36}$$

We must now argue that $\psi_M(p, \phi) = (1 - p^2)^{-1/2} \check{f}_M(p, \phi)$ is a polynomial in p of degree $\leqslant M$. To see this, make a further change of variables in (6.35),

$t = \sqrt{1 - p^2}\, s$. Then

$$\psi_M(p, \phi) = \frac{\check{f}_M(p, \phi)}{\sqrt{1 - p^2}}$$

$$= \int_{-1}^{1} f_M\Big(p \cos\phi - s\sqrt{1 - p^2} \sin\phi,\ p \sin\phi + s\sqrt{1 - p^2} \cos\phi\Big)\, ds. \tag{6.37}$$

Now the integrand is a polynomial in p and $s\sqrt{1 - p^2}$, but odd powers of s do not contribute to the integral. Thus $\psi_M(p, \phi) = \check{f}_M(p, \phi)/\sqrt{1 - p^2}$ is a polynomial in p of degree $\leqslant M$.

Example 11. Suppose $M = 4$ (from x^4), then terms of the form

$$p^4\cos^4\phi, \qquad p^2 s^2(1 - p^2)\cos^2\phi \sin^2\phi, \qquad s^4(1 - p^2)^2\sin^4\phi$$

contribute to the integral (6.37) while terms of the form

$$p^3 s\sqrt{1 - p^2}\cos^3\phi \sin\phi, \qquad p s^3(1 - p^2)^{3/2}\cos\phi \sin^3\phi$$

do not contribute. □

From (6.36) and (6.37),

$$\int_{-1}^{1} p^l \psi_M(p, \phi)\sqrt{1 - p^2}\, dp = 0. \tag{6.38}$$

We further assume that it is possible to select the members of $P_M(D)$ such that the polynomials ψ_M are of exact degree M. Then we observe that the only polynomials of degree M which satisfy (6.38) for all $l < M$ may be constructed by orthogonalization of the set of polynomials

$$\{1, p, p^2, \ldots, p^M\}$$

with respect to the weight function $\sqrt{1 - p^2}$ [Rivlin (1974)]. The resulting orthogonal polynomials are Tchebycheff polynomials of the second kind $U_M(p)$, as discussed in Appendix C.

This concludes the argument for using Tchebycheff polynomials in Radon space. As mentioned earlier, by transforming from $\check{f}$ to $\tilde{f}$ to f, the Zernike polynomials in real space emerge. Also, the properties assumed for the members of $P_M(D)$ can be used to discover the Zernike polynomials [Marr (1974)].

[7.7] ORTHOGONAL FUNCTION EXPANSIONS OVER THE ENTIRE PLANE

Basis Functions

One basic result has been discussed already in [3.8]. Recall that we found a complete set of orthogonal polynomials $\mathscr{L}^{\pm l}_{l+2k}$,

$$\mathscr{L}^{\pm l}_{l+2k}(\lambda x, \lambda y) = \frac{(-1)^k \lambda}{\sqrt{\pi}} \left[\frac{k!}{(l+k)!}\right]^{1/2}$$

$$\times e^{\pm il\theta}\left[\lambda^2(x^2+y^2)\right]^{l/2} L^l_k\left[\lambda^2(x^2+y^2)\right], \quad (7.1)$$

with Radon transform

$$\mathscr{R}\left\{\mathscr{L}^{\pm l}_{l+2k}(\lambda x, \lambda y) e^{-\lambda^2(x^2+y^2)}\right\}$$

$$= \left[\frac{1}{k!(l+k)!}\right]^{1/2} \frac{1}{2^{l+2k}} e^{\pm il\phi} H_{l+2k}(\lambda p) e^{-\lambda^2 p^2}. \quad (7.2)$$

The Laguerre polynomials L^l_k and the Hermite polynomials H_m are discussed in Appendix C. We observe that for $x = r\cos\theta$ and $y = r\sin\theta$,

$$\mathscr{L}^{\pm l}_{l+2k}(\lambda r\cos\theta, \lambda r\sin\theta) = \frac{(-1)^k \lambda^{l+1}}{\sqrt{\pi}} \left[\frac{k!}{(l+k)!}\right]^{1/2} e^{\pm il\theta} r^l L^l_k(\lambda^2 r^2).$$

$$(7.3)$$

The orthogonality integral with respect to the weight function $e^{-\lambda^2(x^2+y^2)}$ in real space is

$$\langle \mathscr{L}^{\pm l}_{l+2k}, \mathscr{L}^{\pm l'}_{l'+2k'} \rangle = \iint_{-\infty}^{\infty} \left(\mathscr{L}^{\pm l}_{l+2k}\right)^* \mathscr{L}^{\pm l'}_{l'+2k'} e^{-\lambda^2(x^2+y^2)}\, dx\, dy$$

$$= \delta_{ll'}\delta_{kk'}. \quad (7.4)$$

This is easy to verify by changing to polar coordinates and making the change of variables $r^2 = t$.

In Radon space, the basis functions

$$h^{\pm l}_{l+2k}(\lambda p, \phi) = N^l_k e^{\pm il\phi} H_{l+2k}(\lambda p), \quad (7.5)$$

with

$$N_k^l = \frac{1}{2^{l+2k}}\left[\frac{1}{k!(l+k)!}\right]^{1/2}, \tag{7.6}$$

satisfy an orthogonality condition with respect to weight function $e^{-\lambda^2 p^2}$,

$$\begin{aligned}\left[h_{l+2k}^{\pm l}, h_{l'+2k'}^{\pm l'}\right] &= \int_0^{2\pi}\int_{-\infty}^{+\infty}\left(h_{l+2k}^{\pm l}\right)^* h_{l'+2k'}^{\pm l'} e^{-\lambda^2 p^2}\, dp\, d\phi \\ &= \frac{2\pi}{\lambda}\delta_{ll'}N_k^l N_{k'}^{l'}\int_{-\infty}^{+\infty} H_{l+2k}(\lambda p)H_{l+2k'}(\lambda p)\, d(\lambda p) \\ &= \frac{2\pi^{3/2}}{\lambda}\left(N_k^l\right)^2 2^{l+2k}(l+2k)!\,\delta_{ll'}\delta_{kk'} \\ &= \frac{2\pi^{3/2}}{\lambda 2^{l+2k}}\frac{(l+2k)!}{k!(l+k)!}\delta_{ll'}\delta_{kk'}.\end{aligned} \tag{7.7}$$

Remark. These same results were suggested by Cormack (1964) and expanded on by Maldonado and co-workers in connection with a study of the relation between the emitted spectral intensity ($\propto \check{f}$) and the internal emission coefficient ($\propto f$) for optically thin light sources (gaseous plasmas) by considerably less elegant methods which did not make use of Radon transforms [Maldonado (1965); Maldonado and Olsen (1966)]. Also see Maldonado, Caron and Olsen (1965) and Olsen, Maldonado, and Duckworth (1968) for applications in plasma physics, and Matulka and Collins (1971), Jagota and Collins (1972), and Kosakoski and Collins (1974) for aerodynamic applications, determination of density fields from holographic interferograms. □

The complex polynomials $\mathscr{L}_{l+2k}^{\pm l}(\lambda x, \lambda y)$ studied in [3.8] are precisely the same as the complex polynomials designated as $U_{l+2k}^{\pm l}(\lambda x, \lambda y)$ by Maldonado and Olsen (1966). To avoid possible confusion with Tchebycheff polynomials, we continue using $\mathscr{L}$ rather than U. Also observe the analogy between the Zernike polynomials on the disk and the polynomials $\mathscr{L}_{l+2k}^{\pm l}$ over the plane.

Series Expansions

In the interest of maintaining some uniformity with conventions already established by Maldonado and co-workers and Collins and co-workers (references given in the preceding remark), the expansion for $f(x, y)$ is the sum of two parts

$$\begin{aligned}f(x, y) = \sum_{k=0}^{\infty}\sum_{l=0}^{\infty}\varepsilon_l\big[&A_{l+2k}^l(\lambda)\mathscr{L}_{l+2k}^l(\lambda x, \lambda y) \\ &+ A_{l+2k}^{-l}(\lambda)\mathscr{L}_{l+2k}^{-l}(\lambda x, \lambda y)\big]e^{-\lambda^2(x^2+y^2)},\end{aligned} \tag{7.8}$$

where $\varepsilon_l = 1$ if $l \neq 0$ and $\varepsilon_0 = \frac{1}{2}$. This form also avoids use of $|l|$, which may appear when the sum on l goes from $-\infty$ to $+\infty$. The expansion coefficients $A^{\pm l}_{l+2k}$ depend on the scale factor λ, but once λ is selected, they are constant. Since a factor of $e^{-\lambda^2(x^2+y^2)}$ is included on the right-hand side of (7.8) it may be convenient in some cases to determine the coefficients for $e^{\lambda^2(x^2+y^2)}f(x, y)$ rather than for $f(x, y)$. By use of (7.4), the coefficients are easy to determine:

$$A^{\pm l}_{l+2k} = \iint_{-\infty}^{+\infty} \mathscr{L}^{\mp l}_{l+2k}(\lambda x, \lambda y) f(x, y) e^{-\lambda^2(x^2+y^2)} dx\, dy. \tag{7.9}$$

Note that ε_l does not need to be included since this formula also holds for $l = 0$.

The Radon transform of $f(x, y)$ as expanded in (7.8) follows directly from [3.8], (7.2), and (7.5):

$$\check{f}(p, \phi) = \sum_{k=0}^{\infty} \sum_{l=0}^{\infty} \varepsilon_l \left[A^l_{l+2k}(\lambda) h^l_{l+2k}(\lambda p) + A^{-l}_{l+2k}(\lambda) h^{-l}_{l+2k}(\lambda p) \right] e^{-\lambda^2 p^2}. \tag{7.10}$$

Using (7.7), the coefficients can be calculated

$$A^{\pm l}_{l+2k} = \frac{\lambda 2^{l+2k}}{2\pi^{3/2}} \frac{k!(l+k)!}{(l+2k)!} \int_0^{2\pi} \int_{-\infty}^{+\infty} h^{\mp l}_{l+2k}(\lambda p) \check{f}(p, \phi) e^{-\lambda^2 p^2}\, dp\, d\phi. \tag{7.11}$$

If the explicit form for h from (7.5) is used, then

$$A^{\pm l}_{l+2k} = \frac{\lambda [k!(l+k)!]^{1/2}}{2\pi^{3/2}(l+2k)!} \int_0^{2\pi} \int_{-\infty}^{\infty} \check{f}(p, \phi) e^{\mp il\phi} H_{l+2k}(\lambda p) e^{-\lambda^2 p^2}\, dp\, d\phi. \tag{7.12}$$

[7.8] OTHER APPROACHES

We have not exhausted the possible series approaches. By partitioning real space into disjoint regions $R_1, R_2, \ldots, R_k$ such that each point (x, y) lies in one and only one of the regions, it is possible to define basis functions

$$b_k(x, y) = \begin{cases} 0 & \text{for } (x, y) \notin R_k \\ \left(\dfrac{1}{A_k} \right)^{1/2} & \text{for } (x, y) \in R_k, \end{cases}$$

where A_k is the area of region R_k. This *gate function set* b_k [see Snyder and Cox (1977)] may be used to obtain a piecewise approximation to $f(x, y)$. Such approximations are commonly used in conjunction with algebraic reconstruction methods. Observe that this set is orthogonal by construction.

Another popular choice is to expand $f(r, \theta)$ in the Bessel function set used by Crowther, Amos, and Klug (1972) and Klug and Crowther (1972). This approach has been reviewed in detail by Snyder and Cox (1977).

Series methods are fundamental when working with hollow or truncated projection data [Lewitt and Bates (1978b)]. Recent work by Buonocore, Brody, and Macovski (1981) points toward the development of more optimal types of basis sets appropriate for limited and/or noisy data. This new approach is especially interesting because it provides a connection between the continuous object to be reconstructed and its discrete representation. The *xy* plane is decomposed into a set of discrete but *overlapping* regions for which algorithms based on continuous projections are computationally feasible.

Chapter 8

More Properties, Applications, and Generalizations

[8.1] INTRODUCTION

Properties and applications of a slightly more technical nature are discussed here. These include some basic theorems and transforms of distributions in [8.2], a discrete version in [8.3], use in picture restoration in [8.4], transforms in geophysics in [8.5], potential scattering in [8.6], and partial differential equations in [8.7]. Some generalizations and other extensions and uses are mentioned in [8.8].

[8.2] CHARACTERIZATION OF THE TRANSFORM

Some Theorems

It may be useful to state certain theorems that help characterize the Radon transform. Although justification for most of these properties appears in previous chapters, the various results were not stated as theorems. Formal proofs are easily available in the paper by Ludwig (1966) and/or the monograph by Helgason (1980). As we shall see, the following general observation emerges: $f \in \mathscr{S}$ has compact support if and only if $\check{f}$ has compact support. (For definitions and notation, see Appendix B.)

Before stating theorems, let us observe that for each $f \in \mathscr{S}$, the integral of the Radon transform,

$$\int_{-\infty}^{\infty} \check{f}(p, \xi) p^k \, dp, \qquad k = 0, 1, 2, \ldots,$$

is a homogeneous polynomial of degree k in $\xi_1, \xi_2, \ldots, \xi_n$. This follows immediately from the manipulations

$$\int_{-\infty}^{\infty} \check{f}(p, \xi) p^k \, dp = \int_{-\infty}^{\infty} dp \, p^k \int d\mathbf{x} \, f(\mathbf{x}) \, \delta(p - \xi \cdot \mathbf{x})$$

$$= \int (\xi \cdot \mathbf{x})^k f(\mathbf{x}) \, d\mathbf{x}.$$

This suggests that we define the space $\mathcal{S}_H(\mathcal{P}^n)$ as the set of all $\psi(p, \xi) \in \mathcal{S}(\mathcal{P}^n)$ such that

$$\int_{-\infty}^{\infty} \psi(p, \xi) p^k \, dp$$

is a homogeneous polynomial in $\xi_1, \xi_2, \ldots, \xi_n$ of degree k. (Recall that $\mathcal{P}^n$ is the space of all hyperplanes in $\mathbb{R}^n$.)

The following theorem is described by Helgason (1980) as the *Schwartz theorem*.

Theorem 1. (The Radon transform $f \to \check{f}$ is a linear one-to-one mapping of $\mathcal{S}(\mathbb{R}^n)$ onto $\mathcal{S}_H(\mathcal{P}^n)$.

The next theorem is known as the *support theorem* [Helgason (1980)].

Theorem 2. Let f be a (once) continuously differentiable function on $\mathbb{R}^n$ such that:

(i) For each integer $k > 0$, $|\mathbf{x}|^k f(\mathbf{x})$ is bounded.
(ii) There exists a constant c such that $\check{f}(p, \xi) = 0$ for $p > c$.

Then $f(\mathbf{x}) = 0$ for $|\mathbf{x}| > c$.

The characterization of the Radon transform of functions $f \in \mathcal{D}(\mathbb{R}^n)$ may be regarded as the *Paley–Wiener theorem* for the Radon transform [Helgason (1980)].

Theorem 3. The Radon transform is a bijection of $\mathcal{D}(\mathbb{R}^n)$ onto $\mathcal{D}_H(\mathcal{P}^n)$, where $\mathcal{D}_H(\mathcal{P}^n) = \mathcal{S}_H(\mathcal{P}^n) \cap \mathcal{D}(\mathcal{P}^n)$.

For an extension and sharper version, see Lax and Phillips (1979).

Ludwig (1966) proves a theorem that summarizes several properties.

Theorem 4. In order for $\psi(p, \xi)$ to be the Radon transform of a function $f \in \mathcal{S}$, it is necessary and sufficient that (a) $\psi \in \mathcal{S}(\mathbb{R}^1 \times S^{n-1})$, (b) ψ be even in p and ξ, and (c) the integral

$$\int_{-\infty}^{\infty} \psi(p, \xi) p^k \, dp$$

be a polynomial of degree $\leqslant k$ in $\xi_1, \xi_2, \ldots, \xi_n$, for all $k \geqslant 0$. Condition (c) is equivalent to the following condition (d): If S_{lm} is a spherical harmonic of degree l, and if $k < l$, then

$$\int_{|\xi|=1} \int_{-\infty}^{\infty} \psi(p, \xi) p^k S_{lm}(\xi) \, dp \, d\xi = 0.$$

The mapping $\psi \to f$ is continuous from $\mathscr{S}$ to $\mathscr{S}$.

Transforms of Distributions

The identity (4.20) in [5.4] can be used to study transforms of distributions. The basic results for the spaces $\mathscr{S}'$ and $\mathscr{D}'$ (defined in Appendix B) as proved by Ludwig (1966) may be summarized in the following fashion:

$$\begin{array}{lll} \text{(i)} & \mathscr{R}: & \mathscr{S}' \leftrightarrow (\Upsilon \mathscr{R} \mathscr{S})' \\ \text{(ii)} & \mathscr{R}: & \mathscr{D}' \leftrightarrow (\Upsilon \mathscr{R} \mathscr{D})' \\ \text{(iii)} & \mathscr{R}^\dagger: & (\mathscr{R} \mathscr{S})' \leftrightarrow \mathscr{S}' \\ \text{(iv)} & \mathscr{R}^\dagger: & (\mathscr{R} \mathscr{D})' \leftrightarrow \mathscr{D}' \end{array}$$

Each mapping is one to one and continuous in both directions. To illustrate more precisely just what these mappings mean, recall from [5.4] for $f(\mathbf{x}) \in \mathscr{S}$ and $\psi(p, \xi) \in \mathscr{S}$ (on $\mathbb{R}^1 \times \mathsf{S}^{n-1}$),

$$\int (f)(\mathscr{R}^\dagger \psi) \, d\mathbf{x} = \int_{|\xi|=1} \int_{-\infty}^{\infty} (\mathscr{R} f)(\psi) \, dp \, d\xi.$$

This identity may be written conveniently as

$$\langle f, \mathscr{R}^\dagger \psi \rangle = [\mathscr{R} f, \psi].$$

Consider (i), for example; suppose $f \in \mathscr{S}'$ and let $\psi = \Upsilon \mathscr{R} h$, where $h \in \mathscr{S}$ is a test function. Then,

$$[\mathscr{R} f, \psi] = [\mathscr{R} f, \Upsilon \mathscr{R} h] = \langle f, \mathscr{R}^\dagger \Upsilon \mathscr{R} h \rangle = \langle f, h \rangle.$$

The right-hand side is a continuous linear functional on $\mathscr{S}$. Conversely, if $\Phi(p, \xi) \in (\Upsilon \mathscr{R} \mathscr{S})'$ such that

$$\langle f, h \rangle = [\Phi, \Upsilon \mathscr{R} h],$$

then

$$[\mathscr{R}f, \psi] = \langle f, \mathscr{R}^\dagger \psi \rangle$$

$$= \left[\Phi, \Upsilon \mathscr{R} \mathscr{R}^\dagger \psi\right] = [\Phi, \psi].$$

Now the right-hand side is a continuous linear functional on $\Upsilon\mathscr{R}\mathscr{S}$. Clearly, Φ is interpreted as the Radon transform of f, and it is possible to go in both directions,

$$\langle f, h \rangle \leftrightarrow [\Phi, \psi].$$

Condition (iii) follows by using the same reasoning as above with the changes $f \in \mathscr{S}$ and $h \in \mathscr{S}'$. To obtain (ii) and (iv), just change from $\mathscr{S}$ to $\mathscr{D}$.

[8.3] A DISCRETE VERSION

As pointed out in Chapter 6, many algorithms for reconstructing a function from knowledge of its projections are available, and slightly less than infinitely more investigators have studied discrete versions and computer implementations of these algorithms (see the references in [6.8]). Here we shall confine the discussion to only one of these methods, that is, a discrete version of the filtered backprojection algorithm. This technique is known to be computationally efficient and accurate; hence its selection to *illustrate* some of the basic ideas. The purpose is to illustrate a basic discrete formalism, not to provide a finished product ready for computer implementation.

Kak, Jakowatz, Baily, and Keller (1977) have given a nice discussion* of the digital implementation of this technique. We shall for the most part follow their outline, but first a brief discussion of the sampling theorem is in order.

Sampling Theorem

The Whittaker–Shannon sampling theorem is discussed in many texts [e.g., Goodman (1968); Hamming (1977); Papoulis (1977)]. Consequently, it should be adequate to refer its proof to one of these standard sources and confine this review to some of the most pertinent results.

Basically, the theorem states that if a continuous function (or signal) $x(t)$ is bandlimited† with bandwidth ν_b, then the function can be uniquely determined from its sampled values, provided the samples are taken at ν_0 samples per unit time with $\nu_0 \geqslant 2\nu_b$. If t is measured in seconds, then the bandwidth ν_b is in

*This work is expanded on by Crawford and Kak (1979), Kak (1979), Rosenfeld and Kak (1982). Also, see Lindgren and Rattey (1981).

†The statement that $x(t)$ is *bandlimited* means that the Fourier transform of $x(t)$ [i.e., $\tilde{x}(\nu)$] vanishes for $|\nu| > \nu_b$.

Hertz. The *maximum* sample spacing Δt is given by $\Delta t = \tau = 1/2\nu_b$. The frequency $1/\tau = 2\nu_b$ is known as the *Nyquist* sampling rate. The expression for recovering $x(t)$ from its samples is given by

$$x(t) = \sum_{k=-\infty}^{\infty} x\left(\frac{k}{2\nu_b}\right) \frac{\sin\left[2\pi\nu_b\left(t - \frac{k}{2\nu_b}\right)\right]}{2\pi\left(t - \frac{k}{2\nu_b}\right)}. \tag{3.1}$$

A Discrete Inversion Algorithm

For convenience let $\check{f}_\phi(p) = \check{f}(p, \phi)$ represent the projections (Radon transform) of $f(x, y)$ and designate the Fourier transform of $\check{f}_\phi(p)$ by

$$F_\phi(q) = \int_{-\infty}^{\infty} \check{f}_\phi(p) e^{-i2\pi qp}\, dp. \tag{3.2}$$

From [6.4], the function $f(x, y)$ can be reconstructed from

$$f(x, y) = \int_0^\pi Q_\phi(x\cos\phi + y\sin\phi)\, d\phi, \tag{3.3}$$

where

$$Q_\phi(s) = \int_{-\infty}^{\infty} F_\phi(q)|q| e^{i2\pi sq}\, dq. \tag{3.4}$$

Observe that since p represents a length, the variable q is a reciprocal length that may be called spatial frequency. Like p, the variable s also has units of length. Although in principle the integral in (3.4) is over all q, in actual practice for all $|q| > W > 0$, the contribution to the integral is negligible. This means that for practical purposes the function $\check{f}_\phi(p)$ may be considered bandlimited with bandwidth W. Thus, by the sampling theorem (with p corresponding to time and q corresponding to frequency),

$$\check{f}_\phi(p) = \sum_{l=-\infty}^{\infty} \check{f}_\phi\left(\frac{l}{2W}\right) \frac{\sin 2\pi W\left(p - \frac{l}{2W}\right)}{2\pi W\left(p - \frac{l}{2W}\right)}. \tag{3.5}$$

By Fourier transforming (3.5) on the p variable, we obtain

$$F_\phi(q) = \frac{1}{2W} \sum_{l=-\infty}^{\infty} \check{f}_\phi\left(\frac{l}{2W}\right) e^{-i2\pi q(l/2W)} \Pi\left(\frac{q}{2W}\right), \tag{3.6}$$

where Bracewell (1978) defines

$$\Pi(x) = \begin{cases} 1, & |x| < \frac{1}{2} \\ 0, & |x| > \frac{1}{2}. \end{cases} \tag{3.7}$$

It is necessary to assume* that the projections can be represented adequately by $N + 1$ samples of $\check{f}_\phi$, where N is arbitrarily selected as an even (possibly quite large) integer. Then (3.6) reduces to

$$F_\phi(q) = \frac{1}{2W} \sum_{l=-N/2}^{N/2} \check{f}_\phi\left(\frac{l}{2W}\right) e^{-i2\pi q(l/2W)} \Pi\left(\frac{q}{2W}\right). \tag{3.8}$$

Clearly, from (3.7) and (3.8), $F_\phi(q)$ vanishes for $|q| > W$. In the interval $-W < q < W$, suppose $F_\phi(q)$ is evaluated at a set of equally spaced points defined by

$$q = m\frac{2W}{N} \qquad \text{with} \quad m = -\frac{N}{2}, \ldots, 0, \ldots, \frac{N}{2}.$$

This yields

$$F_\phi\left(m\frac{2W}{N}\right) = \frac{1}{2W} \sum_{l=-N/2}^{N/2} \check{f}_\phi\left(\frac{l}{2W}\right) e^{-i2\pi(ml/N)}, \tag{3.9}$$

which is in the form of a discrete Fourier transform (DFT) that can be evaluated by use of fast Fourier transform (FFT) methods [Brigham (1974)].

The next step is to obtain an expression for the discrete evaluation of $Q_\phi(s)$ from (3.4). In view of the fact that $F_\phi(q)$ has already been assumed band-limited, the discrete version of (3.3) follows immediately:

$$\begin{aligned} Q_\phi(s) &= \int_{-W}^{W} F_\phi(q)|q|e^{i2\pi sq}\, dq \\ &\simeq \frac{2W}{N} \sum_{m=-N/2}^{N/2} F_\phi\left(m\frac{2W}{N}\right)\left|m\frac{2W}{N}\right| e^{i2\pi sm(2W/N)}. \end{aligned}$$

We evaluate $Q_\phi(s)$ at those values of p where the projections are sampled:

$$Q_\phi\left(\frac{k}{2W}\right) \simeq \left(\frac{2W}{N}\right) \sum_{m=-N/2}^{N/2} F_\phi\left(m\frac{2W}{N}\right)\left|m\frac{2W}{N}\right| e^{i2\pi(mk/N)}, \tag{3.10}$$

*Assumption of finite order.

where $k = -N/2, \ldots, 0, \ldots, N/2$. Thus Q_ϕ is approximated by an inverse DFT. When noise is taken into consideration, improved results are usually obtained when $F_\phi[m(2W/N)]|m(2W/N)|$ is multiplied by a window function [Crawford and Kak (1979)]. This tends to suppress high frequencies, which usually are associated with observation noise. Thus (3.10) might be modified to read

$$Q_\phi\left(\frac{k}{2W}\right) = \frac{2W}{N} \sum_{m=-N/2}^{N/2} F_\phi\left(m\frac{2W}{N}\right)\left|m\frac{2W}{N}\right| H\left(m\frac{2W}{N}\right) e^{i2\pi(mk/N)}, \tag{3.11}$$

where $H[m(2W/N)]$ represents the window function, for example, a Hamming window [Hamming (1977)]. This represents a Fourier domain evaluation of Q_ϕ. It is also possible to evaluate Q_ϕ in the spatial (signal) domain. If F_ϕ in (3.11) is replaced by its expansion in (3.9), then by the convolution theorem for discrete Fourier transforms [Brigham (1974)]

$$Q_\phi\left(\frac{k}{2W}\right) = \sum_j \check{f}_\phi\left(\frac{k-j}{2W}\right) h\left(\frac{j}{2W}\right), \tag{3.12}$$

where $h(j/2W)$ is the inverse DFT of $|m(2W/N)|H[m(2W/N)]$.

The final step is to evaluate $f(x, y)$ from knowlege of Q_ϕ by a discrete version of (3.3),

$$f(x, y) \simeq \frac{\pi}{K} \sum_{l=1}^{K} Q_{\phi_l}(x \cos \phi_l + y \sin \phi_l).$$

The angles ϕ_l with $l = 1, 2, \ldots K$ are those angles for which the projections $\check{f}_\phi(p)$ are known.

Computer Implementation

It must be emphasized that the purpose of the preceding discussion is to illustrate some of the basic formulas associated with discrete versions of the continuous formalism. There are several technical difficulties with using the above formulas. The simultaneous assumptions of finite bandwidth and finite order are not strictly valid. Certain artifacts are introduced by the fact that (3.12) is not a periodic convolution. There are problems of interpolation. Of course, the ever present difficulties associated with sampling effects and less than ideal methods of data collection remain. There are ways to cope with these and other difficulties, many of which are discussed by Crawford and Kak (1979), Kak (1979), and Rosenfeld and Kak (1982) using essentially the same notation and formalism as used here. In connection with sampling effects, see also Joseph, Spital, and Stockham (1980); for a treatment of some fundamental restrictions relative to data collection with finite probes and detectors, see Verly and Bracewell (1979). Many other references are contained in [6.8]. Finally, computer codes that implement the inversion algorithm discussed here in both the spatial and Fourier domains are available [Huesman, Gullberg, Greenberg, and Budinger (1977); Herman and Rowland (1977)].

[8.4] PICTURE RESTORATION

Rosenfeld and Kak (1982) have illustrated how the Radon transform is involved in certain a posteriori determinations of the point spread function (PSF). Under noise-free and shift-invariant conditions [described in detail by Rosenfeld and Kak (1982)], the ideal picture $f(x, y)$ and the corresponding degraded picture $g(x, y)$ may be related by

$$g(x, y) = \int\int h(x - x', y - y') f(x', y')\, dx'\, dy.$$

The integral is over the picture, and h is the degradation function. Since the degradation of a point source $f(x', y') = \delta(x' - \alpha)\, \delta(y' - \beta)$ is given by

$$g(x, y) = h(x - \alpha, y - \beta),$$

the function h is identified as the point spread function (PSF). Here the degraded image of a point is independent of the position of the point except for translations.

It may not be possible to determine $h(x, y)$ analytically if the reasons for degradation are too complex. The alternative is to use the degraded picture itself to estimate the PSF. If there are good resons to believe that the original picture contains a sharp point, then the image of that point in the degraded picture is the PSF. Consider the image of a faint star in an astronomical photograph, for example.

Now, if there are reasons to expect that the original scene contains sharp lines, it is desirable to have a method for estimating $h(x, y)$ from the images of these lines. If the lines source is parallel to the x axis in the original scene, then Rosenfeld and Kak (1982) show that the image of such a line source $h_L(y)$ is related to the PSF by

$$h_L(y) = \int_{-\infty}^{\infty} h(x, y)\, dx.$$

Clearly, if the line is at some arbitrary angle with respect to the x axis and the equation of the line is given by $p = x \cos \phi + y \sin \phi$, then the image of the line is just the Radon transform of the PSF,

$$\check{h}(p, \phi) = \int\int h(x, y)\, \delta(p - x \cos \phi - y \sin \phi)\, dx\, dy.$$

For circularly symmetric point spread functions, only one radial line is needed to determine the PSF. For further discussion of this problem in terms of transfer functions, see Rosenfeld and Kak (1982).

[8.5] TRANSFORMATIONS IN GEOPHYSICS

Record Sections and Slant Stacks

The application considered here apparently is being rapidly accepted as a standard method for analysis of record sections* in geophysics: Phinney, Chowdhury, and Frazer (1981) (PCF in the following discussion). We adopt the coordinate conventions of PCF associated with a seismic record section. A transient point source is excited† at the origin at time $t = 0$, and sensors distributed along the x axis (at $z = 0$) record the signals $u(x, t)$. The coordinate z is positive downward and x is horizontal along the surface (earth or water). The propagating wavefield is described by $u(x, z, t)$, which is assumed continuous and twice differentiable in all of its arguments.

Ultimately, of course, the aim is to obtain structural information about the medium in the xz plane by some sort of inversion procedure, velocity versus depth information for example. Phinney, Chowdhury, and Frazer (1981) have given a convincing argument that an important first step in such an endeavor is to use the Radon transform of the record section $\check{u}$ rather than the record section itself. They point out several mathematical and physical reasons for this.

The desired transformation is from the variables x and t to variables p (ray parameter) and τ (intercept time). The new scalar function $\check{u}(p, \tau)$ may be referred to as the slowness representation, slant stack, or plane wave decomposition. We shall not delve into a discussion of travel time curves and interrelationships among the variables x, t, p, and τ; however, the paper by McMechan, and Ottolini (1980) is highly recommended in this regard.‡ They also discuss the Radon transform relation between the seismic profile and the slant stack. An informative review is also given by Robinson (1982).

As mentioned in [1.9], the pioneer work that brought the Radon transform to prominence in geophysics was by Chapman (1978), who emphasized its importance in a study of synthesizing seismograms for earth models that depend only on depth. It is evident from the work by Chapman, PCF, and others that due to the particular way in which the natural parameters x, t, p, and τ enter into the analysis of seismic data as well as in theoretical models,

*Depending on the usage and author, this may be referred to as a seismogram, seismic profile, or the wavefield observed along $z = 0$.

†Some sort of "shot" to generate a compressional wave.

‡Briefly, p is the slope and τ the intercept of a line tangent to an arbitrary point (x, t) on a travel time curve. The units are reciprocal velocity for p and time for τ.

the Radon transform equations have a slightly different appearance than in the case where the variable parameters were x, y, p, and ϕ. [Here p is being used two different ways, but it seems desirable to risk some confusion to maintain contact with the common usage of p to represent slowness (reciprocal velocity) in the geophysics literature.] Consequently, a reformulation of the main equations to conform with current usage in geophysics is in order.

Reformulation*

The main difference comes from the way a line is parametrized. Rather than write the equation in normal form

$$p = x \cos \phi + y \sin \phi, \tag{5.1}$$

express the equation in slope-intercept form

$$t = px + \tau. \tag{5.2}$$

Thus, in place of $f(x, y)$, we have $u(x, t)$, and in place of $\check{f}(p, \phi)$, we have $\check{u}(p, \tau)$. Note the different interpretation for the parameter p. In (5.1), p is the distance from the origin to the line in the xy plane, and in (5.2) p is the slope of the line in the xt plane. The Radon transform equation has the form

$$\begin{aligned} \check{u}(p, \tau) &= \int\int u(x, t)\, \delta(t - px - \tau)\, dx\, dt \\ &= \int u(x, \tau + px)\, dx. \end{aligned} \tag{5.3}$$

Once again we bow to convention and utilize the asymmetric form of the Fourier transform. The two-dimensional transform of $u(x, t)$ is given by†

$$\tilde{u}(k, \omega) = \int\int u(x, t) e^{-i(\omega t - kx)}\, dx\, dt, \tag{5.4}$$

and the inverse is

$$u(x, t) = \frac{1}{(2\pi)^2} \int\int \tilde{u}(k, \omega) e^{i(\omega t - kx)}\, dk\, d\omega. \tag{5.5}$$

The inverse Radon transform can be found in the usual fashion by following the path indicated in Fig. 4.2.

*All integrals in this subsection are from $-\infty$ to ∞.
†Here, the symbol ω is used to represent a scalar.

From (5.4), with $k = \omega p$,

$$\begin{aligned}\tilde{u}(\omega p, \omega) &= \int dx \int dt\, u(x, t) \int d\tau\, e^{-i\omega\tau}\, \delta(\tau - t + px) \\ &= \int dx \int d\tau\, e^{-i\omega\tau} \int dt\, u(x, t)\, \delta(\tau - t + px) \\ &= \int d\tau\, e^{-i\omega\tau} \int dx\, u(x, \tau + px) \\ &= \int \check{u}(p, \tau) e^{-i\omega\tau}\, d\tau. \end{aligned} \tag{5.6}$$

This just expresses the relation

$$\mathscr{F}_2 u = \mathscr{F}_1 \check{u}.$$

It is desired to find u in terms of $\check{u}$. From (5.5),

$$\begin{aligned} u(x, t) &= \frac{1}{(2\pi)^2} \int d\omega \int dk \int dp\, \tilde{u}(k, \omega) e^{i(\omega t - kx)}\, \delta\!\left(\frac{k}{\omega} - p\right) \\ &= \frac{1}{(2\pi)^2} \int d\omega \int dp \int dk |\omega| \tilde{u}(k, \omega) e^{i(\omega t - kx)}\, \delta(k - \omega p) \\ &= \frac{1}{(2\pi)^2} \int\!\!\int |\omega| \tilde{u}(\omega p, \omega) e^{i\omega(t - px)}\, d\omega\, dp. \end{aligned} \tag{5.7}$$

Then by use of (5.6),

$$u(x, t) = \frac{1}{(2\pi)^2} \int dp \int d\tau\, \check{u}(p, \tau) \int d\omega |\omega| e^{i\omega(t - px - \tau)}. \tag{5.8}$$

We have encountered similar expressions in earlier sections, for example, [5.5]. There are various ways to proceed. One approach is to observe that the transform on ω gives [Lighthill (1958)] $-2/(t - px - \tau)^2$. Thus

$$\begin{aligned} u(x, t) &= \frac{-1}{2\pi^2} \int dp \int d\tau \frac{\check{u}(p, \tau)}{(\tau - t + px)^2} \\ &= \frac{-1}{2\pi^2} \int dp \int d\tau \frac{\check{u}(p, \tau - px)}{(\tau - t)^2}, \end{aligned} \tag{5.9}$$

where the last step was obtained by making a change of variables $\tau \to \tau - px$.

An integration by parts yields one of the familiar forms

$$u(x,t) = \frac{-1}{2\pi}\int dp\,\frac{1}{\pi}\int d\tau\,\frac{\frac{\partial \check{u}}{\partial \tau}(p,\tau - px)}{\tau - t}, \tag{5.10}$$

in terms of the Hilbert transform. Recall that the Hilbert transform of a function f is given by a principal value integral. See [B.11] and Sneddon (1972).

$$f_H(t) = \mathcal{H}\,[f(\tau);t] = \frac{1}{\pi}\int \frac{f(\tau)}{\tau - t}\,d\tau,$$

and in terms of derivatives

$$\frac{d}{dt}f_H(t) = \frac{1}{\pi}\int \frac{\frac{df}{d\tau}(\tau)}{\tau - t}\,d\tau. \tag{5.11}$$

It follows that $u(x,t)$ can be expressed as

$$u(x,t) = \frac{-1}{2\pi}\int dp\,\frac{d}{dt}\,\frac{1}{\pi}\int d\tau\,\frac{\check{u}(p,\tau - px)}{\tau - t}$$

$$= \frac{-1}{2\pi}\int dp\,\frac{d}{dt}\,\mathcal{H}\,[\check{u}(p,\tau - px);t]. \tag{5.12}$$

By designating

$$u^{\dagger}(p,t - px) = \frac{d}{dt}\,\mathcal{H}\,[\check{u}(p,\tau - px);t], \tag{5.13}$$

it follows that the inverse transform may be written in the form

$$u(x,t) = \frac{-1}{2\pi}\int u^{\dagger}(p,t - px)\,dp$$

$$= \frac{-1}{2\pi}\int\!\!\int u^{\dagger}(p,t')\,\delta(t' - t + px)\,dt'\,dp. \tag{5.14}$$

Remark. There seems to be some confusion in the geophysics literature regarding which equation [(5.3) or (5.14)] represents the Radon transform and which one represents the inverse Radon transform. The approach here agrees with that taken by Phinney, Chowdhury, and Frazer (1981). □

One-dimensional Models

It must be emphasized that utilization of the Radon transform is only a first, last, or intermediate step in the very involved problems of deducing structural information about the earth from artificial source seismic surveys or from theoretical seismogram studies in which theoretical record sections may be compared with observed record sections. The important point is that for one-dimensional models in which density and elastic constants depend only on depth, the Radon transform and the inverse Radon transform serve as useful, if not essential, tools in the analysis.* For further discussion, we refer the reader to the geophysics literature cited in [1.10], in particular to the following sources: Chapman (1978), Dey-Sarkar and Chapman (1978), McMechan and Ottolini (1980), Clayton and McMechan (1981), Phinney, Chowdhury, and Frazer (1981), Stoffa, Buhl, Diebold, and Wenzel (1981), and Robinson (1982).

[8.6] THE INTEGRAL EQUATIONS OF POTENTIAL SCATTERING

The study of elastic scattering using the Schrödinger equation leads to the integral equation [page 153 of Roman (1965)]

$$\psi_k^+(\mathbf{r}) = e^{ikz} - \frac{1}{4\pi}\int \frac{e^{ik|\mathbf{r}-\mathbf{r}'|}}{|\mathbf{r}-\mathbf{r}'|} U(\mathbf{r}')\psi_k^+(\mathbf{r}')\, d^3\mathbf{r}' \tag{6.1}$$

for the wave function ψ. A particle of momentum k (in natural units $\hbar = c = 1$) traveling in the $+z$ direction encounters a potential $V(\mathbf{r})$ [proportional to $U(\mathbf{r})$] and is scattered in a direction specified by the position vector $\mathbf{r}$. The wave function for the scattered particle is $\psi_k^+(\mathbf{r})$. In a great many cases, it is adequate to work with this equation in the asymptotic region and assume that the potential falls of faster with r than $1/r$. This leads to a scattering amplitude [page 154 of Roman (1965)]

$$f_k(\theta, \phi) = \frac{-1}{4\pi}\int e^{-ik\mathbf{n}\cdot\mathbf{x}} U(\mathbf{x})\psi_k^+(\mathbf{x})\, d\mathbf{x}, \qquad \mathbf{n} = \frac{\mathbf{r}}{r}. \tag{6.2}$$

As usual, $\mathbf{x} = (x, y, z)$ and the integral is over all $\mathbb{R}^3$. The significance of $f_k(\theta, \phi)$ is that its magnitude squared f^*f is proportional to the probability for scattering in a direction specified by the polar angle θ and azimuthal angle ϕ. Note that since the incident direction is along $+z$, the angle θ measures the deviation from the incident direction. If U is a central potential, dependent only on r, then every angle ϕ is equally likely and f depends only on θ. At this point, we do not make that assumption, however.

Equation (6.2) is deceptively simple since to use it ψ_k^+ must first be determined from (6.1). Consequently an approximation and/or iterative scheme is in order. Here we consider only one such approximation: the *first Born*

*As mentioned earlier this is a strong point made by PCF. It was also emphasized to the author by private communication with S. Coen; see also, S. Coen (1981a, b) and references therein.

approximation. This approximation consists of making the replacement

$$\psi_k^+(\mathbf{x}) \to e^{ikz}.$$

For further convenience, we write

$$kz = k\boldsymbol{\kappa} \cdot \mathbf{x}$$

in terms of the unit vector $\boldsymbol{\kappa}$, which points in the $+z$ direction. With these replacements, (6.2) becomes

$$f_k(\theta, \phi) = \frac{-1}{4\pi} \int e^{-ik\mathbf{n}\cdot\mathbf{x}} e^{ik\boldsymbol{\kappa}\cdot\mathbf{x}} U(\mathbf{x})\, d\mathbf{x}. \tag{6.3}$$

We want to relate this equation to the Radon transform of the potential.

Let $\xi = (\mathbf{n} - \boldsymbol{\kappa})/s$, where

$$\begin{aligned} s = |\mathbf{n} - \boldsymbol{\kappa}| &= [(\mathbf{n} - \boldsymbol{\kappa}) \cdot (\mathbf{n} - \boldsymbol{\kappa})]^{1/2} \\ &= [2(1 - \cos\theta)]^{1/2} \\ &= 2\sin\frac{\theta}{2}. \end{aligned}$$

Then, by definition,

$$\check{U}\left(\frac{p}{s}, \xi\right) = \int U(\mathbf{x})\, \delta\left(\frac{p}{s} - \xi \cdot \mathbf{x}\right) d\mathbf{x},$$

and from [3.2]

$$|s|\check{U}(p, s\xi) = |s| \int U(\mathbf{x})\, \delta(p + \boldsymbol{\kappa} \cdot \mathbf{x} - \mathbf{n} \cdot \mathbf{x})\, d\mathbf{x}.$$

Now the (nonsymmetric) Fourier transform of $\check{U}$ with respect to the p variable yields

$$\begin{aligned} \tilde{\check{U}}(k, s\xi) &= \int_{-\infty}^{\infty} dp\, e^{-ipk} \int d\mathbf{x}\, U(\mathbf{x})\, \delta(p + \boldsymbol{\kappa} \cdot \mathbf{x} - \mathbf{n} \cdot \mathbf{x}) \\ &= \int e^{-ik\mathbf{n}\cdot\mathbf{x}} e^{ik\boldsymbol{\kappa}\cdot\mathbf{x}} U(\mathbf{x})\, d\mathbf{x}. \end{aligned} \tag{6.4}$$

Thus we see that $-4\pi f_k(\theta, \phi)$ is to be identified with $\tilde{\check{U}}(k, s\xi)$, and by use of the inverse Fourier transform

$$\begin{aligned} \check{U}(p, s\xi) &= \frac{1}{2\pi} \int_{-\infty}^{\infty} \tilde{\check{U}}(k, s\xi) e^{ikp}\, dk \\ &= -2 \int_{-\infty}^{\infty} f_k(\theta, \phi) e^{ikp}\, dk. \end{aligned} \tag{6.5}$$

Similar types of equations emerge in connection with classical scattering of waves. For a nice discussion of a way the Radon transform can be utilized in these situations, see the article by Norton and Linzer (1981) on reflectivity imaging in three dimensions.

[8.7] PARTIAL DIFFERENTIAL EQUATIONS

Preliminary Reminders

We observed in [3.6] that if L is a linear operator with constant coefficients then

$$\mathscr{R}\left\{\mathsf{L}\left(\frac{\partial}{\partial x}, \frac{\partial}{\partial y}, \frac{\partial}{\partial z}\right) f(x, y, z)\right\}$$

$$= \mathsf{L}\left(\xi_1 \frac{\partial}{\partial p}, \xi_2 \frac{\partial}{\partial p}, \xi_3 \frac{\partial}{\partial p}\right) \check{f}(p, \xi). \tag{7.1}$$

To be specific, the discussion is for $\mathbb{R}^3$; however, the approach is quite general [John (1955); Ludwig (1966)]. When time is included as an additional variable, then L has the form

$$\mathsf{L} = \mathsf{L}\left(\frac{\partial}{\partial x}, \ldots, \frac{\partial}{\partial t}\right),$$

and the Radon transform of $\mathsf{L}u$, where $u = u(x, y, z; t) = u(\mathbf{x}; t)$ is given by

$$\mathscr{R}\{\mathsf{L}u\} = \mathsf{L}\left(\xi_1 \frac{\partial}{\partial p}, \ldots; \frac{\partial}{\partial t}\right) \check{u}(p, \xi; t).$$

Example 1. Suppose $\mathsf{L} = \partial^2/\partial x^2 + (\partial/\partial x)(\partial/\partial t) + (\partial/\partial y)(\partial/\partial z) + \partial^2/\partial t^2$, then

$$\mathscr{R}\{\mathsf{L}u\} = \left(\xi_1^2 \frac{\partial^2}{\partial p^2} + \xi_1 \frac{\partial}{\partial p}\frac{\partial}{\partial t} + \xi_2\xi_3 \frac{\partial^2}{\partial p^2} + \frac{\partial^2}{\partial t^2}\right) \check{u}(p, \xi; t),$$

with

$$\check{u}(p, \xi; t) = \int u(\mathbf{x}; t)\, \delta(p - \xi \cdot \mathbf{x})\, d\mathbf{x}.$$

Recall that $\mathscr{R}$ and $\partial/\partial t$ commute. □

The Cauchy Problem

To be entirely specific, we consider the initial value problem for hyperbolic homogeneous equations with constant coefficients. The Cauchy problem to be considered is that of finding a solution $u(\mathbf{x}; t)$ of the equation

$$\mathsf{L}\left(\frac{\partial}{\partial x}, \frac{\partial}{\partial y}, \frac{\partial}{\partial z}; \frac{\partial}{\partial t}\right)u(x, y, z; t) = 0, \tag{7.2}$$

subject to the initial conditions for $t = 0$

$$\left(\frac{\partial}{\partial t}\right)^l u(\mathbf{x}; t)\bigg|_{t=0} = 0 \quad \text{for} \quad 0 \leqslant l \leqslant m - 2, \tag{7.3a}$$

$$\left(\frac{\partial}{\partial t}\right)^{m-1} u(\mathbf{x}; t)\bigg|_{t=0} = f(\mathbf{x}). \tag{7.3b}$$

Here m is the degree of the polynomial operator L.

Remark. Once this problem has been solved, the result may be used to obtain solutions of the more general Cauchy problem, where the partial differential equation is inhomogeneous and the initial value data more general [Chapter VI of Courant and Hilbert (1962)]. □

Now apply the Radon transform operator to equation (7.2) to obtain the reduced equation

$$\mathsf{L}\left(\xi_1\frac{\partial}{\partial p}, \xi_2\frac{\partial}{\partial p}, \xi_2\frac{\partial}{\partial p}; \frac{\partial}{\partial t}\right)\check{u}(p, \xi; t) = 0. \tag{7.4}$$

Since $\mathscr{R}$ and $\partial/\partial t$ commute, the initial conditions for the reduced equation are

$$\left(\frac{\partial}{\partial t}\right)^l \check{u}(p, \xi; t)\bigg|_{t=0} = 0; \qquad 0 \leqslant l \leqslant m - 2, \tag{7.5a}$$

$$\left(\frac{\partial}{\partial t}\right)^{m-1} \check{u}(p, \xi; t)\bigg|_{t=0} = \check{f}(p, \xi). \tag{7.5b}$$

The equation for $\check{u}$ is one-dimensional in the spatial variable p and may be solved by standard methods for treating partial differential equations with one spatial variable and time as the two independent variables. The components of ξ are treated as parameters when solving for $\check{u}$.

Once $\check{u}$ is determined, the solution $u(\mathbf{x}; t)$ may be computed by use of the inversion formula; for $\mathbb{R}^3$, the result is, from [5.4],

$$u(\mathbf{x}; t) = \mathscr{R}^\dagger\bar{\check{u}}(s, \xi; t)$$

$$= \int_{|\xi|=1} \bar{\check{u}}(\xi \cdot \mathbf{x}, \xi; t)\, d\xi, \tag{7.6}$$

where

$$\bar{\check{u}}(s, \xi; t) = \frac{-1}{8\pi} \frac{\partial^2}{\partial p^2} \check{u}(p, \xi; t)\bigg|_{p=s}. \tag{7.7}$$

We consider a simple example to help fix the basic ideas more firmly.

Example 2. Consider the wave equation (with wave velocity c)

$$\left(\frac{\partial^2}{\partial x^2} + \frac{\partial^2}{\partial y^2} + \frac{\partial^2}{\partial z^2} - \frac{1}{c^2}\frac{\partial^2}{\partial t^2}\right) u(\mathbf{x}; t) = 0$$

with initial conditions

$$u(\mathbf{x}; 0) = 0$$

$$u_t(\mathbf{x}; 0) = e^{-x^2-y^2-z^2}.$$

The reduced equation is

$$\left(\frac{\partial^2}{\partial p^2} - \frac{1}{c^2}\frac{\partial^2}{\partial t^2}\right) \check{u}(p, \xi; t) = 0$$

with initial conditions

$$\check{u}(p, \xi; 0) = 0$$

$$\check{u}_t(p, \xi; 0) = \pi e^{-p^2}.$$

By standard techniques, the solution is

$$\check{u}(p, \xi; t) = \frac{\pi}{2c}\int_{p-ct}^{p+ct} e^{-s^2}\, ds.$$

Note that

$$\frac{\partial^2 \check{u}}{\partial p^2} = \frac{1}{c^2}\frac{\partial^2 \check{u}}{\partial t^2} = \frac{\pi}{c}\left[(p - ct)e^{-(p-ct)^2} - (p + ct)e^{-(p+ct)^2}\right].$$

Thus $\bar{\check{u}}$ can be found immediately,

$$\bar{\check{u}}(s, \xi; t) = \frac{-1}{8c}\left[(s - ct)e^{-(s-ct)^2} - (s - ct)e^{-(s+ct)^2}\right],$$

and $u(\mathbf{x}; t)$ can be computed from (7.6). □

For further theory of the Radon transform with a view toward its use in solving differential equations see John (1955), Ludwig (1966), Lax and Phillips (1979), Helgason (1980), and references quoted in these sources.

[8.8] GENERALIZATIONS AND OTHER USES

Attenuated Radon Transform

A study of the single photon ECT problem in [1.3] leads to a natural generalization of the Radon transform. Recall that if attenuation of the medium is neglected, the radionuclide source distribution function $f(\mathbf{x}) = f(x, y)$ and the measured projection data may be related by the Radon transform $\check{f}$. On the other hand, if the attenuating properties of the medium are not negligible, as is often the situation in nuclear medicine problems, then the connection between the projection data and the source distribution function is considerably more complicated.

Suppose a photon is emitted at position $\mathbf{x} = (x, y)$ in Fig. 8.1. If the photon travels along the line defined by

$$p = \mathbf{x}' \cdot \xi = x' \cos \phi + y' \cos \phi$$

from the point (x, y) to the detector, then the linear attenuation coefficient $\mu(x', y')$ is effective over the solid portion of the line L where

$$(\mathbf{x}' - \mathbf{x}) \cdot \xi^{\perp} \geqslant 0.$$

As usual, ξ is a unit vector

$$\xi = (\cos \phi, \sin \phi)$$

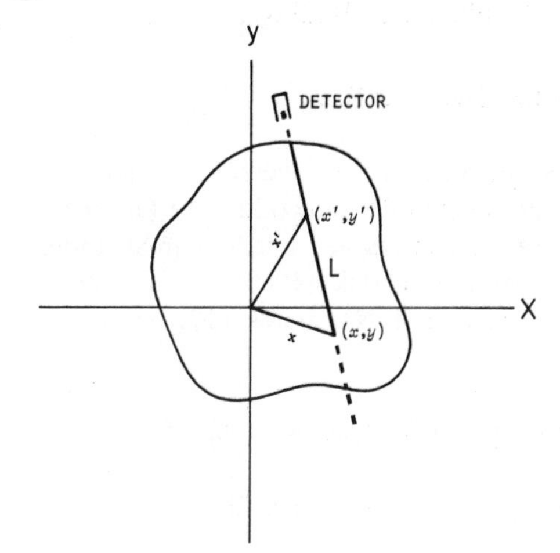

Figure 8.1 A photon is emitted at $\mathbf{x} = (x, y)$ and travels along line L to the detector.

from the origin along a perpendicular to the line L and

$$\xi^{\perp} = (-\sin\phi, \cos\phi)$$

points along L toward the detector.

Under these conditions, the distribution function f is modified by an exponential attenuation factor of the form

$$\mathscr{A}(x, y; p, \phi) = \exp\left[-\int_{L'} \mu(x', y')\, dl'\right],$$

where the line integral is over only the solid portion of L in Fig. 8.1. This can be expressed in a more transparent form with $\mu(\mathbf{x}') = \mu(x', y')$:

$$\mathscr{A}(x, y; p, \phi) = \exp\left(-\int_{\mathbf{x}'\cdot\xi^{\perp} \geqslant \mathbf{x}\cdot\xi^{\perp}} \mu(\mathbf{x}')\, \delta(p - \xi\cdot\mathbf{x}')\, d\mathbf{x}'\right).$$

Finally, the projections are related to f and $\mathscr{A}$ by*

$$\check{f}_{\mu}(p, \xi) = \iint_{-\infty}^{\infty} \mathscr{A}(x, y; p, \phi) f(x, y)\, \delta(p - x\cos\phi - y\sin\phi)\, dx\, dy. \tag{8.1}$$

Note that if μ is negligible, the attenuation factor goes to unity and $\check{f}_{\mu}$ is just the Radon transform $\check{f}$ of the distribution f.

The transformation (8.1) is known as the *attenuated* or *exponential* Radon transform. It has been studied by several authors; the work by Gullberg (1979b, 1980) is especially recommended. Other important contributions include those of Tretiak and Metz (1980) and Natterer (1979). For other references, see [1.10] and the review by Gullberg and Budinger (1981).

Inversion Formula for Divergent Rays

It is not essential that the line integrals $\check{f}$ be parallel. An inversion formula has been developed for diverging rays of fan shape in connection with fan-beam projections in X-ray TCT work. Several references are cited in [6.8] under "Fan-Beam Methods"; in particular, see Narparstek (1980) for a concise statement of the basic results, and Herman and Naparstek (1977) for full detail.

*This can also be expressed using vectors ξ and $\xi^{\perp}$ in the form [Natterer (1979)]

$$\check{f}_{\mu}(p, \xi) = \int_{-\infty}^{\infty} f(p\xi + t\xi^{\perp}) \exp\left[-\int_{t}^{\infty} \mu(p\xi + s\xi^{\perp})\, ds\right] dt.$$

Other Extensions and Uses

Several authors have extended the Radon transform to include various non-Euclidean spaces. For a discussion of these generalizations see the last part of Helgason (1980), Lax and Phillips (1979), and earlier work cited in these sources. Further generalizations and additional results on characterization of classical and spherical Radon transforms may be found in the work by Cormack and Quinto (1980) and Quinto (1980a, b, 1981a, b).

Guillemin and Sternberg (1977) utilized the Radon transform in their study of the geometric aspects of distributions. They show how the transform is useful in decomposing a distribution into a superposition of simpler types of distributions.

In the area of theoretical statistics, Cavaretta, Micchelli, and Sharma (1980a, b) have utilized Radon inversion to extend univariate interpolation procedures to higher dimensions.

Several discussions of the ill-posed nature of the inverse problem are contained among the various contributions to the Delaware conference edited by Nashed (1982). Questions of uniqueness and resolution are discussed by Katz (1978, 1979) also.

An interesting application that may have physical importance is suggested by Newton (1978). He points out that although the existence of nonlinear evolution equations with soliton solutions are of great physical interest, most of the equations actually utilized have been confined to one spatial dimension. Newton uses Radon transform methods to generate an integro-differential evolution in three spatial variables from a known one-dimensional equation. The new equation has soliton solutions that vanish at large distances in all directions. His method, using the inverse Radon transform, can be applied to other equations to generate integro-differential equations in higher dimensions that have soliton solutions from one-dimensional differential equations already known to have soliton solutions. Although the method is straightforward, problems with physical interpretations and applications remain.

Other applications and further generalizations are contained in the proceedings of the Oberwolfach conference edited by Herman and Natterer (1981). Some other sources for more technical and abstract work on the Radon transform are contained under the heading "Mathematical and Generalizations" in [1.10]. The more recent articles in this section are mainly related to extensions and generalizations, whereas the earlier ones often contain original material regarding the transform on Euclidean space.

Appendix A

Translation of Radon's 1917 Paper*

ON THE DETERMINATION OF FUNCTIONS FROM THEIR INTEGRALS ALONG CERTAIN MANIFOLDS

If one integrates a function of two variables x, y—a *point-function* $f(P)$ in the plane—that satisfies suitable regularity conditions, along an arbitrary straight line g, then the values $F(g)$ of these integrals define a *line-function*. The problem that is solved in part A of this paper is the inversion of this functional transformation. That is, answers to the following questions are given: Is every line-function that satisfies suitable regularity conditions obtainable by this process? If this is the case, is the point-function f then uniquely determined by F and how can it be found?

The problem of finding a line-function $F(g)$ from the mean values over its points $f(P)$, which is in a sense the dual problem, is solved in part B.

Finally, in part C, certain generalizations that arise particularly from considering non-Euclidian manifolds as well as higher-dimensional spaces are briefly discussed.

Interesting in themselves, the treatment of these problems is gaining even more interest because of the fact that there are numerous relations between this subject and the theory of the logarithmic and the Newtonian potential. These will be pointed out in the appropriate places.

A. DETERMINATION OF A POINT-FUNCTION IN THE PLANE FROM ITS INTEGRALS ALONG STRAIGHT LINES

1. Let $f(x, y)$ be a real function defined for all real points $p = [x, y]$ that satisfies the following regularity conditions:

(a_1) $f(x, y)$ is continuous.

(b_1) The following double integral, which is to be taken over the whole plane, is convergent:

$$\iint \frac{|f(x, y)|}{\sqrt{x^2 + y^2}} \, dx \, dy.$$

*Translated by R. Lohner, School of Mathematics, Georgia Institute of Technology, Atlanta, GA 30332.

(c_1) For an arbitrary point $P = [x, y]$ and any $r \geqslant 0$, let

$$\bar{f}_P(r) = \frac{1}{2\pi}\int_0^{2\pi} f(x + r\cos\phi,\ y + r\sin\phi)\, d\phi.$$

Then for every point P,

$$\lim_{r\to\infty} \bar{f}_P(r) = 0.$$

Thus the following theorems hold.

Theorem I. The integral of f along the straight line g with the equation $x\cos\phi + y\sin\phi = p$, given by

(I)

$$F(p,\phi) = F(-p, \phi + \pi) = \int_{-\infty}^{+\infty} f(p\cos\phi - s\sin\phi,\ p\sin\phi + s\cos\phi)\, ds$$

is "in general" well-defined. This means that on any circle those points that have tangent lines for which F does not exist form a set of linear measure zero.

Theorem II. If the mean value of $F(p, \phi)$ is formed for the tangent lines of the circle with center $P = [x, y]$ and radius q:

$$\bar{F}_P(q) = \frac{1}{2\pi}\int_0^{2\pi} F(x\cos\phi + y\sin\phi + q, \phi)\, d\phi, \tag{II}$$

then this integral is absolutely convergent for all P, q.

Theorem III. The value of f is completely determined by F and can be computed as follows:

$$f(P) = -\frac{1}{\pi}\int_0^{\infty} \frac{d\bar{F}_P(q)}{q}. \tag{III}$$

Here the integral is to be understood in the Stieltjes sense and it can also be defined by the formula:

$$f(P) = \frac{1}{\pi}\lim_{\varepsilon\to 0}\left(\frac{\bar{F}_P(\varepsilon)}{\varepsilon} - \int_{\varepsilon}^{\infty} \frac{\bar{F}_P(q)}{q^2}\, dq\right). \tag{III$'$}$$

Before starting with the proof of these theorems, we note that conditions a_1–c_1 are invariant under rigid motions of the plane. Thus we can always consider the point $[0, 0]$ to represent an arbitrary point of the plane.

Now the double integral

$$\iint_{x^2+y^2>q^2} \frac{f(x, y)}{\sqrt{x^2 + y^2 - q^2}} dx\, dy \tag{1}$$

is seen to converge absolutely. Using the transformation

$$x = q\cos\phi - s\sin\phi, \qquad y = q\sin\phi + s\cos\phi,$$

it becomes

$$\int_0^{2\pi} d\phi \int_0^{\infty} f(q\cos\phi - s\sin\phi, q\sin\phi + s\cos\phi)\, ds$$

$$= \int_0^{2\pi} d\phi \int_{-\infty}^{0} f(q\cos\phi - s\sin\phi, q\sin\phi + s\cos\phi)\, ds,$$

so that its value can also be expressed as

$$\frac{1}{2}\int_0^{2\pi} d\phi \int_{-\infty}^{+\infty} f(q\cos\phi - s\sin\phi, q\sin\phi + s\cos\phi)\, ds = \frac{1}{2}\int_0^{2\pi} F(q, \phi)\, d\phi.$$

From well-known properties of absolute convergent double integrals, theorems I and II follow.

In order to derive formula (III), one can choose the following path: Introducing polar coordinates in (1) yields

$$\int_q^{\infty} dr \int_0^{2\pi} \frac{f(r\cos\phi, r\sin\phi)}{\sqrt{r^2 - q^2}} d\phi$$

or, using the mean value notation from c_1:

$$2\pi \int_q^{\infty} \frac{\bar{f}_0(r)\, dr}{\sqrt{r^2 - q^2}}.$$

Comparing this with the value of (1) obtained before,

$$\bar{F}_0(q) = 2\int_q^{\infty} \frac{\bar{f}_0(r)\, dr}{\sqrt{r^2 - q^2}}. \tag{2}$$

Introducing the variables $r^2 = v$, $q^2 = u$, this integral equation of the first kind can easily be solved by the well-known method of Abel, which yields formula (III) for

$$\bar{f}_0(0) = f(0, 0).$$

However, it seems to be hard to derive this without placing further restrictions on f; therefore, we prefer a direct verification.

To prove the equality of the expressions (III) and (III′), it first must be shown that

$$\lim_{q\to\infty} \frac{\bar{F}_0(q)}{q} = 0.$$

Because of (2),

$$\left|\frac{\bar{F}_0(q)}{q}\right| \leqslant \frac{2}{q}\left|\int_q^{2q} \frac{\bar{f}_0(r) r\, dr}{\sqrt{r^2 - q^2}}\right| + \frac{2}{q}\left|\int_{2q}^{\infty} \frac{\bar{f}_0(r)\, dr}{\sqrt{r^2 - q^2}}\right|$$

$$\leqslant 2\sqrt{3}\,|\bar{f}_0(t)| + \frac{4}{\sqrt{3}}\int_{2q}^{\infty} |\bar{f}_0(r)|dr \qquad (q < t < 2q)$$

and this converges to zero as $q \to \infty$ because of b_1 and c_1.

Introducing (2), the right-hand side of (III′) is transformed into

$$\frac{2}{\pi}\lim_{\varepsilon\to 0}\left(\frac{1}{\varepsilon}\int_{\varepsilon}^{\infty} \frac{r\bar{f}_0(r)}{\sqrt{r^2 - \varepsilon^2}} dr - \int_{\varepsilon}^{\infty} \frac{dq}{q^2}\int_q^{\infty} \frac{r\bar{f}_0(r)}{\sqrt{r^2 - q^2}} dr\right).$$

If the order of integration is interchanged in the second integral, one can integrate with respect to q and see that this integral is an absolute convergent double integral that justifies the interchange. Moreover, one finds the value

$$\frac{2}{\pi}\lim_{\varepsilon\to 0}\int_{\varepsilon}^{\infty} \frac{\bar{f}_0(r)}{r\sqrt{r^2 - \varepsilon^2}} dr$$

for the preceding expression which yields, in fact, $\bar{f}_0(0) = f(0, 0)$, as can be shown without difficulty.

2. Let $F(p, \phi) = F(-p, \phi + \pi)$ be a line-function satisfying the following regularity conditions:

(a_2) F and its derivatives F_p, F_{pp}, F_{ppp}, F_{ϕ}, $F_{p\phi}$, $F_{pp\phi}$ are continuous for all $[p, \phi]$.

(b_2) F, F_{ϕ}, pF_p, $pF_{p\phi}$ and pF_{pp} converge to zero uniformly in ϕ as $p \to \infty$.

(c_2) The integrals

$$\int_0^{\infty} F_{pp} \ln p\, dp, \int_0^{\infty} F_{ppp}\, p \ln p\, dp, \int_0^{\infty} F_{pp\phi}\, p \ln p\, dp$$

converge absolutely and uniformly in ϕ.

Then we can prove the following theorem.

Theorem IV. If $f(P)$ is formed according to (III) or (III′), then it satisfies conditions a_1, b_1, c_1 and its integrals along straight lines yield the given $F(p, \phi)$. Due to theorem III, it is the only such function.

Introducing polar coordinates, we get

$$f(\rho\cos\psi, \rho\sin\psi) = -\frac{1}{2\pi^2}\int_0^\infty \frac{dp}{p}\int_0^{2\pi} F_p(\rho\cos\omega + p, \omega + \psi)\, d\omega$$

$$= \frac{1}{2\pi^2}\int_0^\infty \ln p\, dp \int_0^{2\pi} F_{pp}(p + \rho\cos\omega, \omega + \psi)\, d\omega$$

since

$$\int_0^{2\pi} F_p(\rho\cos\omega + p, \omega + \psi)\, d\omega = \int_0^{2\pi} F_p(\rho\cos\omega, \omega + \psi)\, d\omega + \int_0^{2\pi} d\omega \int_0^p F_{pp}(\rho\cos\omega + t, \omega + \psi)\, dt$$

and the first term is equal to zero because of $F(p, \phi) = F(-p, \phi + \pi)$. Thus the product of the integral with $\ln p$ converges to zero as $p \to 0$. From the same property of F, it also follows that

$$(3)\quad f(\rho\cos\psi, \rho\sin\psi) = \frac{1}{2\pi^2}\int_0^\pi d\omega \int_{-\infty}^{+\infty} F_{pp}(p, \omega + \psi)\ln|p - \rho\cos\omega|\, dp.$$

Now it suffices to show

$$(4)\qquad \int_{-\infty}^{+\infty} f(\rho, 0)\, d\rho = F\left(0, \frac{\pi}{2}\right),$$

since the conditions a_2–c_2 are invariant under rigid motions. We let

$$F(p, \phi) = F\left(p, \frac{\pi}{2}\right) + \cos\phi \cdot G(p, \phi).$$

G satisfies regularity conditions that can be easily specified. According to this decomposition, $f(\rho, 0)$ is split into two parts $f_1(\rho)$ and $f_2(\rho)$ that have to be investigated separately. Because of

$$\int_0^\pi \ln|p - \rho\cos\omega|\, d\omega = \begin{cases} \pi\ln\dfrac{|p| + \sqrt{p^2 - \rho^2}}{2}, & |p| > |\rho| \\[2ex] \pi\ln\dfrac{|\rho|}{2}, & |p| \leqslant |\rho| \end{cases}$$

it follows that

$$f_1(\rho) = \frac{1}{2\pi^2}\int_0^{\pi} d\omega \int_{-\infty}^{+\infty} F_{pp}\left(p, \frac{\pi}{2}\right)\ln|p - \rho\cos\omega|\, dp$$

$$= \frac{1}{2\pi}\int_{|\rho|}^{\infty} F_{pp}\left(p, \frac{\pi}{2}\right)\ln\frac{|p| + \sqrt{p^2 - \rho^2}}{|\rho|}\, dp$$

$$+ \frac{1}{2\pi}\int_{-\infty}^{-|\rho|} F_{pp}\left(p, \frac{\pi}{2}\right)\ln\frac{|p| + \sqrt{p^2 - \rho^2}}{|\rho|}\, dp.$$

Now, this is absolutely integrable from $-\infty$ to $+\infty$ with respect to ρ, which can be seen from interchanging the order of integration. The value of the integral is

$$\int_{-\infty}^{+\infty} f_1(\rho)\, d\rho = \frac{1}{2\pi}\int_{-\infty}^{+\infty} F_{pp}\left(p, \frac{\pi}{2}\right)\int_{-|p|}^{+|p|}\ln\frac{|p| + \sqrt{p^2 - \rho^2}}{|\rho|}\, d\rho\, dp$$

$$= \frac{1}{2}\int_{-\infty}^{+\infty} F_{pp}\left(p, \frac{\pi}{2}\right)|p|\, dp = F\left(0, \frac{\pi}{2}\right).$$

As to $f_2(\rho)$, we will show that it is also absolutely integrable and yields zero when integrated from $-\infty$ to $+\infty$.

We can write $f_2(\rho)$ as follows:

$$f_2(\rho) = \frac{1}{2\pi^2}\int_0^{\pi} d\omega \int_{-\infty}^{+\infty} G_{pp}(p, \omega)\ln|p - \rho\cos\omega| \cdot \cos\omega\, d\omega$$

$$= \frac{1}{2\pi^2}\int_0^{\pi} d\omega \int_{-\infty}^{+\infty} G_{pp}(p, \omega)\left[\ln\left|\frac{p - \rho\cos\omega}{\rho\cos\omega}\right|\cos\omega + \frac{\rho p\cos^2\omega}{1 + \rho^2\cos^2\omega}\right] dp$$

since the integral of the additional terms is zero and in this form integration with respect to ρ leads to an *absolutely* convergent threefold integral. This is so because of

$$\int_{-\infty}^{+\infty}\left|\cos\omega\ln\left|\frac{p - \rho\cos\omega}{\rho\cos\omega}\right| + \frac{\rho p\cos^2\omega}{1 + \rho^2\cos^2\omega}\right| d\rho$$

$$= |p|\int_{-\infty}^{+\infty}\left|\ln\left|1 - \frac{1}{\tau}\right| + \frac{p^2\tau}{1 + p^2\tau^2}\right| d\tau = \lambda(p)$$

with

$$\lim_{|p|\to\infty}\frac{\lambda(p)}{|p|\ln|p|} = 2.$$

The integration with respect to ρ yields the value

$$\int_{-\infty}^{+\infty} f_2(\rho)\, d\rho = 0,$$

which completes the proof of (4).

Now it remains to show that f satisfies the conditions a_1–c_1.

The continuity follows from the representation (3) because of the assumptions a_2–c_2. Condition b_1 is also satisfied since

$$\int_{-\infty}^{+\infty} |f(\rho\cos\psi, \rho\sin\psi)|\, d\rho$$

is integrable with respect to ψ, as is easily seen. To show that c_1 holds, we form

$$\begin{aligned}
\bar{f}_0(\rho) &= \frac{1}{2\pi}\int_0^{2\pi} f(\rho\cos\psi, \rho\sin\psi)\, d\psi \\
&= \frac{1}{4\pi^3}\int_0^{\pi} d\omega \int_0^{2\pi} d\psi \int_{-\infty}^{+\infty} F_{pp}(p,\psi)\ln|p - \rho\cos\omega|\, dp \\
&= \frac{1}{4\pi^2}\int_0^{2\pi} d\psi \left[\int_{-\infty}^{-|\rho|} F_{pp}(p,\psi)\ln\frac{|p| + \sqrt{p^2-\rho^2}}{2}\, dp\right. \\
&\quad + \int_{|\rho|}^{+\infty} F_{pp}(p,\psi)\ln\frac{|p| + \sqrt{p^2-\rho^2}}{2}\, dp \\
&\quad \left. + F_p(\rho,\psi)\ln\frac{\rho}{2} - F_p(\rho,\psi)\ln\frac{\rho}{2}\right],
\end{aligned}$$

from which the validity of c_1 can be seen. This completes the proof of theorem IV.

B. DETERMINATION OF A LINE-FUNCTION FROM ITS POINT MEAN VALUES

3. Let $F(p,\phi) = F(-p, \phi+\pi)$ be a line-function satisfying the following regularity conditions:

(a_3) F, F_ϕ, F_p are continuous, $|F_\phi| < M$ for all p, ϕ.

(b_3) $F_p \ln|p|$ is convergent to zero uniformly in ϕ as $p \to \infty$.

(c_3) $\int_{-\infty}^{+\infty} |F_p|\ln|p|\, dp$ is uniformly convergent in ϕ.

Again these conditions are invariant under rigid motions. We form the point mean value of $F(p, \phi)$ for $P = [x, y]$:

$$f(x, y) = \frac{1}{\pi}\int_{-\pi/2}^{+\pi/2} F(x\cos\phi + y\sin\phi, \phi)\, d\phi. \tag{5}$$

Then the following theorem holds.

Theorem V. F is uniquely determined by specifying f; that is

$$F\left(0, \frac{\pi}{2}\right) = -\frac{1}{2\pi}\int_{-\infty}^{+\infty}\frac{dx}{x}\int_{-\infty}^{+\infty} f_x(x, y)\, dy, \tag{V}$$

where the Cauchy principal value is to be taken for the integral with respect to x. The value of F for any other straight line can be determined from this formula by means of a suitable rigid motion.

To prove this, we first deduce from (5) that

$$\int_{-A}^{B} f_x(x, y)\, dy = \frac{1}{\pi}\int_{-\pi/2}^{+\pi/2} d\phi \int_{-A}^{B} F_p(x\cos\phi + y\sin\phi, \phi)\cos\phi\, dy, \tag{6}$$

where A, B are two positive constants.

Now, as we have done already earlier, we let

$$F(p, \phi) = F(p, 0) + \sin\phi\, G(p, \phi),$$

where $G(p, \phi)$ is bounded in the domain of integration and has the limit zero as $p \to \infty$. From

$$\int_{-A}^{B} G_p(x\cos\phi + y\sin\phi, \phi)\cos\phi\sin\phi\, dy$$

$$= \left[G(x\cos\phi + B\sin\phi, \phi) - G(x\cos\phi - A\sin\phi, \phi)\right]\cos\phi$$

it follows that the second term of (6) tends to zero as $A \to \infty$, $B \to \infty$ thus leaving only the first one to be investigated. Performing the analogous integration, one sees that in this first term the integral with respect to ϕ also tends to zero as $A \to \infty$, $B \to \infty$ if the integration is carried out over an interval which does *not* contain $\phi = 0$. Therefore, it remains to consider

$$\lim_{\substack{A\to\infty \\ B\to\infty}} \frac{1}{\pi}\int_{-\varepsilon}^{+\varepsilon} d\phi \int_{-A}^{B} F_p(x\cos\phi + y\sin\phi, 0)\cos\phi\, dy, \qquad 0 < \varepsilon < \frac{\pi}{2}.$$

This integral can be written as

$$\frac{1}{\pi}\int_{-\varepsilon}^{+\varepsilon} d\phi \int_{x\cos\phi - A\sin\phi}^{x\cos\phi + B\sin\phi} F_p(p, 0)\cot\phi\, dp.$$

Then, assuming A and B sufficiently large and interchanging the order of integration, after some computations one obtains the value

$$\frac{1}{\pi}\int_{x\cos\varepsilon - B\sin\varepsilon}^{x\cos\varepsilon + B\sin\varepsilon} \ln\frac{(B^2+x^2)\sin\varepsilon}{\left|Bp - x\sqrt{B^2+x^2-p^2}\right|} F_p(p,0)\,dp$$

$$+\frac{1}{\pi}\int_{x\cos\varepsilon - A\sin\varepsilon}^{x\cos\varepsilon + A\sin\varepsilon} \ln\frac{(A^2+x^2)\sin\varepsilon}{\left|Ap - x\sqrt{A^2+x^2-p^2}\right|} F_p(p,0)\,dp.$$

It is sufficient to determine the limit of the second integral as $A \to \infty$. We write it as follows:

$$\frac{1}{\pi}\ln(A\sin\varepsilon)\left[F(x\cos\varepsilon + A\sin\varepsilon, 0) - F(x\cos\varepsilon - A\sin\varepsilon, 0)\right]$$

$$+\frac{1}{\pi}\int_{x\cos\varepsilon - A\sin\varepsilon}^{x\cos\varepsilon + A\sin\varepsilon} \ln\frac{1}{|p-x|} F_p(p,0)\,dp$$

$$+\frac{1}{\pi}\int_{x\cos\varepsilon - A\sin\varepsilon}^{x\cos\varepsilon + A\sin\varepsilon} \ln\frac{\left|Ap + x\sqrt{A^2+x^2-p^2}\right|}{A|p+x|} F_p(p,0)\,dp.$$

Since in the last integral the logarithm tends to zero *uniformly* as $A \to \infty$, the limit follows:

$$-\frac{1}{\pi}\int_{-\infty}^{+\infty} F_p(p,0)\ln|p-x|\,dp$$

which leads to the limit of (6):

$$\int_{-\infty}^{+\infty} f_x(x,y)\,dy = -\frac{2}{\pi}\int_{-\infty}^{+\infty} F_p(p,0)\ln|p-x|\,dp.$$

It should be noted here that the latter expression represents the boundary values of the imaginary part of a regular analytic function in the upper half plane whose real part has the boundary values $2F(x,0)$.

If we now form

$$-\int_{-\infty}^{+\infty}\frac{dx}{x}\int_{-\infty}^{+\infty} f_x(x,y)\,dy = \frac{2}{\pi}\int_0^{\infty}\frac{dx}{x}\int_{-\infty}^{+\infty} F_p(p,0)\ln\left|\frac{p-x}{p+x}\right|dx$$

in the spirit of formula (V), then this double integral is absolutely convergent and leads directly to formula (V) since

$$\int_0^{\infty}\ln\left|\frac{p-x}{p+x}\right|\frac{dx}{x} = -\frac{\pi^2}{2}\operatorname{sgn} p.$$

4. Now let f be a point-function with the following regularity properties:

(a_4) f and its derivatives up to the the second order are continuous.

(b_4) The expressions $f(x, y)$, $\sqrt{x^2 + y^2}\ln(x^2 + y^2)f_x(x, y)$, $\sqrt{x^2 + y^2}\ln(x^2 + y^2)f_y(x, y)$ approach zero as $x^2 + y^2 \to \infty$.

(c_4) The integrals

$$\int_{-\infty}^{+\infty}\int_{-\infty}^{+\infty} D_1 f \frac{\ln(x^2 + y^2)}{\sqrt{x^2 + y^2}}\,dx\,dy$$

and

$$\int_{-\infty}^{+\infty}\int_{-\infty}^{+\infty} D_2 f \ln(x^2 + y^2)\,dx\,dy,$$

where $D_1 f$ means any first and $D_2 f$ any second derivative, are absolutely convergent.

Again these conditions are invariant under rigid motions. Then the following theorem holds.

Theorem VI. The line-function formed from f according to (V) has the point mean values $f(x, y)$.

It is sufficient to show the proof for the origin. For an arbitrary straight line through the origin, (V) yields after an integration by parts:

$$F(0, \phi) = \frac{1}{2\pi}\int_{-\infty}^{+\infty}\int_{-\infty}^{+\infty}\left[f_{xx}\cos^2\phi + 2f_{xy}\sin\phi\cos\phi + f_{yy}\sin^2\phi\right]$$

$$\cdot\ln|x\cos\phi + y\sin\phi|\,dx\,dy$$

or, after introducing polar coordinates ρ, ψ:

$$F(0, \phi) = \frac{1}{2\pi}\int_0^{\infty}\rho\,d\rho\int_0^{2\pi}\left[\frac{\partial^2 f}{\partial\rho^2}\cos^2(\phi - \psi)\right.$$

$$+2\frac{\partial^2 f}{\partial\rho\,\partial\psi}\frac{\sin(\phi - \psi)\cos(\phi - \psi)}{\rho} + \frac{\partial^2 f}{\partial\psi^2}\frac{\sin^2(\phi - \psi)}{\rho^2}$$

$$+\frac{\partial f}{\partial\rho}\frac{\sin^2(\phi - \psi)}{\rho}$$

$$\left.-2\frac{\partial f}{\partial\psi}\frac{\sin(\phi - \psi)\cos(\phi - \psi)}{\rho^2}\right]\ln|\rho\cos(\phi - \psi)|\,d\psi.$$

In order to form the point mean value for $[0,0]$, the integration with respect to ϕ from 0 to 2π can be carried out under the double integral, and then one has to divide by 2π. The term containing $\partial^2 f/\partial\psi^2$ that appears during this computation cancels when integrating with respect to ψ and there remains

$$\frac{1}{2\pi}\int_0^{2\pi} d\psi \int_0^{\infty}\left[\frac{1}{4}\left(\rho\frac{\partial^2 f}{\partial\rho^2}-\frac{\partial f}{\partial\rho}\right)+\tfrac{1}{2}\ln\frac{\rho}{2}\frac{\partial}{\partial\rho}\left(\rho\frac{\partial f}{\partial\rho}\right)\right]d\rho,$$

which indeed reduces to $f(0,0)$.

In order to show the uniqueness of F, it remains to show that the conditions a_3–c_3 are satisfied, which makes it obviously necessary to place further restrictions on f.

5. Here the following remark, for which I am indebted to Mr. W. Blaschke, who also posed the problem, should be made: Both problems treated here are closely related to the theory of the Newtonian potential. That is, if we consider the transition from a point-function f to its mean values F along straight lines as a linear functional transformation

$$F = Rf$$

and similarly the transition from a line-function F to its point mean values v

$$v = BF,$$

then it is natural to consider the composed transformation $H = BR$ defined by

$$v = Hf = B[Rf] = BRf.$$

It can now be readily seen that Hf is nothing but the Newtonian potential in the points of the plane that is covered with a mass of density $(1/\pi)f$. According to a remark made by G. Herglotz, this can be used to construct the inverse of the transformation H; this leads to

$$f(P) = H^{-1}v = -\frac{1}{2}\int_0^{\infty}\frac{d\bar{v}_P(r)}{r} = -\frac{1}{4\pi}\int_{-\infty}^{+\infty}\int_{-\infty}^{+\infty}\frac{\Delta v(x',y')}{r_{PP'}}dx'\,dy',$$

where $\bar{v}_P$ is a notation for a mean value analogous to the previously introduced notations and Δ is the Laplacian operator.

Now we could think of performing the inversion of R and H, which was done directly in 1–3 by means of

$$R^{-1} = H^{-1}B \qquad \text{and} \quad B^{-1} = RH^{-1}.$$

In fact, I first found the inversion formula (IV) in this way. However, carrying out this thought in a strict manner seems to be more difficult than the direct verification, and it even fails in the non-Euclidian cases, which will be soon discussed.

Finally, we remark that the regularity conditions assumed in parts A and B are of course by no means the most general ones. This can be shown with simple examples.

C. GENERALIZATIONS

6. A far-reaching generalization of the problem treated in part A could be formulated as follows: Let S be a surface on which a line-element ds is defined by any means, and a twice infinite family of curves C is given on S. Then, a point-function on the surface is to be determined from the integrals $\int f\,ds$ along the curves C.

The nearest specialization is obtained by taking a non-Euclidian plane for S, the corresponding line-element for ds, and the corresponding straight lines for the curves C. In the elliptic case, the problem can be carried over to the geometry on a sphere. Interpreting in a well-known fashion a diametrical pair of points on the sphere as a point in the elliptic plane, there results the problem of the determination of an even function on the sphere (i.e., a function with the same value in diametrical points) from its integrals along the great circles. Minkowski was the first to deal with this problem in principle*) and he solved it by expansions in terms of spherical functions. Later P. Funk computed Minkowski's solution and he has shown how to obtain this solution from the Abel integral equation.† This is the method to which I owe the solution of problem A. Funk's solution is analogous to (III) with the exception that the sinus of the spherical radius appears in the denominator and to the integral there is added the value of F at the pole of the corresponding great circle divided by π. In the hyperbolic plane, the solution of the problem is analogous to (III) too:

$$f(P) = -\frac{1}{\pi}\int_0^\infty \frac{d\bar{F}_p(q)}{\sinh q}$$

(here the measure of curvature is assumed to be $= -1$). This can be shown to be in total agreement with the derivation of (III) indicated in 1.

In both cases, the question analogous to B can be posed also. In the elliptic geometry, nothing new results because of the absolute polarity, and in the hyperbolic case a solution analogous to (V) does not seem to exist.

Gesammelle Abhandlungen **II**, pp. 277ff.

†*Math. Ann.*, **74**, pp. 283–288.

A second specialization results if (in the Euclidian or in the non-Euclidian geometry) the circles with constant radius are taken for the curves C. Here Minkowski's method using spherical functions can be applied on the sphere and so the problem can be solved to a certain degree. However, it is interesting that in this case the uniqueness of the solution can be lost. The reason for this is that for certain radii ρ defined by the zeros of the Legendre polynomials of even order there exist even functions on the sphere that do not vanish identically, but whose integrals along any circle with spherical radius ρ are zero. In the Euclidian case, the spherical functions are replaced by the integral theorem of the Bessel functions. Here there are always functions which do not vanish identically but whose integrals along any circle with fixed radius yield zero. If this radius is one then these functions are (in polar coordinates)

$$J_n(x_\nu \rho)\cos n\phi, \qquad J_n(x_\nu \rho)\sin n\phi,$$

and linear combinations, where x_ν is a zero of J_0. In the hyperbolic case, the Bessel functions are replaced by so-called conical functions for which the corresponding integral theorem has been proven by Weyl.* The results are analogous to the Euclidian case.

7. The results in parts A and B can be generalized in another direction by passing on to higher-dimensional spaces. In a Euclidian space $\mathbb{R}^n$, one can try to determine a point-function $f(p) = f(x_1, x_2, \ldots, x_n)$ from its integrals $F(\alpha_1, \ldots, \alpha_n, p)$ over all hyperplanes $\alpha_1 x_1 + \cdots + \alpha_n x_n = p$, $(\alpha_1^2 + \cdots + \alpha_n^2 = 1)$. Following a procedure analogous to that applied in 1, we form the mean value $\bar{F}_0(q)$ of F over the tangent-planes of the sphere with center $[0, 0, \ldots, 0]$ and radius q. It is given by the $(n-1)$-fold integral:

$$\bar{F}_0(q) = \frac{1}{\Omega_n} \int F(\alpha, q)\, d\omega,$$

where $d\omega$ is the surface element and $\Omega_n = (2\pi^{n/2})/(\Gamma(n/2))$ is the surface area of the n dimensional sphere $\alpha_1^2 + \cdots + \alpha_n^2 = 1$.

$\bar{F}_0$ can be represented as an n-fold integral over f:

$$\bar{F}_0(q) = \frac{\Omega_{n-1}}{\Omega_n} \iint_{x_1^2 + \cdots x_n^2 > q^2} f(x_1, x_2, \ldots, x_n) \times \frac{\left(x_1^2 + \cdots + x_n^2 - q^2\right)^{(n-3)/2}}{\left(x_1^2 + \cdots + x_n^2\right)^{(n-2)/2}} dx_1 \cdots dx_n \tag{7}$$

Gött. Nachr., 1910, p. 454.

or, using an already often used mean value notation:

$$\bar{F}_0(q) = \Omega_{n-1} \int_q^{\infty} \bar{f}_0(q)(r^2 - q^2)^{(n-3)/2} r\, dr.$$

This formula is analogous to (1) and has corresponding consequences. The substitution $r^2 = v$, $q^2 = u$ leads to the integral equation

$$\Phi(u) = \frac{\Omega_{n-1}}{2} \int_u^{\infty} \phi(v)(v - u)^{(n-3)/2}\, dv.$$

If n is *even*, we get the same equation as (2) by differentiating $((n/2) - 1)$ times, and from this,

$$\phi(0) = f(0, 0, \ldots, 0)$$

can be found. Thus, for a given F, the formation of F differentiations and one integration is necessary. If n is *odd*, then this integration is omitted, since we now get from differentiating $((n - 1)/2)$ times:

$$\phi(0) = \frac{2(-1)^{(n-1)/2}}{\Omega_{n-1}\left(\frac{n-3}{2}\right)!} \Phi^{((n-3)/2)}(0).$$

The three-dimensional case is particularly simple, but this case can also be treated using a method analogous to 5 that yields very elegant results. From (7), the point mean value of F for $q = 0$ follows:

$$\bar{F}_0 = \frac{1}{2} \iiint \frac{f(x, y, z)}{\sqrt{x^2 + y^2 + z^2}}\, dx\, dy\, dz.$$

This equation can be considered the Newtonian potential of the space covered with a mass of density $\frac{1}{2}f$. Therefore, it follows that

$$f(x, y, z) = -\frac{1}{2\pi} \Delta \bar{F},$$

where $\bar{F}$ stands for the point mean value of F.

Here also the problem analogous to B can be solved. Using the method indicated in 5, one finds for a plane-function F with known point mean values f that

$$F(E) = -\frac{1}{2\pi} \iint \Delta f\, d\sigma,$$

where $d\sigma$ is the surface area element of the plane E. Δ is the Laplacian operator for the three-dimensional space, and the integration is to be taken over the whole plane E.

Appendix B

Generalized Functions

[B.1] INTRODUCTION

In this appendix we state and discuss briefly some of the foundations and concepts from the theory of distributions and generalized functions. The terms *distributions* and *generalized functions* are used interchangeably; in [B.8], there are some comments regarding a distinction. In view of the current availability of a large number of excellent sources* for this material,[1-19] it has been included mainly for the purposes of establishing notation, recording formulas, stating certain basic results, indicating relevant source material, and for preserving the overall continuity of discussion.

[B.2] PRELIMINARY REMARKS AND NOTATION

Let $\mathbf{x} = (x_1, x_2, \ldots, x_n)$ be a vector in real n-dimensional Euclidean space $\mathbb{R}^n$. The inner product of two vectors $\mathbf{x}$ and $\mathbf{y}$ in $\mathbb{R}^n$ is written as

$$\mathbf{x} \cdot \mathbf{y} = \langle \mathbf{x}, \mathbf{y} \rangle = \sum_{j=1}^{n} x_j y_j,$$

and the magnitude of $\mathbf{x}$ is $|\mathbf{x}| = (\mathbf{x} \cdot \mathbf{x})^{1/2}$. A complex-valued function of n real variables is written as $F(x_1, x_2, \ldots, x_n)$ or, more simply, as $F(\mathbf{x})$. In most cases, $F(\mathbf{x})$ will be real-valued; however, since a discussion of complex-valued functions automatically includes the real-valued functions, it is not necessary to bother with a distinction.

Let $\mathscr{K}$ be the closure[20] of the set of points $\mathbf{x} \in \mathbb{R}^n$ for which $F(\mathbf{x}) \neq 0$. Then $\mathscr{K}$ is the *support* of F, and if the support $\mathscr{K}$ is bounded, then it is *compact*. By the Heine–Borel theorem,[20] we may use "closed and bounded" and "compact" interchangeably. Rephrasing, the support of a function in $\mathbb{R}^n$ is the smallest closed subset in $\mathbb{R}^n$ outside of which the function vanishes.

If $F(\mathbf{x})$ is infinitely differentiable, it is said to be of class C^∞. The C^∞ function $F(\mathbf{x})$ is called *rapidly decreasing* if F and all its derivatives decrease for $|\mathbf{x}| \to \infty$ more rapidly than any inverse power of $|\mathbf{x}|$.

*References appear at the end of the appendix.

Remark 1. In detail, this means that $|\mathbf{x}^l D^m F| < C(l, m)$, $l = \sum_i l_i$, $m = \sum_i m_i$, where $(l_1, l_2, \ldots, l_n)$ and $(m_1, m_2, \ldots, m_n)$ are n-tuples of nonnegative integers and $C(l, m)$ is a constant that depends on l, m, and F. The symbol $\mathbf{x}^l$ means $x_1^{l_1} x_2^{l_2} \cdots x_n^{l_n}$ and $D^m F$ is the derivative

$$D^m F = \frac{\partial^m}{\partial x_1^{m_1} \partial x_2^{m_2} \cdots \partial x_n^{m_n}} F. \qquad \square$$

There is a need to consider the integral, over all $\mathbb{R}^n$, of the product of two functions, say F and G. This is expressed in the usual fashion,

$$\langle G, F \rangle = \int G^*(\mathbf{x}) F(\mathbf{x})\, d\mathbf{x}, \tag{2.1}$$

where * means complex conjugation, and $d\mathbf{x}$ is shorthand for $dx_1\, dx_2 \cdots dx_n$. The "value" of the integral is in general some complex number $\langle \cdot\, , \cdot \rangle$. Of course, if G is a real-valued function, then the right-hand side of (2.1) becomes $\int G(\mathbf{x}) F(\mathbf{x})\, d\mathbf{x}$.

Remark 2. Note that

$$\langle G, F \rangle^* = \langle F, G \rangle. \qquad \square \tag{2.2}$$

Remark 3. A weight function may be included as a generalization of (2.1),

$$\langle G, F \rangle = \int G^*(\mathbf{x}) F(\mathbf{x}) w(\mathbf{x})\, d\mathbf{x}. \qquad \square \tag{2.3}$$

[B.3] SPACES OF TEST FUNCTIONS

The Space $\mathcal{D}_{\mathcal{K}}$. The space of C^∞ functions on $\mathbb{R}^n$ with compact support $\mathcal{K} \in \mathbb{R}^n$ is designated by $\mathcal{D}_{\mathcal{K}}$.

The Space $\mathcal{D}$. The space of all C^∞ functions on $\mathbb{R}^n$ with compact support is designated by $\mathcal{D}$.

The Space $\mathcal{S}$. The space of rapidly decreasing C^∞ functions on $\mathbb{R}^n$ is designated by $\mathcal{S}$.

The Space $\mathcal{E}$. The space of all C^∞ functions on $\mathbb{R}^n$ with arbitrary support is designated by $\mathcal{E}$.

Observe that these spaces are linear vector spaces over the field of complex numbers and the inclusion is

$$\mathcal{D} \subset \mathcal{S} \subset \mathcal{E}.$$

The term test function often refers to a function from $\mathcal{D}$ or $\mathcal{S}$.

[B.4] CONVERGENCE IN THE SPACES OF TEST FUNCTIONS

The Space $\mathscr{D}_{\mathscr{K}}$. Let $\{F_j\}$ represent a nonempty sequence of functions in $\mathscr{D}_{\mathscr{K}}$. When we write $F_j \to F \in \mathscr{D}_{\mathscr{K}}$ (or $\lim_{j\to\infty} F_j = F$), it is understood that the sequence $\{F_j - F\}$ converges uniformly to zero on the compact set $\mathscr{K} \in \mathbb{R}^n$; furthermore, each sequence of derivatives of arbitrary order converges uniformly on $\mathscr{K}$ to the corresponding derivative of F, $\{D^m F_j - D^m F\} \to 0$ for every fixed $m \geqslant 0$.

The Space $\mathscr{D}$. Convergence in the sense of $\mathscr{D}$ is defined the same way as for $\mathscr{D}_{\mathscr{K}}$ except that one must first specify that all supports of the functions in the sequence $\{F_j\}$ are contained in some fixed compact region $\mathscr{K} \in \mathbb{R}^n$.

The Space $\mathscr{S}$. Let $\{F_j\}$ be a sequence of functions in $\mathscr{S}$. This sequence is said to converge to zero if and only if: (i) The F_j and the $D^m F_j$ converge to zero uniformly on every compact subset $\mathscr{K}$ of $\mathbb{R}^n$; and (ii) the constants $C(l, m)$ in the expression $|\mathbf{x}^l D^m F_j| < C(l, m)$ are independent of j for all j. For notation, see Remark 1 in [B.2]. The sequence $\{F_j\}$ is said to converge to $F \in \mathscr{S}$ if $\{F_j - F\}$ converges to zero. This may be indicated by $F_j \to F$ or $\lim_{j\to\infty} F_j = F$ and called convergence in the sense of $\mathscr{S}$.

The Space $\mathscr{E}$. Let $\{F_j\}$ be a sequence of functions in $\mathscr{E}$. This sequence is said to converge to zero if and only if for each $m \geqslant 0$, the sequence converges uniformly to zero on every compact subset of $\mathbb{R}^n$. The sequence is said to converge to $F \in \mathscr{E}$ if $\{F_j - F\}$ converges to zero. Convergence of this type is called convergence in the sense of $\mathscr{E}$.

Remark 1. It is important to realize that an inclusion also applies for convergence. If $\mathfrak{CS}$ means "convergence in the sense of," then

$$\mathfrak{CS}(\mathscr{D}) \Rightarrow \mathfrak{CS}(\mathscr{S}) \Rightarrow \mathfrak{CS}(\mathscr{E}). \qquad \square$$

[B.5] LINEAR FUNCTIONALS

A *functional* defined on a linear space $\mathfrak{L}$ of functions is a mapping or rule that assigns a complex number to each member of $\mathfrak{L}$. That is, the domain of the functional is $\mathfrak{L}$, and the range lies in the field of complex numbers. In the following discussion, $\mathfrak{L}$ may be replaced by $\mathscr{D}$, $\mathscr{S}$, or $\mathscr{E}$.

A *linear functional* T on the space $\mathfrak{L}$ satisfies

$$\langle T, \alpha F_1 + \beta F_2 \rangle = \alpha \langle T, F_1 \rangle + \beta \langle T, F_2 \rangle \tag{5.1}$$

for any $F_1, F_2 \in \mathfrak{L}$ and complex numbers α, β. The set of all linear functionals on a linear space $\mathfrak{L}$ forms a linear space known as the *dual* space $\mathfrak{L}'$.

Remark 1. When we say that $\langle T, F \rangle$ forms a linear functional on $\mathfrak{L}$, the implication is that F is any member of $\mathfrak{L}$. In other words, it is as if F is the variable that has $\mathfrak{L}$ as its domain. To emphasize this we could write $T(F)$ rather than $\langle T, F \rangle$. It is customary to refer to either T or $\langle T, F \rangle$ as the linear functional, depending on context and convenience. □

We shall have special use for *continuous* linear functionals. Consider a sequence of functions $\{F_j\}$ from some linear space $\mathfrak{L}$. A linear functional is continuous if and only if

$$\lim_{j \to \infty} \langle T, F_j \rangle = \left\langle T, \lim_{j \to \infty} F_j \right\rangle. \tag{5.2}$$

To understand precisely what this means, it is necessary to specify the meaning of $\lim_{j \to \infty} F_j$. This is done for various spaces of interest in [B.4].

[B.6] DISTRIBUTIONS

We are now all set for the following definitions.[1]

Definition 1. A continuous linear functional on the vector space $\mathscr{D}$ is called a *distribution*.

Definition 2. The space of all distributions is designated by $\mathscr{D}'$.

Definition 3. A continuous linear functional on the vector space $\mathscr{S}$ is called a *tempered distribution*.

Definition 4. The space of all tempered distributions is designated by $\mathscr{S}'$.
Also, designate $\mathscr{E}'$ as the space of all continuous linear functionals defined on $\mathscr{E}$. The space $\mathscr{E}'$ is the space of distributions with compact support.
The inclusion for these distributions is: $\mathscr{E}' \subset \mathscr{S}' \subset \mathscr{D}'$.

Remark 1. Definitions of equality, addition, and scalar multiplication in these spaces are the obvious ones. If $\mathfrak{L}'$ represents any one of the spaces $\mathscr{D}'$, $\mathscr{S}'$, or $\mathscr{E}'$,

(i) $T_1 \in \mathfrak{L}'$ and $T_2 \in \mathfrak{L}'$ are said to be equal if and only if $\langle T_1, F \rangle = \langle T_2, F \rangle$ for all $F \in \mathfrak{L}$.
(ii) $(T_1 + T_2) \in \mathfrak{L}'$ is defined by $\langle T_1 + T_2, F \rangle = \langle T_1, F \rangle + \langle T_2, F \rangle$ for all $F \in \mathfrak{L}$.
(iii) $\alpha T \in \mathfrak{L}'$, where α is complex, is defined by $\langle \alpha T, F \rangle = \alpha^* \langle T, F \rangle$ for all $F \in \mathfrak{L}$. □

The most straightforward way to define convergence in these spaces is in terms of what is known as *weak convergence*. Let $\mathfrak{L}'$ represent $\mathscr{D}'$, $\mathscr{S}'$, or $\mathscr{E}'$. The sequence of generalized functions $\{T_j\} \in \mathfrak{L}'$ is said to converge to the generalized function $T \in \mathfrak{L}'$ when

$$\lim_{j\to\infty} \langle T_j, F\rangle = \langle T, F\rangle \qquad \text{for all } F \in \mathfrak{L}. \tag{6.1}$$

We shall have more to say about convergent sequences of generalized functions and distributions in following sections. For a summary of results, see Table B1, after Marchand (1962) with permission.

[B.7] SOME MEMBERS OF $\mathscr{D}$ AND $\mathscr{D}'$

In the interest of simplicity, we confine the current discussion to real valued functions defined on $\mathbb{R}^1$ rather than $\mathbb{R}^n$. This is not a serious lack of generality since the extension from $\mathbb{R}^1$ to $\mathbb{R}^n$ may be made in a natural fashion.[1]

Table B1. Spaces and Convergence

Space	Convergence[a]
$\mathscr{D}$	$F_j \to 0 \Leftrightarrow D^m F_j \overset{\text{unif}}{\to} 0$
C^∞ functions of compact support	$\forall$ fixed $m \geqslant 0$, $\mathbb{S}\{F_j\} \in \mathscr{K}$
$\mathscr{S}$	$F_j \to 0 \Leftrightarrow \mathbf{x}^l D^m F_j \overset{\text{unif}}{\to} 0$
Rapidly decreasing C^∞ functions	$\forall$ fixed $l, m \geqslant 0$, $\quad \forall \mathscr{K} \in \mathbb{R}^n$, $\quad \mathbb{S}\{F_j\} \in \mathscr{K}$
$\mathscr{E}$	$F_j \to 0 \Leftrightarrow D^m F_j \overset{\text{unif}}{\to} 0$
C^∞ functions	$\forall$ fixed $m \geqslant 0$, $\quad \forall \mathscr{K} \in \mathbb{R}^n$, $\quad \mathbb{S}\{F_j\} \in \mathscr{K}$
$\mathscr{D}'$	$\lim_{j\to\infty} \langle T_j, F\rangle = \langle T, F\rangle$
Distributions	$\forall F \in \mathscr{D}$; $\quad \{T_j\}, T \in \mathscr{D}'$
$\mathscr{S}'$	$\lim_{j\to\infty} \langle T_j, F\rangle = \langle T, F\rangle$
Tempered distributions	$\forall F \in \mathscr{S}$; $\quad \{T_j\}, T \in \mathscr{S}'$
$\mathscr{E}'$	$\lim_{j\to\infty} \langle T_j, F\rangle = \langle T, F\rangle$
Distributions with compact support	$\forall F \in \mathscr{E}$; $\quad \{T_j\}, T \in \mathscr{E}'$

[a] If $(F_j - F) \to 0$, then we say $F_j \to F$. $\mathscr{K}$ represents a fixed compact support in $\mathbb{R}^n$. $\mathbb{S}\{F_j\}$ means the supports of the functions in the sequence. The symbol $\overset{\text{unif}}{\to}$ means "converges uniformly in $\mathbb{R}^n$ to."

Consider the function $F \in \mathscr{D}$, $x \in \mathbb{R}^1$,

$$F(x) = \begin{cases} e^{-1/(1-x^2)}, & |x| < 1 \\ 0, & |x| \geqslant 1. \end{cases} \tag{7.1}$$

Its graph is shown in Fig. B.1.

Remark 1. This is the standard example of a function in $\mathscr{D}$. It may be compared with $F(x) = e^{-x^2}$, which is a standard example of a function in $\mathscr{S}$. □

Consider also, the sequence of rectangular pulse functions defined by

$$S_k(x) = \begin{cases} 0, & x < 0 \\ k, & 0 \leqslant x \leqslant \dfrac{1}{k} \\ 0, & x > \dfrac{1}{k} \end{cases} \tag{7.2}$$

for integer values of k. Clearly, the sequence $\{S_k\}$ does not have a well defined

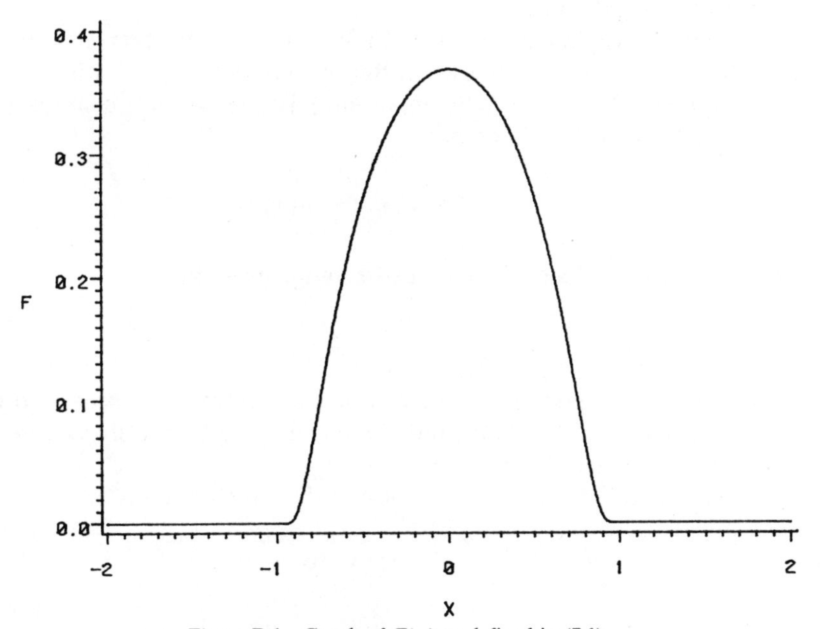

Figure B.1 Graph of $F(x)$ as defined in (7.1).

limit (as $k \to \infty$) in the elementary calculus sense of a limit. Form the integral

$$\langle S_k, F \rangle = \int_{-\infty}^{\infty} S_k(x)F(x)\,dx = k\int_0^{1/k} F(x)\,dx. \tag{7.3}$$

The value of this integral is a complex number (real in this case) represented by $\langle S_k, F \rangle$ for the F given in (7.1) or for *any* $F \in \mathscr{D}$. For any sequence $\{F_j\} \in \mathscr{D}$ that converges to $F \in \mathscr{D}$, we have

$$\lim_{j \to \infty} \langle S_k, F_j \rangle = \langle S_k, F \rangle.$$

Hence $\langle S_k, F \rangle \forall F \in \mathscr{D}$ forms a continuous linear functional on $\mathscr{D}$. This yields the following example.

Example 1. The pulse function S_k defined in (7.2) is a distribution. □

Example 2. Any function $T(x)$ that is integrable in either the Riemann or Lebesgue sense over every finite interval $[a, b]$ of $\mathbb{R}^1$ (i.e., *locally integrable*) is a distribution,

$$\langle T, F \rangle = \int_{-\infty}^{\infty} T(x)F(x)\,dx, \qquad \forall F \in \mathscr{D}.$$ □

The distributions in the preceding examples are *regular* distributions. In fact, the distributions described in Example 2 can be used to *define* a *regular distribution*. We now turn our attention to distributions that are not regular; these are called *singular* distributions.

Let us go back to $S_k(x)$ defined by (7.2). We have already observed that $\{S_k\}$ does not have a limit in the usual sense. But observe that $\langle S_k, F \rangle$ with $F \in \mathscr{D}$ does have a limit. (This is crucially important.) To see this, just consider that by use of the Mean Value Theorem[21]

$$\langle S_k, F \rangle = \frac{k}{k} F(\eta_k) = F(\eta_k),$$

where $0 < \eta_k < 1/k$. Hence the limit exists and is given by

$$\lim_{k \to \infty} \langle S_k, F \rangle = F(0). \tag{7.4}$$

This points out the important deviation from the elementary approach to the study of the behavior of functions; rather than studying S_k directly as $k \to \infty$, we study the functional $\langle S_k, F \rangle$ as $k \to \infty$.

The usual way one writes (7.4) is by use of the δ distribution

$$\langle \delta, F \rangle = \int_{-\infty}^{\infty} \delta(x)F(x)\,dx = F(0). \tag{7.5}$$

There is no problem with this equation if we realize that the middle part is purely formal and all we really need is the continuous linear functional

$$\langle \delta, F \rangle = F(0), \qquad \forall F \in \mathscr{D}.$$

With this in mind, we can attach meaning to the distribution defined by

$$\delta = \lim_{k \to \infty} S_k. \tag{7.6}$$

Remark 2. The δ distribution obtained here is identical to the δ function or impulse function, which now carries the name of Dirac as a tribute to his skillful use of this "function" (actually it is not a function but a distribution) in his classic treatise on quantum mechanics,[22] and in his 1927 paper on quantum dynamics.[23] The Dirac δ function, written in terms of $x \in \mathbb{R}^1$, is defined by

$$\delta(x) = 0, \qquad \forall x \neq 0, \qquad \int_{-\infty}^{\infty} \delta(x)\, dx = 1,$$

and the *sifting* property it has on functions,

$$\int_{-\infty}^{\infty} F(x)\delta(x-a)\, dx = F(a).$$

The expression $\delta(x)$ and the preceding formulas are now familiar to scientists, engineers, and mathematicians worldwide. In fact, the early success with the formal use of the δ function played a major role in the eventual development of a mathematically rigorous theory of distributions and generalized functions. An especially interesting treatment and history of the δ function may be found in Chapter V of Van der Pol and Bremmer,[24] and a concise discussion of its formal properties is available in the article by Sneddon.[25] These properties are also discussed in almost all recent texts on quantum mechanics and the analysis of linear systems. For that reason, we shall feel free to use these properties as they are needed. At certain times we shall make use of the theory of distributions to justify some particular use of the δ function, and at other times, we shall simply use it in the usual heuristic fashion so familiar to physicists and engineers. However, when we do follow the latter approach, it is always with the comforting realization that our formal manipulations can be made rigorous in the framework of distribution theory and the theory of generalized functions. □

The distribution δ is an example of a *singular* distribution. One of the beautiful by-products of this functional approach is that both regular singular distributions may be treated formally in exactly the same way; indeed, some authors do not even bother making a distinction. Hence, when we write

$$\langle T, F \rangle = \int_{-\infty}^{\infty} T(x)F(x)\,dx, \tag{7.7}$$

we do not care whether T is a regular or singular distribution. But, if T is singular it is understood that the right side of (7.7) is just formal. More will be said about this when we discuss differentiation of distributions.

Remark 3. There is nothing unique about (7.6). Many other sequences yield δ (see [B.10]). The δ distribution is not the only singular distribution of physical importance (see [B.11]). □

[B.8] OTHER METHODS

The concepts already discussed and those to be discussed in the remaining sections only touch the surface in the development of the theory of distributions, carefully developed in detail by Laurent Schwartz.[1] To go deeper and even to justify rigorously several of the statements already made, would require considerable background in topology and functional analysis. For this reason, the reader not already familiar with an in-depth treatment of distributions may wish to make use of certain references at the end of this appendix. The little books by Marchand[4] and De Jager[3] might be good to start with.

In addition to the functional method of Schwartz,[1] there are other approaches to the study of distributions and generalized functions, such as the method of sequences[12–15] and the method of convolution quotients.[26,27] (For additional directions to the literature see the preface of Zemanian.[16]) It is especially appropriate to mention one of these here. This is the method pioneered by Temple[12] and Lighthill,[13] whereby certain sequences of ordinary functions are used to represent generalized functions. This approach, although perhaps not as elegant mathematically, is in some respects more physically appealing especially when working with functions of a single variable so that easy-to-visualize graphs can be drawn. Actually, we have already touched on this approach, and we shall find further use for some of the concepts associated with this method in following sections.

It is interesting to note that although Schwartz is responsible for the major developments of distribution theory (from 1945–49), there were important foundations already being laid in the late twenties and during the thirties. The work of Sobolev[28] is of special importance in this respect. For a discussion of these matters and a guide to some of the important early papers, the reader may wish to see the first few pages of the first chapter of De Jager.[3]

It should be pointed out that the various approaches are, for all practical purposes, mathematically equivalent and they can all be understood in terms of the Schwartz theory.[2,16] For this reason, various aspects of the different approaches are becoming entwined in the literature, especially where applications are involved. This appears to be useful and we also follow this trend. In keeping with this theme, this section ends with a quote from Lighthill.[13] On the dedication page, he writes, "To Paul Dirac, who saw that it must be true, Laurent Schwartz, who proved it, and George Temple, who showed how simple it could be made."

[B.9] GENERALIZED FUNCTIONS FROM SEQUENCES

Let $\mathfrak{L}$ represent a linear space of test functions defined on $\mathbb{R}^1$. This space is to remain unspecified further for now. Consider a *regular sequence* of ordinary functions, that is, functions $T_j(x)$ which (i) are continuous, (ii) have continuous derivatives of all orders, and (iii) are such that $\lim_{j\to\infty}\langle T_j, F\rangle$ exists for all $F \in \mathfrak{L}$. It turns out that different regular sequences can lead to the same value for $\lim_{j\to\infty}\langle T_j, F\rangle$; several examples appear in [B.10]. Hence it is necessary to define *equivalent regular sequences*. Different regular sequences $\{T_j\}$ and $\{T_j'\}$ are said to be *equivalent* if and only if

$$\lim_{j\to\infty}\langle T_j, F\rangle = \lim_{j\to\infty}\langle T_j', F\rangle, \qquad \forall F \in \mathfrak{L}.$$

With this background, we can define the *generalized function* $T(x)$ as a class of equivalent regular sequences such that

$$\langle T, F\rangle = \lim_{j\to\infty}\langle T_j, F\rangle = \lim_{j\to\infty}\int_{-\infty}^{\infty} T_j(x)F(x)\,dx. \tag{9.1}$$

Note that here the integral on the right-hand side is not just symbolic. This is one reason for the popularity of this approach pioneered by Temple[12] and Lighthill.[13]

Remark 1. Equation (9.1) should be compared with (6.1). Clearly, the approach sketched here is encompassed by the Schwartz theory; suppose $\mathfrak{L}$ is identified with $\mathscr{S}$, for example. □

Remark 2. Now we can point to a slight conceptual difference between distributions and generalized functions. Technically, distributions live in $\mathscr{D}'$, and generalized functions are defined as an equivalence class of regular sequences. It should be noted that even this distinction is not made uniformly throughout the literature. □

Remark 3. This approach to the study of generalized functions tends to place the distributions and test functions on an equal footing, since the test functions themselves can always be set up as the limit of one or more regular sequences. This is a strong point made by Lighthill[13] in his discussion of "good" functions. □

[B.10] SEQUENCES

Suppose we consider the three sequences $\{S_k^{(1)}\}$, $\{S_k^{(2)}\}$, and $\{S_k^{(3)}\}$ where

$$S_k^{(1)}(x) = \begin{cases} k, & 0 \leqslant x < \dfrac{1}{2k} \\ 0, & x > \dfrac{1}{2k} \\ S_k^{(1)}(-x), & x < 0 \end{cases} \tag{10.1a}$$

$$S_k^{(2)}(x) = \begin{cases} 2k^2 x & 0 \leqslant x \leqslant \dfrac{1}{2k} \\ 2k^2\left(\dfrac{1}{k} - x\right) & \dfrac{1}{2k} < x \leqslant \dfrac{1}{k} \\ 0 & x > \dfrac{1}{k} \\ S_k^{(2)}(-x) & x < 0 \end{cases} \tag{10.1b}$$

$$S_k^{(3)}(x) = \begin{cases} -k, & 0 \leqslant x < \dfrac{1}{2k} \\ 2k, & \dfrac{1}{2k} < x < \dfrac{1}{k} \\ 0, & x > \dfrac{1}{k} \\ S_k^{(3)}(-x), & x < 0. \end{cases} \tag{10.1c}$$

Graphs of these sequences are shown in Fig. B.2. Each sequence has the property that

$$\lim_{k \to \infty} S_k^{(j)}(x) = 0, \qquad \forall x \neq 0; \tag{10.2a}$$

also,

$$\lim_{k \to \infty} \int_{-\infty}^{\infty} S_k^{(j)}(x) F(x)\, dx = F(0), \tag{10.2b}$$

for all $F \in \mathscr{D}$ and for $j = 1, 2, 3$.

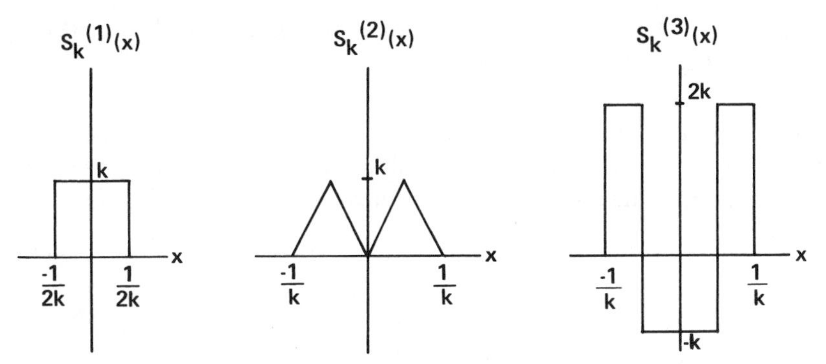

Figure B.2 Graphs of $S_k^{(j)}(x)$ as given in (10.1).

Remark 1. Repeated application of the Mean Value Theorem is perhaps the easiest approach to verify (10.2), see [B.7]. There are instances, such as with $S_k^{(3)}$ for example, when a more general approach to delta-convergent sequences may be more satisfactory. See the method used by Gel'fand and Shilov[8] in Chapter 1. □

Observe that, although each of the sequences (10.1) converges to $\delta(x)$ in the sense that (10.2) holds, for $x = 0$, there is very different behavior as $k \to \infty$,

$$\lim_{k \to \infty} S_k^{(1)}(0) = \infty, \qquad \lim_{k \to \infty} S_k^{(2)}(0) = 0, \qquad \lim_{k \to \infty} S_k^{(3)}(0) = -\infty.$$

Sequences that converge in the fashion indicated in (10.2) are called *equivalent* δ sequences, and the distributions obtained as $k \to \infty$ are said to be *equal* as indicated in Remark 1 of [B.6].

There are two important general results to be emphasized from the preceding example:

(i) One may not assign a value to the singular distribution $\delta(x)$ at $x = 0$.
(ii) Many different sequences may converge to the same distribution (or generalized function).

Here we have seen how a distribution comes from a certain type of sequence of ordinary functions in a natural way. Although we do not wish to go into this in more detail than has already been discussed in [B.8], it can be shown that one may always view a distribution or generalized function as an equivalence class of what are called fundamental sequences. In other words, one may think of a distribution or generalized function as the common limit of all of its fundamental sequences. □

Remark 2. For in-depth discussions of fundamental sequences, see Korevaar,[14] Lighthill,[13] or Temple.[12] □

We now turn our attention to various equivalent δ sequences. The approach will be to give several examples of possible ways one may represent $\delta(x)$ as a limit of an approximating sequence of functions $S(x; a)$ or $S(x; \varepsilon)$ such that

$$\lim_{a\to\infty} S(x; a) = 0, \qquad \forall x \neq 0, \tag{10.3a}$$

and

$$\lim_{a\to\infty} \langle S(x; a), F(x)\rangle = \langle \delta, F\rangle = F(0), \tag{10.3b}$$

or

$$\lim_{\varepsilon\to 0} S(x; \varepsilon) = 0, \qquad \forall x \neq 0, \tag{10.4a}$$

and

$$\lim_{\varepsilon\to 0} \langle S(x; \varepsilon), F(x)\rangle = \langle \delta, F\rangle = F(0). \tag{10.4b}$$

Remark 3. Observe that always $\int_{-\infty}^{\infty} S(x; a)\, dx = \int_{-\infty}^{\infty} S(x; \varepsilon)\, dx = 1$, independent of a or ε. □

Three examples of δ sequences have already been given in this section; others follow.

Example 1.

$$S(x; a) = \frac{a}{\sqrt{\pi}} e^{-a^2x^2}$$

$$\delta(x) = \lim_{a\to\infty} \frac{a}{\sqrt{\pi}} e^{-a^2x^2}. \qquad \square$$

Remark 4. Note that by an appropriate change of variable, it is possible to obtain another δ sequence. For example, suppose $a \to 1/2\sqrt{\varepsilon}$, then

$$\delta(x) = \lim_{\varepsilon\to +0} \frac{1}{2\sqrt{\pi\varepsilon}} e^{-x^2/4\varepsilon}. \qquad \square$$

Example 2.

$$\delta(x) = \lim_{\varepsilon\to +0} \frac{1}{\pi} \frac{\varepsilon}{x^2 + \varepsilon^2}. \qquad \square$$

In the next example the more general approach to delta-convergent sequences mentioned in Remark 1 is appropriate.

Example 3.

$$\delta(x) = \lim_{a\to\infty} \frac{\sin ax}{\pi x}. \qquad \square$$

If we observe that

$$\frac{1}{2\pi}\int_{-a}^{a} e^{ikx}\,dk = \frac{\sin ax}{\pi x},$$

then we see the basis for the famous result of the next example.

Example 4.

$$\delta(x) = \lim_{a\to\infty} \frac{1}{2\pi}\int_{-a}^{a} e^{ikx}\,dk = \frac{1}{2\pi}\int_{-\infty}^{\infty} e^{ikx}\,dk. \qquad \square$$

Remark 5. It should be clear that x may be replaced by $x - x_0$ to obtain $\delta(x - x_0)$. Thus, from Example 4, we have

$$\delta(x - x_0) = \frac{1}{2\pi}\int_{-\infty}^{\infty} e^{ik(x-x_0)}\,dk.$$

Equivalently,

$$\delta(x - x_0) = \int_{-\infty}^{\infty} e^{\pm i2\pi k(x-x_0)}\,dk. \qquad \square$$

The preceding examples are fairly standard. Several other classical examples are discussed by Van der Pol and Bremmer[24] and by Korevaar.[14]

Remark 6. Since $\mathscr{D} \subset \mathscr{S}$ and $\delta \in \mathscr{D}$, it follows that δ is also a tempered distribution. $\square$

[B.11] SOME OTHER IMPORTANT SINGULAR DISTRIBUTIONS

Thus far the only singular distribution mentioned is the δ distribution. Another very useful singular distribution is the *principle value* distribution $\mathcal{P}(1/x)$ on the interval $(-\infty, \infty)$. For $F \in \mathscr{D}$, this distribution is defined by

$$\left\langle \mathscr{P}\frac{1}{x}, F \right\rangle = \lim_{\varepsilon \to +0} \left[\int_{-\infty}^{-\varepsilon} \frac{F(x)}{x} dx + \int_{+\varepsilon}^{\infty} \frac{F(x)}{x} dx \right]$$

$$= \mathscr{P} \int_{-\infty}^{\infty} \frac{F(x)}{x} dx, \tag{11.1}$$

where $\mathscr{P}$ indicates that the Cauchy principle value is to be taken.

Remark 1. Equation (11.1) may be written as $\langle \mathscr{P}(1/x), F \rangle = \int_{-\infty}^{\infty} F(x)/x \, dx$, where the principle value is understood. □

Two other examples are the Heisenberg delta functions, which are of great use in scattering theory (e.g., see Chapter 7 in Newton[29]),

$$\delta^{+}(x) = \tfrac{1}{2}\delta(x) - \frac{1}{2\pi i}\mathscr{P}\frac{1}{x} = -\lim_{\varepsilon \to +0} \frac{1}{2\pi i} \frac{1}{x + i\varepsilon}, \tag{11.2a}$$

$$\delta^{-}(x) = \tfrac{1}{2}\delta(x) + \frac{1}{2\pi i}\mathscr{P}\frac{1}{x} = +\lim_{\varepsilon \to +0} \frac{1}{2\pi i} \frac{1}{x - i\varepsilon}. \tag{11.2b}$$

Remark 2. We must delay the justification of the second equality in these equations until the end of [B.12]. □

Observe that upon solving (11.2) for $\mathscr{P}(1/x)$, we obtain

$$\mathscr{P}\frac{1}{x} = \pi i(\delta^{-} - \delta^{+}). \tag{11.3}$$

If x is replaced by $x - x_0$ in (11.2b), the result is

$$\mathscr{P}\frac{1}{x - x_0} = \lim_{\varepsilon \to +0} \frac{1}{x - x_0 - i\varepsilon} - i\pi\delta(x - x_0). \tag{11.4}$$

When this distribution is used with $F \in \mathscr{D}$, we obtain the useful result (see Appendix 4 in Roman[30])

$$\mathscr{P}\int_a^b \frac{F(x)}{x - x_0} dx = \lim_{\varepsilon \to +0} \int_a^b \frac{F(x)}{x - x_0 - i\varepsilon} dx - i\pi F(0), \tag{11.5}$$

where $a \leqslant x_0 \leqslant b$.

It is useful to introduce the Hilbert transform[31,32] and write the Hilbert transform in terms of a convolution:

$$F_H(t) = \mathcal{H}[F(x); t] = \frac{1}{\pi}\mathcal{P}\int_{-\infty}^{\infty} \frac{F(x)}{x-t}\,dx$$

$$= \frac{-1}{\pi}\mathcal{P}\left(\frac{1}{t}\right) * F(t) = i(\delta^+ - \delta^-) * F. \tag{11.6}$$

From (11.6) and the derivative theorem for Hilbert transforms,

$$F_H'(t) = \frac{d}{dt}\mathcal{H}[F(x); t] = \mathcal{H}[F'(x); t],$$

all the following forms are equivalent.

$$\mathcal{H}\left[\frac{dF(x)}{dx}; t\right] = \frac{1}{\pi}\mathcal{P}\int_{-\infty}^{\infty} \frac{\frac{dF}{dx}(x)}{x-t}\,dx$$

$$= \frac{-1}{\pi}\frac{dF}{dt}(t) * \mathcal{P}\left(\frac{1}{t}\right)$$

$$= \frac{-1}{\pi}\frac{d}{dt}\left[F(t) * \mathcal{P}\left(\frac{1}{t}\right)\right]$$

$$= \frac{-1}{\pi}F(t) * \frac{d}{dt}\mathcal{P}\left(\frac{1}{t}\right)$$

$$= \frac{1}{\pi}\int_{-\infty}^{\infty} \frac{F(x)}{(x-t)^2}\,dx. \tag{11.7}$$

Another way to generate singular distributions is by differentiating singular (and, in some cases, regular) distributions. This will be seen in the following section. In addition to the singular distributions mentioned here, many others are important. For a discussion of some of them, see Zemanian[16] and Bremermann.[2]

[B.12] DERIVATIVES OF DISTRIBUTIONS

The Dirac δ function is often presented as the derivative of the Heaviside, or unit step, function

$$u(x) = \begin{cases} 0, & x < 0 \\ \frac{1}{2}, & x = 0 \\ 1, & x > 0. \end{cases} \tag{12.1}$$

Let us investigate this possibility by formal manipulations of distributions.

It is easy to see that u is a (regular) distribution $\langle \mathrm{u}, F \rangle$ when applied to all $F \in \mathscr{D}$. Consider the integration by parts of the distribution $\langle \mathrm{u}, F' \rangle$:

$$\int_{-\infty}^{\infty} \mathrm{u}(x) F'(x)\, dx = \mathrm{u}(x) F(x) \Big|_{-\infty}^{+\infty} - \int_{-\infty}^{\infty} \mathrm{u}'(x) F(x)\, dx, \quad (12.2)$$

or, since the first term on the right-hand side vanishes,

$$\langle \mathrm{u}', F \rangle = -\langle \mathrm{u}, F' \rangle. \quad (12.3)$$

Indeed, this equation may be taken as the *definition* of a derivative of a distribution.[1] Upon looking at $\langle \mathrm{u}, F' \rangle$ in a different way, we obtain

$$\langle \mathrm{u}, F' \rangle = \int_{0}^{\infty} F'\, dx = F(x) \Big|_{0}^{+\infty} = -F(0), \quad (12.4)$$

where explicit use has been made of (12.1) and the fact that F vanishes at ∞. Now, we have already established that $\langle \delta, F \rangle = F(0)$, which yields

$$\langle \delta, F \rangle = \langle \mathrm{u}', F \rangle, \qquad \forall F \in \mathscr{D}, \quad (12.5)$$

and by equality of distributions we have the formal result

$$\delta(x) = \frac{d}{dx} \mathrm{u}(x). \quad (12.6)$$

Remark 1. Note that since $\mathrm{u}(x) + \mathrm{u}(-x) = 1$ it follows that

$$\delta(x) + \frac{d}{dx} \mathrm{u}(-x) = 0 \quad \text{and} \quad \frac{d}{dx} \mathrm{u}(-x) = -\delta(x). \qquad \square$$

Considering the preceding discussion, it is clear that the appropriate definition of the distributional or generalized derivative of the distribution T is

$$\langle T', F \rangle = -\langle T, F' \rangle. \quad (12.7)$$

By repeated partial integrations, the general result

$$\left\langle \left(\frac{d}{dx} \right)^{k} T, F \right\rangle = (-1)^{k} \left\langle T, \left(\frac{d}{dx} \right)^{k} F \right\rangle \quad (12.8)$$

is obtained for the kth derivative.

Example 1.

$$\langle \delta', F \rangle = -\langle \delta, F' \rangle = -F'(0)$$

and

$$\langle \delta'(x-a), F \rangle = -F'(a). \qquad \square$$

Remark 2. If F is defined on $\mathbb{R}^n$ rather than $\mathbb{R}^1$, then (12.7) generalizes in the obvious way to $\langle \partial T/\partial x_j, F \rangle = -\langle T, \partial F/\partial x_j \rangle$ or, in the general case, $\langle D^m T, F \rangle = (-1)^m \langle T, D^m F \rangle$, were the notation is that of Remark 1 in [B.2]. $\square$

In the previous section, we mentioned that $\mathcal{P}(1/x)$ is a singular distribution. This distribution comes from differentiating the distribution $\log|x|$,

$$\mathcal{P}\frac{1}{x} = \frac{d}{dx}\log|x|. \tag{12.9}$$

The proper interpretation of this equation is, of course,

$$\left\langle \mathcal{P}\frac{1}{x}, F \right\rangle = \left\langle \frac{d}{dx}\log|x|, F \right\rangle, \qquad \forall F \in \mathcal{D}$$

Now, let us define

$$\log(x \pm i0) = \lim_{y \to \pm 0} \log(x + iy),$$

and recall that the complex number $z = x + iy$ may always be written as $z = re^{i\theta}$, where $r = |z|$ and $\theta = \arg z$. Also, keep in mind that the branch cut for $\log z$ is along the negative real axis. With these things in mind, upon examining the limit, we find

$$\begin{aligned}\log(x \pm i0) &= \lim_{y \to \pm 0} \{\log|x + iy| + i\arg(x + iy)\} \\ &= \log|x| \pm i\pi \mathrm{u}(-x).\end{aligned}$$

By differentiation

$$\frac{1}{x \pm i0} = \lim_{y \to \pm 0} \frac{1}{x + iy} = \mathcal{P}\frac{1}{x} \mp i\pi\delta(x), \tag{12.10}$$

and by making some obvious notational changes in this equation, we obtain (11.2), as promised in Remark 2 of [B.11].

Finally, we consider the derivative of a distribution T, which depends on a parameter a in addition to the independent variable x. If $T(x; a)$ represents the distribution, then

$$\frac{dT}{da} = \lim_{h \to 0} \frac{T(x; a + h) - T(x; a)}{h}, \tag{12.11}$$

provided the limit exists. As usual, this is to be interpreted as

$$\left\langle \frac{dT}{da}, F \right\rangle = \lim_{h \to 0} \frac{\langle T(x; a + h), F \rangle - \langle T(x; a), F \rangle}{h}.$$

To be specific, suppose T is $\delta(x - a)$ with both x and a in $\mathbb{R}^1$, then

$$\left\langle \frac{d}{da}\delta(x - a), F \right\rangle = \lim_{h \to 0} \frac{\langle \delta(x - a - h), F \rangle - \langle \delta(x - a), F \rangle}{h}$$

$$= \lim_{h \to 0} \frac{F(a + h) - F(a)}{h} = \left.\frac{dF}{dx}\right|_{x=a} = F'(a).$$

However, we also know that

$$\left\langle \frac{d}{dx}\delta(x - a), F \right\rangle = -\langle \delta(x - a), F' \rangle = -F'(a),$$

and this leads to the conclusion that

$$\frac{d}{da}\delta(x - a) = -\frac{d}{dx}\delta(x - a). \tag{12.12}$$

If $\mathbf{x}$ and $\mathbf{a}$ come from $\mathbb{R}^n$, the preceding results generalize to

$$\frac{\partial}{\partial a_j}\delta(\mathbf{x} - \mathbf{a}) = -\frac{\partial}{\partial x_j}\delta(\mathbf{x} - \mathbf{a}), \tag{12.13}$$

and if the argument of the δ function is $p - \xi \cdot \mathbf{x}$, where $p \in \mathbb{R}^1$ and $\xi, \mathbf{x} \in \mathbb{R}^n$, we obtain

$$\frac{\partial}{\partial \xi_j}\delta(p - \xi \cdot \mathbf{x}) = -x_j \frac{\partial}{\partial p}\delta(p - \xi \cdot \mathbf{x}). \tag{12.14}$$

[B.13] FOURIER TRANSFORMS

When the Fourier transform is generalized to include transforms of distributions, some very useful and important relationships emerge. For a rigorous treatment of this topic see Zemanian[16] and Schwartz.[1] One approach is to consider some distribution $T \in \mathscr{S}'$ and test function $F \in \mathscr{S}$ and *define* $\tilde{T} = \mathscr{F}T$ in terms of the Parseval relation

$$\langle T, F \rangle = \langle \tilde{T}, \tilde{F} \rangle. \tag{13.1}$$

Formally, in terms of an integral, this is

$$\int_{-\infty}^{\infty} T^*(t) F(t)\, dt = \int_{-\infty}^{\infty} \tilde{T}^*(x) \tilde{F}(x)\, dx, \tag{13.2}$$

where * means complex conjugation and

$$\tilde{F}(x) = \int_{-\infty}^{\infty} e^{-i2\pi xt} F(t)\, dt,$$

$$F(t) = \int_{-\infty}^{\infty} e^{i2\pi xt} \tilde{F}(x)\, dx.$$

By identifying T with $\delta(t - a)$, we obtain the Fourier transform pair

$$\delta(t - a) \leftrightarrow e^{-i2\pi ax}, \tag{13.3}$$

since

$$\langle \delta(t - a), F \rangle = F(a) = \int_{-\infty}^{\infty} e^{i2\pi ax} \tilde{F}(x)\, dx$$

holds for all test functions F. Of course, the pair may be written in the equivalent form

$$e^{i2\pi at} \leftrightarrow \delta(x - a). \tag{13.4}$$

When used in (13.2), this gives $\tilde{F}(a) = \tilde{F}(a)$.

Now meaning can be given to the transform of a periodic function. For example, consider

$$\begin{aligned} \mathscr{F}(\cos 2\pi x_0 t) &= \tfrac{1}{2} \mathscr{F}(e^{i2\pi x_0 t} + e^{-i2\pi x_0 t}) \\ &= \tfrac{1}{2}\delta(x - x_0) + \tfrac{1}{2}\delta(x + x_0). \end{aligned}$$

Another important transform pair follows from this more general approach[31]:

$$\operatorname{sgn} t \leftrightarrow \frac{1}{i\pi} \mathscr{P}\left(\frac{1}{x}\right). \tag{13.5}$$

Also,

$$\mathscr{P}\left(\frac{1}{t}\right) \leftrightarrow -i\pi \operatorname{sgn} x, \tag{13.6}$$

where, as usual,

$$\operatorname{sgn} t = \begin{cases} 1, & t > 0 \\ -1, & t < 0. \end{cases}$$

Remark 1. The $\mathcal{P}$ in (13.5) and (13.6) may be omitted when confusion does not result. □

The derivative theorem for $F \in \mathcal{S}$ is[31]

$$\mathcal{F}F'(t) = i2\pi x\tilde{F}(x).$$

It follows that a similar result holds for $T \in \mathcal{S}'$,

$$\begin{aligned} \langle \mathcal{F}T', \tilde{F} \rangle &= \langle T', F \rangle = -\langle T, F' \rangle \\ &= -\langle \tilde{T}, \tilde{F}' \rangle = -\langle \tilde{T}(x), i2\pi x\tilde{F}(x) \rangle \\ &= \langle i2\pi x\tilde{T}(x), \tilde{F}(x) \rangle. \end{aligned}$$

This must hold for arbitrary F, and we conclude that

$$\mathcal{F}T'(t) = i2\pi x\tilde{T}(x). \tag{13.7}$$

Example 1. Suppose $T(t) = |t|$, then $T'(t) = \operatorname{sgn} t$ and $\tilde{T}(x) = \mathcal{F}(|t|)$. By use of (13.5) and (13.7) we have

$$\mathcal{F}(\operatorname{sgn} t) = \frac{1}{i\pi x} = i2\pi x\mathcal{F}(|t|).$$

Thus,

$$\mathcal{F}(|t|) = -1/2\pi^2 x^2.$$

□

Example 2. Note that the unit step function can be expressed as

$$\mathrm{u}(t) = \tfrac{1}{2} + \tfrac{1}{2}\operatorname{sgn} t.$$

Therefore, from (13.3) and (13.5),

$$\tilde{\mathrm{u}}(x) = \tfrac{1}{2}\delta(x) + \frac{1}{2\pi i}\mathcal{P}\left(\frac{1}{x}\right),$$

and by use of (11.2),

$$\tilde{\mathrm{u}}(x) = \delta^{-}(x).$$

□

[B.14] REFERENCES FOR APPENDIX B

1. Schwartz (1957, 1959, 1966)
2. Bremermann (1965)
3. De Jager (1969)
4. Marchand (1962)
5. Halperin (1952)
6. Roos (1969)
7. Erdélyi (1961)
8. Gel'fand and Shilov (1964)
9. Zadeh and Desoer (1963)
10. Challifour (1972)
11. Liverman (1964)
12. Temple (1953, 1955)
13. Lighthill (1958)
14. Korevaar (1968)
15. Mikusiński (1948)
16. Zemanian (1965)
17. Friedman (1963)
18. Friedman (1969)
19. Stakgold (1967, 1968)
20. Weir (1973)
21. Taylor (1955)
22. Dirac (1927)
23. Dirac (1930, 1958)
24. Van der Pol and Bremmer (1955)
25. Sneddon (1955)
26. Erdélyi (1962)
27. Mikusiński (1959)
28. Sobolev (1936)
29. Newton (1966)
30. Roman (1965)
31. Bracewell (1978)
32. Sneddon (1972)

Appendix C

Special Functions

[C.1] INTRODUCTION

In this appendix, formulas are recorded for easy reference. None of these results are new; the references used are listed in [C.6].[1-6] The indices l, m, and n represent nonnegative integers and the indices α and ν may be complex. If the argument of a function is not explicitly shown, it is assumed to be x. Primes or D may be used to represent derivatives with respect to the argument.

[C.2] TCHEBYCHEFF POLYNOMIALS

First Kind $T_l(x)$

$$T_l(x) = \cos(l \operatorname{arc}\cos x), \qquad 0 < x < 1.$$

$$T_l(x) = \cosh(l \operatorname{arc}\cosh x), \qquad 1 < x < \infty.$$

$$T_l(x) = \tfrac{1}{2}\left[\left(x + \sqrt{x^2 - 1}\right)^l + \left(x - \sqrt{x^2 - 1}\right)^l\right], \qquad 0 < x < \infty.$$

$$T_l(-x) = (-1)^l T_l(x), \qquad T_l(1) = 1, \qquad T_l(0) = \cos\frac{l\pi}{2}.$$

Generating Function

$$\frac{1 - z^2}{1 - 2xz + z^2} = 1 + 2\sum_{l=1}^{\infty} T_l(x) z^l, \qquad \begin{array}{c} -1 < x < 1 \\ |z| < 1. \end{array}$$

Recurrence and Derivatives

$$T_{l+1} = 2xT_l - T_{l-1}.$$

$$(1 - x^2)T_l' = lT_{l-1} - lxT_l.$$

$$(1 - x^2)T_l'' - xT_l' + l^2 T_l = 0.$$

Orthogonality

$$\int_{-1}^{1} T_l(x) T_m(x)(1 - x^2)^{-1/2}\, dx = \begin{cases} 0, & l \neq m \\ \dfrac{\pi}{2}, & l = m \neq 0 \\ \pi, & l = m = 0. \end{cases}$$

First Few

$$T_0 = 1.$$

$$T_1 = x.$$

$$T_2 = 2x^2 - 1.$$

$$T_3 = 4x^3 - 3x.$$

$$T_4 = 8x^4 - 8x^2 + 1.$$

Second Kind $U_l(x)$

$$U_{l-1}(x) = \frac{\sin(l \arccos x)}{(1 - x^2)^{1/2}}, \qquad 0 < x < 1$$

$$U_{l-1}(x) = \frac{\sinh(l \arccos hx)}{(x^2 - 1)^{1/2}}, \qquad 1 < x < \infty$$

$$U_{l-1}(x) = \frac{\left(x + \sqrt{x^2 - 1}\right)^l - \left(x - \sqrt{x^2 - 1}\right)^l}{2(x^2 - 1)^{1/2}}, \qquad 0 < x < \infty,\ x \neq 1$$

$$U_l(-x) = (-1)^l U_l(x), \qquad U_l(1) = l + 1, \qquad U_l(0) = \cos\frac{l\pi}{2}.$$

Generating Function

$$\frac{1}{1 - 2xz + z^2} = \sum_{l=0}^{\infty} U_l(x) z^l, \qquad \begin{matrix} -1 < x < 1 \\ |z| < 1. \end{matrix}$$

Recurrence and Derivatives

$$U_{l+1} = 2xU_l - U_{l-1}.$$

$$(1 - x^2) U_l' = (l + 1) U_{l-1} - lxU_l.$$

$$(1 - x^2) U_l'' - 3xU_l' + l(l + 2) U_l = 0.$$

Orthogonality

$$\int_{-1}^{1} U_l(x)U_m(x)(1-x^2)^{1/2}\,dx = \frac{\pi}{2}\delta_{lm}.$$

First Few

$$U_0 = 1.$$

$$U_1 = 2x.$$

$$U_2 = 4x^2 - 1.$$

$$U_3 = 8x^3 - 4x.$$

$$U_4 = 16x^4 - 12x^2 + 1.$$

Miscellaneous Relationships

$$U_{l-1} = \frac{1}{l}T_l', \qquad l \geqslant 1.$$

$$T_l = U_l - xU_{l-1}, \quad l \geqslant 1.$$

$$(1-x^2)U_l = xT_{l+1} - T_{l+2}.$$

$$\frac{T_l}{(x^2-1)^{1/2}} - U_{l-1} = \frac{\left(x-\sqrt{x^2-1}\right)^l}{(x^2-1)^{1/2}} = \frac{\left(x+\sqrt{x^2-1}\right)^{-l}}{(x^2-1)^{1/2}}, \qquad x \neq 1.$$

[C.3] HERMITE POLYNOMIALS $H_l(x)$

Generating Function

$$e^{2xt-t^2} = \sum_{l=0}^{\infty} \frac{H_l(x)t^l}{l!}.$$

Special Values

$$H_l(x) = (-1)^l H_l(-x), \quad H_{2l}(0) = (-1)^l\frac{(2l)!}{l!}, \qquad H_{2l+1}(0) = 0.$$

Recurrence and Derivatives

$$H_{l+1} = 2xH_l - 2lH_{l-1}.$$

$$H_l' = 2lH_{l-1}.$$

$$H_l'' - 2xH_l' + 2lH_l = 0.$$

Orthogonality

$$\int_{-\infty}^{\infty} H_l(x) H_m(x) e^{-x^2}\, dx = \sqrt{\pi}\, 2^l l! \delta_{lm}.$$

Rodrigues Formula

$$H_l(x) = (-1)^l e^{x^2} \left(\frac{d}{dx}\right)^l e^{-x^2}.$$

$$\left(-\frac{d}{dx}\right)^m e^{-x^2} H_n(x) = e^{-x^2} H_{m+n}(x).$$

First Few

$$H_0 = 1.$$

$$H_1 = 2x.$$

$$H_2 = 4x^2 - 2.$$

$$H_3 = 8x^3 - 12x.$$

$$H_4 = 16x^4 - 48x^2 + 12.$$

$$H_5 = 32x^5 - 160x^3 + 120x.$$

Expansions

$$x^0 = H_0.$$

$$x^1 = \tfrac{1}{2} H_1.$$

$$x^2 = \tfrac{1}{4}(2H_0 + H_2).$$

$$x^3 = \tfrac{1}{8}(H_3 + 6H_1).$$

$$x^4 = \tfrac{1}{16}(H_4 + 12H_2 + 12H_0).$$

$$x^5 = \tfrac{1}{32}(H_5 + 20H_3 + 60H_1).$$

[C.4] LAGUERRE POLYNOMIALS $L_n^\alpha(x)$

Rodrigues Formula

$$L_n^\alpha(x) = \frac{x^{-\alpha} e^x}{n!} \left(\frac{d}{dx}\right)^n [e^{-x} x^{n+\alpha}].$$

First Few

$$L_0^\alpha = 1.$$

$$L_1^\alpha = 1 + \alpha - x.$$

$$L_2^\alpha = \tfrac{1}{2}(1+\alpha)(2+\alpha) - (2+\alpha)x + \tfrac{1}{2}x^2.$$

$$L_3^\alpha = \tfrac{1}{6}(1+\alpha)(2+\alpha)(3+\alpha) - \tfrac{1}{2}(2+\alpha)(3+\alpha)x + \tfrac{1}{2}(3+\alpha)x^2 - \tfrac{1}{6}x^3.$$

Recurrence and Derivatives

$$nL_n^\alpha = (2n - 1 + \alpha - x)L_{n-1}^\alpha - (n - 1 + \alpha)L_{n-2}^\alpha.$$

$$L_n^\alpha = L_{n-1}^\alpha + L_n^{\alpha-1}.$$

$$(n - x)L_n^\alpha = (\alpha + n)L_{n-1}^\alpha - xL_n^{\alpha+1}.$$

$$(1 + \alpha + n)L_n^\alpha = (n + 1)L_{n+1}^\alpha + xL_n^{\alpha+1}.$$

$$\frac{d}{dx}L_n^\alpha = -L_{n-1}^{\alpha+1}.$$

Orthogonality

$$\int_0^\infty x^\alpha e^{-x} L_n^\alpha(x) L_m^\alpha(x)\,dx = \begin{cases} 0, \quad m \neq n, & \mathrm{Re}(\alpha) > -1 \\ \dfrac{\Gamma(1+\alpha+n)}{n!}, \quad m = n, & \mathrm{Re}(\alpha) > -1. \end{cases}$$

Miscellaneous

$$H_{2m}(x) = (-1)^m 2^{2m} m! L_m^{-1/2}(x^2).$$

$$H_{2m+1}(x) = (-1)^m 2^{2m+1} m! x L_m^{1/2}(x^2).$$

[C.5] GEGENBAUER POLYNOMIALS $C_l^\nu(x)$

Generating Function

$$(1 - 2xt + t^2)^{-\nu} = \sum_{l=0}^{\infty} C_l^\nu(x) t^l, \qquad \nu \neq 0.$$

Special Values

$$C_0^0(1) = 1.$$

$$C_l^0(1) = \frac{2}{l}.$$

$$C_l^0(x) = \lim_{\nu \to 0} \frac{1}{\nu} C_l^\nu(x).$$

$$C_{2l+1}^\nu(0) = 0.$$

$$C_{2l}^\nu(0) = \frac{(-1)^l}{l!} \frac{\Gamma(l+\nu)}{\Gamma(\nu)}.$$

$$C_l^\nu(1) = \frac{\Gamma(2\nu + l)}{l!\Gamma(2\nu)}.$$

$$C_l^\nu(-x) = (-1)^l C_l^\nu(x).$$

First Few

$$C_0^\nu = 1.$$

$$C_1^\nu = 2\nu x.$$

$$C_2^\nu = 2\nu(\nu + 1)x^2 - \nu.$$

$$C_3^\nu = \tfrac{4}{3}\nu(\nu + 1)(\nu + 2)x^3 - 2\nu(\nu + 1)x.$$

$$C_4^\nu = \tfrac{2}{3}\nu(\nu + 1)(\nu + 2)(\nu + 3)x^4 - 2\nu(\nu + 1)(\nu + 2)x^2 + \tfrac{1}{2}\nu(\nu + 1).$$

Recurrence and Derivatives

$$(l+1)C_{l+1}^\nu = 2(l+\nu)xC_l^\nu - (l + 2\nu - 1)C_{l-1}^\nu.$$

$$2\nu(1 - x^2)C_{l-1}^{\nu+1} = (2\nu + l - 1)C_{l-1}^\nu - lxC_l^\nu,$$

$$= (l + 2\nu)xC_l^\nu - (l+1)C_{l+1}^\nu.$$

$$DC_l^\nu = 2\nu C_{l-1}^{\nu+1}, \qquad D \equiv \frac{d}{dx}.$$

$$(1 - x^2)DC_l^\nu = (l + 2\nu - 1)C_{l-1}^\nu - lxC_l^\nu,$$

$$= (l + 2\nu)xC_l^\nu - (l+1)C_{l+1}^\nu.$$

$$(1 - x^2)D^2C_l^\nu - (2\nu + 1)xDC_l^\nu + l(l + 2\nu)C_l^\nu = 0.$$

Orthogonality

$$\int_{-1}^{1} C_l^\nu(x) C_m^\nu(x)(1-x^2)^{\nu-1/2}\,dx = \frac{\pi 2^{1-2\nu}\Gamma(l+2\nu)}{l!(\nu+l)\Gamma(\nu)\Gamma(\nu)}\delta_{lm}.$$

Miscellaneous

$$T_l(x) = \frac{l}{2} C_l^0(x).$$

$$U_l(x) = C_l^1(x).$$

$$C_l^\nu(x) = \frac{\Gamma(2\nu+l)}{\Gamma(2\nu)} \cdot \frac{\Gamma(\nu+\frac{1}{2})}{\Gamma(\nu+\frac{1}{2}+l)} P_l^{(\nu-1/2,\,\nu-1/2)}(x),$$

where $P_l^{(\alpha,\,\alpha)}(x)$ is a Jacobi polynomial.

$$C_l^{1/2}(x) = P_l(x),$$

where $P_l(x)$ is a Legendre polynomial.

[C.6] REFERENCES

1. Abramowitz and Stegun (1964)
2. Arfken (1970)
3. Erdélyi, Magnus, Oberhettinger, and Tricomi (1953b)
4. Rainville (1960)
5. Rivlin (1974)
6. Szegö (1939)

References

Abramowitz, M. and I. A. Stegun, Eds. (1964). *Handbook of Mathematical Functions with Formulas, Graphs, and Mathematical Tables*, National Bureau of Standards Applied Mathematics Series 55, U.S. Government Printing Office, Washington, DC.

Aksnes, K. and F. A. Franklin (1976). "Mutual phenomena of the Galilean satellites in 1973. III. Final results from 91 light curves," *Astron. J.*, **81**, 464–481.

Aksnes, K. and F. A. Franklin (1978a). "Mutual phenomena of Jupiter's five inner satellites in 1979," *Icarus*, **34**, 188–193.

Aksnes, K. and F. A. Franklin (1978b). "Mutual phenomena of Saturn's satellites in 1979–1980," *Icarus*, **34**, 194–207.

Albers, S. C. (1979). "Mutual occultations of planets: 1557 to 2230," *Sky and Telescope*, **57**, 220–222.

Altschuler, M. D. (1979). "Reconstruction of the global-scale three-dimensional solar corona," in *Image Reconstruction From Projections*, G. T. Herman, Ed., Vol. 32 of *Topics in Applied Physics*, Springer-Verlag, New York, pp. 105–145.

Altschuler, M. D., G. T. Herman, and A. Lent (1978). "Fully three-dimensional image reconstruction from cone-beam sources," in *Proceedings of the 1978 IEEE Computer Society Conference on Pattern Recognition and Image Processing*, May 31–June 2 (1978), Chicago, Illinois, IEEE, New York, pp. 194–199.

Alvarez, R. E. and A. Macovski (1976). "Energy-selective reconstructions in x-ray computerized tomography," *Phys. Med. Biol.*, **21**, 733–744.

Andrew, E. R. (1969). *Nuclear Magnetic Resonance*, Cambridge University, Cambridge, England.

Andrew, E. R. (1976) "Zeugmatography," in *Proceedings of the Fourth Ampere International Summer School*, Pula, Yugoslavia, R. Blinc and G. Lahajnar, Ed.. University of Ljubljana, Ljubljana, pp. 1–39.

Andrew, E. R. (1980a). "Nuclear magnetic resonance imaging: the multiple sensitive point method," *IEEE Trans. Nucl. Sci.*, **NS-27**(3), 1232–1238.

Andrew, E. R. (1980b). "NMR imaging of intact biological systems," *Philos. Trans. R. Soc. London Ser. B*, **289**, 471–481.

Arfken, G. (1970). *Mathematical Methods for Physicists*, 2nd ed., Academic, New York.

Artzy, E., T. Elfving, and G. T. Herman (1979). "Quadratic optimization for image reconstruction II," *Comput. Graph. Image Proc.*, **11**, 242–261.

Asanuma, T., Ed. (1979). *Flow Visualization*, McGraw-Hill, New York.

Baars, J. W. M., J. F. van der Brugge, J. L. Casse, J. P. Hamaker, L. H. Sondaar, J. J. Visser, and K. J. Wellington (1973). "The synthesis radio telescope at Westerbork," *Proc. IEEE*, **61**, 1258–1266.

Baba, N. and K. Murata (1977). "Filtering for image reconstruction from projections," *J. Opt. Soc. Am.*, **67**, 662–668.

Baily, N. A. (1979). "The fluoroscopic image as input for computed tomography," *IEEE Trans. Nucl. Sci.*, **NS-26**(2), 2707–2709.

Baker, K. D. and G. D. Sullivan (1980). "Multiple bandpass filters in image processing," *IEE Proc.*, **127**, Part E., 173–184.

Ball, J. S., S. A. Johnson, and F. Stenger (1980). "Explicit inversion of the Helmholtz equation for ultrasound insonification and spherical detection," in Vol. 9 of *Acoustical Imaging*, K. Y. Wang, Ed. Plenum, New York and London, pp. 451–461.

Ballard, D. H. (1981). "Generalizing the Hough transform to detect arbitrary shapes," *Pattern Recognition*, **13**, 111–122.

Bang, T. (1951). "A solution of the 'plank problem,'" *Proc. Am. Math. Soc.*, **2**, 990–992.

Barakat, R. (1964). "Solution of an Abel integral equation for band-limited functions by means of sampling theorems," *J. Math. and Phys., Cambridge Mass.*, **43**, 325–331.

Barnes, T. G., D. S. Evans, and T. J. Moffett (1978). "Stellar angular diameters and visual surface brightness. III. An improved definition of the relationship," *Mon. Not. R. Astron. Soc.*, **183**, 285–304.

Barr, W. L. (1962). "Method for computing the radial distribution of emitters in a cylindrical source," *J. Opt. Soc. Am.*, **52**, 885–888.

Barrett, H. H. and W. Swindell (1977). "Analog reconstruction methods for transaxial tomography," *Proc. IEEE*, **65**, 89–107.

Barrett, H. H. and W. Swindell (1981). *Radiological Imaging; The Theory of Image Formation, Detection, and Processing*, Vols. 1 and 2, Academic, New York.

Barrett, H. H., S. K. Gordon, and R. S. Hershell (1976). "Statistical limitations in transaxial tomography," *Comput. Biol. Med.*, **6**, 307–323.

Barton, J. P. (1978). "Feasibility of neutron radiography for large bundles of fast reactor fuel." Tech. Report IRT 6247-004, Instrumentation Research Technology Corp., San Diego, CA.

Bates, R. H. T. and G. C. McKinnon (1979). "Towards improving images in ultrasonic transmission tomography," *Australas. Phys. Sci. Med.*, **2–3**, 134–140.

Bates, R. H. T. and T. M. Peters (1971). "Towards improvements in tomography," *N. Z. J. Sci.*, **14**, 883–896.

Beattie, J. W. (1975). "Tomographic reconstruction from fan beam geometry using Radon's integration method," *IEEE Trans. Nucl. Sci.*, **NS-22**(1), 359–363.

Bendel, P., C.-M. Lai, and P. C. Lauterbur (1980). "^{31}P spectroscopic zeugmatography of phosphorus metabolites," *J. Magn. Reson.*, **38**, 343–356.

Bender, R., S. H. Bellman, and R. Gordon (1970). "ART and the ribosome: A preliminary report on the three-dimensional structure of individual ribosomes determined by an algebraic reconstruction technique," *J. Theor. Biol.*, **29**, 483–487.

Béné, G.-J., B. Borcard, E. Hiltbrand, and P. Magnin (1980). "In situ identification of human physiological fluids by nuclear magnetism in the Earth's field," *Philos. Trans. R. Soc. London Ser. B*, **289**, 501–502.

Bennett, F. D., W. C. Carter, and V. E. Bergdolt (1952). "Interferometric analysis of air flow about projectiles in free flight," *J. Appl. Phys.*, **23**, 453–469.

Berry, M. V. and D. F. Gibbs (1970). "The interpretation of optical projections," *Proc. R. Soc. London Ser. A*, **314**, 143–152.

Bhatia, A. B. and E. Wolf (1954). "On the circle polynomials of Zernike and related orthogonal sets," *Proc. Cambridge Philos. Soc.*, **50**, 40–48.

Birkhoff, G. D. (1940). "On drawings composed of uniform straight lines," *J. de Matématiques*, **19**, 221–236.

Bleistein, N. and J. K. Cohen (1979). "Inverse methods for reflector mapping and sound speed profiling," in *Ocean Acoustics*, J. A. DeSanto, Ed., Vol. 8 of *Topics in Current Physics*, Springer-Verlag, Berlin, Heidelberg, and New York, pp. 225–242.

Bloch, F. (1978). "The early days of NMR," VIII International Conference on Magnetic Resonance in Biological Systems, Nara, Japan (unpublished lecture).

Bloch, F., W. W. Hansen, and M. E. Packard (1946). "The nuclear induction experiment," *Phys. Rev.*, **70**, 474–485.

Bockasten, K. (1961). "Transformation of observed radiances into radial distribution of the emission of a plasma," *J. Opt. Soc. Am.*, **51**, 943–947.

Bockwinkel, H. B. A. (1906). "Over de voortplanting van licht in een twee-assig kristal rondom een middelpunt van trilling." *K. Akad. van Wet. Amsterdam Versl. Natuurk.*, **14**(2), 636–651.

Boerner, W.-M. (1980). "Development of physical optics inverse scattering techniques using Radon projection theory," in *Mathematical Methods and Applications of Scattering Theory*, J. A. DeSanto, A. W. Sàenz, and W. W. Zachary, Eds., Vol. 130 of *Lecture Notes in Physics*, Springer-Verlag, Berlin, Heidelberg, New York, pp. 301–307.

Bohm, C., L. Eriksson, M. Bergstrom, J. Litton, R. Sundman and M. Singh (1978). "A computer assisted ringdetector positron camera system for reconstruction tomography of the brain," *IEEE Trans. Nucl. Sci.*, **NS-25**(1), 624–637.

Born, M. and E. Wolf (1975). *Principles of Optics*, 5th ed., Pergamon, New York.

Bottomley, P. A. (1979). "A comparative evaluation of proton NMR imaging results," *J. Magn. Reson.*, **36**, 121–127.

Bovée, W. M. M., J. H. N. Creyghton, K. W. Getreuer, D. Korbee, S. Lobregt, J. Smidt, R. A. Wind, J. Lindeman, L. Smid, and H. Posthuma (1980). "NMR relaxation and images of human breast tumors *in vitro*," *Philos. Trans. R. Soc. London Ser. B*, **289**, 535–538.

Bracewell, R. N. (1956a). "Strip integration in radio astronomy," *Aust. J. Phys.*, **9**, 198–217.

Bracewell, R. N. (1956b). "Two-dimensional aerial smoothing in radio astronomy," *Aust. J. Phys.*, **9**, 297–314.

Bracewell, R. N. (1958). "Restoration in the presence of errors," *Proc. IRE*, **46**, 106–111.

Bracewell, R. N. (1974). "Three-dimensional reconstruction: An overview," Keynote Address, in *Techniques of Three-Dimensional Reconstruction*, R. B. Marr, Ed., Proceedings of an international workshop held at Brookhaven National Laboratory, July 16–19 (1974), Applied Mathematics Department, BNL 20425, Brookhaven National Laboratory, Upton, NY, pp. 3–14.

Bracewell, R. N. (1977). "Correction for collimator width (restoration) in reconstructive x-ray tomography," *J. Comput. Assisted Tomog.*, **1**, 6–15.

Bracewell, R. N. (1978). *The Fourier Transform and Its Applications*, 2nd ed., McGraw-Hill, New York.

Bracewell, R. N. (1979). "Image reconstruction in radio astronomy," in *Image Reconstruction from Projections*, G. T. Herman, Ed., Vol 32 of *Topics in Applied Physics*, Springer-Verlag, New York, pp. 81–104.

Bracewell, R. N., and A. C. Riddle (1967). "Inversion of fan-beam scans in radio astronomy," *Astrophys. J.*, **150**, 427–434.

Bracewell, R. N. and J. A. Roberts (1954). "Aerial smoothing in radio astronomy," *Aust. J. Phys.*, **7**, 615–640.

Bracewell, R. N., R. S. Colvin, K. M. Price, and A. R. Thompson (1971). "Stanford's high resolution radio interferometer," *Sky and Telescope*, **42**, 4–9.

Bracewell, R. N., R. G. Colvin, L. R. D'Addario, C. J. Grebenkemper, K. M. Price, and A. R. Thompson (1973). "The Stanford five-element radiotelescope," *Proc. IEEE*, **61**, 1249–1257.

Bradley, J. W. (1968). "Density determination from axisymmetric interferograms," *AIAA J.*, **6**, 1190–1192.

Bragg, W. H. (1915). "IX Bakerian Lecture—X rays and crystal structure," *Philos. Trans. R. Soc. London*, **A215**, 253–274.

Bremermann, H. (1965). *Distributions, Complex Variables, and Fourier Transforms,* Addison-Wesley, Reading, PA.

Brigham, E. O. (1974). *The Fast Fourier Transform*, Prentice-Hall, Englewood Cliffs, NJ.

Brinkman, R. T. and R. L. Millis (1973). "Mutual phenomena of Jupiter's satellites in 1973–74," *Sky and Telescope*, **45**, 93–95.

Brooker, H. R. and W. S. Hinshaw (1978). "Thin-section NMR imaging," *J. Magn. Reson.*, **30**, 129–131.

Brooks, R. A. (1977). "A quantitative theory of the Hounsfield unit and its application to dual energy scanning," *J. Comput. Assisted Tomog.*, **1**, 487–493.

Brooks, R. A. and G. Di Chiro (1975). "Theory of image reconstruction in computed tomography." *Radiology*, **117**, 561–572.

Brooks, R. A. and G. Di Chiro (1976a). "Principles of computer assisted tomography (CAT) in radiographic and radioisotopic imaging," *Phys. Med. Biol.*, **21**, 689–732.

Brooks, R. A. and G. Di Chiro (1976b). "Beam hardening in x-ray reconstructive tomography," *Phys. Med. Biol.*, **21**, 390–398.

Brooks, R. A. and G. Di Chiro (1976c). "Statistical limitations in x-ray reconstructive tomography," *Med. Phys.*, **3**, 237–240.

Brooks, R. A. and G. H. Weiss (1976). "Interpolation problems in image reconstruction," *Proc. Soc. Photo Opt. Instrum. Eng.*, **96**, 313–319.

Brooks, R. A., G. H. Glover, A. J. Talbert, R. L. Eisner, and F. A. DiBianca (1979). "Aliasing: A source of streaks in computed tomograms," *J. Comput. Assisted Tomog.*, **3**, 511–518.

Brooks, R. A., M. R. Keller, C. M. O'Connor, and W. T. Sheridan (1980). "Progress toward quantitative computed tomography," *IEEE Trans. Nucl. Sci.*, **NS-27**(3), 1121–1127.

Brooks, R. A. V. J. Sank, W. S. Friauf, S. B. Leighton, H. E. Cascio, and G. Di Chiro (1981). "Design considerations for positron emission tomography," *IEEE Trans. Biomed. Eng.*, **BME-28**, 158–177.

Brooks, R. A., V. J. Sank, A. J. Talbert, and G. Di Chiro (1979). "Sampling requirements and detector motion for positron emission tomography," *IEEE Trans. Nucl. Sci.*, **NS-26**(2), 2760–2763.

Brooks, R. A., G. H. Weiss, and A. J. Talbert (1978). "A new approach to interpolation in computed tomography," *J. Comput. Assisted Tomog.*, **2**, 577–585.

Brouw, W. N. (1975). "Aperture synthesis," in Vol. 14 of *Methods of Computational Physics*, B. Alder, S. Fernbach, and M. Rotenberg, Eds. Academic, New York, pp. 131–175.

Brownell, G. L. and S. Cochavi (1978). "Transverse section imaging with carbon-11 labeled carbon monoxide," *J. Comput. Assisted Tomog.*, **2**, 533–538.

Brownell, G. L., G. A. Correia, and R. G. Zamenhof (1978). "Positron instrumentation," in Vol. 5 of *Recent Advances in Nuclear Medicine*, J. H. Lawrence and T. F. Budinger, Eds., Grune & Stratton, New York, pp. 1–49.

Brunner, P. and R. R. Ernst (1979). "Sensitivity and performance time in NMR imaging," *J. Magn. Reson.*, **33**, 83–106.

Budinger, T. F. (1979a). "Physiology and physics of nuclear cardiology," in *Nuclear Cardiology*, J. T. Willerson, Ed. Cardiovascular Clinics, Davis, Philadelphia, pp. 9–78.

Budinger, T. F. (1979b). "Thresholds for physiological effects due to RF and magnetic fields used in NMR imaging," *IEEE Trans. Nucl. Sci.*, **NS-26**(2), 2821–2825.

Budinger, T. F., and G. T. Gullberg (1974). "Three-dimensional reconstruction in nuclear medicine emission imaging," *IEEE Trans. Nucl. Sci.*, **NS-21**(3), 2–20.

Budinger, T. F. and G. T. Gullberg (1975). "Reconstruction by two-dimensional filtering of simple superposition transverse image," in *Image Processing for 2-D and 3-D Reconstructions from Projections: Theory and Practice in Medicine and the Physical Sciences*, Technical Digest,

August 4–7 (1975), Stanford University, Optical Society of America, Washington, DC, pp. ThA9-1–ThA9-4.

Budinger, T. F. and G. T. Gullberg (1977). "Transverse section reconstruction of gamma-ray emitting radionuclides in patients," in *Reconstruction Tomography in Diagnostic Radiology and Nuclear Medicine*, M. M. Ter-Pogossian et al., Eds., University Park Press, Baltimore, MD, pp. 315–342.

Budinger, T. F., S. E. Derenzo, G. T. Gullberg, W. L. Greenberg, and R. H. Huesman (1977). "Emission computer assisted tomography with single-photon and positron annihilation photon emitters," *J. Comput. Assisted Tomog.*, **1**, 131–145.

Budinger, T. F., S. E. Derenzo, G. T. Gullberg, and R. H. Huesman (1979). "Trends and prospects for circular ring positron cameras," *IEEE Trans. Nucl. Sci.*, **NS-26**(2), 2742–2745.

Budinger, T. F., G. T. Gullberg, and R. H. Huesman (1979). "Emission computed tomography," in *Image Reconstruction from Projections*, G. T. Herman, Ed., Vol. 32 of *Topics in Applied Physics*, Springer-Verlag, New York, pp. 147–246.

Buhl, D. (1973). "Molecular clouds in the galaxy," *Proc. IEEE*, **61**, 1198–1204.

Buonocore, M. H., W. R. Brody, and A. Macovski (1981). "A natural pixel decomposition for two-dimensional image reconstruction," *IEEE Trans. Biomed. Eng.*, **BME-28**, 69–78.

Bureau, F. J. (1955). "Divergent integrals and partial differential equations," *Comm. Pure Appl. Math.*, **8**, 143–202.

Byer, R. L. and L. A. Shepp (1979). "Two-dimensional remote air-pollution monitoring via tomography," *Optics Letters*, **4**, 75–77.

Carson, P. L., D. E. Dick, G. A. Thieme, M. L. Dick, E. J. Bayly, T. V. Oughton, G. L. Dubuque, and H. P. Bay (1978). "Initial investigation of computed tomography for breast imaging with focused ultrasound beams," in Vol. 4 of *Ultrasound in Medicine*, D. White and E. A. Lyons, Eds., Plenum, New York, pp. 319–322.

Cavaretta, A. S., Jr., C. A. Micchelli, and A. Sharma (1980a). "Multivariate interpolation and the Radon transform," *Math. Z.*, **174**, 263–279.

Cavaretta, A. S., Jr., C. A. Micchelli, and A. Sharma (1980b). "Multivariate interpolation and the Radon transform, Part II, some further examples," in *Quantitative Approximation*, R. A. DeVore and K. Scherer, Eds., Academic, New York, pp. 49–62.

Censor, Y., D. E. Gustafson, A. Lent, and H. Tuy (1979). "A new approach to the emission computerized tomography problem: Simultaneous calculation of attenuation and activity coefficients," *IEEE Trans. Nucl. Sci.*, **NS-26**(2), 2775–2779.

Cha, S. and C. M. Vest (1979). "Interferometry and reconstruction of strongly refracting asymmetric-refractive-index fields," *Optics Letters*, **4**, 311–313.

Challifour, J. L. (1972). *Generalized Functions and Fourier Analysis*, Benjamin, New York.

Chang, L. T. (1978). "A method for attenuation correction in radionuclide computed tomography," *IEEE Trans. Nucl. Sci.*, **NS-25**(1), 638–643.

Chang, L. T. (1979). "Attenuation correction and incomplete projection in single photon emission computed tomography," *IEEE Trans. Nucl. Sci.*, **NS-26**(2), 2780–2789.

Chapman, C. H. (1977). "Body waves in seismology," in *Modern Problems in Elastic Wave Propagation*, J. Miklowitz and J. D. Achenbach, Eds., Wiley-Interscience, New York, pp. 477–498.

Chapman, C. H. (1978). "A new method for computing synthetic seismograms," *Geophys. J. R. Astron. Soc.*, **54**, 481–518.

Chapman, C. H. (1979). "On impulsive wave propagation in a spherically symmetric model," *Geophys. J. R. Astron. Soc.*, **58**, 229–234.

Chapman, C. H. (1981a). "Long-period corrections to body waves: Theory." *Geophys. J. R. Astron. Soc.*, **64**, 321–372.

Chapman, C. H. (1981b). "Generalized Radon transforms and slant stacks," *Geophys. J. R. Astron. Soc.*, **66**, 455–460.

Chen, Y. W. (1978). "Solution of the wave equation in R^{n+1} with data on a characteristic cone," *J. Math. Anal. Appl.*, **64**, 223–249.

Chesler, D. A. (1973). "Positron tomography and three-dimensional reconstruction techniques," in *Tomographic Imaging in Nuclear Medicine*, G. S. Freedman, Ed., Society of Nuclear Medicine, New York, pp. 176–183.

Chesler, D. A. and S. J. Riederer (1975). "Ripple suppression during reconstruction in transverse tomography," *Phys. Med. Biol.*, **20**, 632–636.

Chesler, D. A., S. J. Riederer, and N. J. Pelc (1977). "Noise due to photon counting statistics in computed x-ray tomography," *J. Comput. Assisted Tomog.*, **1**, 64–74.

Chiu, M. Y., H. H. Barrett, R. G. Simpson, C. Chou, J. W. Arendt, and G. R. Gindi (1979). "Three-dimensional radiographic imaging with a restricted view angle," *J. Opt. Soc. Am.*, **69**, 1323–1333.

Chiu, M. Y., H. H. Barrett, and R. G. Simpson (1980). "Three-dimensional reconstruction from planar projections," *J. Opt. Soc. Am.*, **70**, 755–762.

Cho, Z. H. (1974). "General views on 3-D image reconstruction and computerized transverse axial tomography," *IEEE Trans. Nucl. Sci.*, **NS-21**(3), 44–71.

Cho, Z. H., I. Ahn, C. Bohm, and G. Huth (1974). "Computerized image reconstruction methods with multiple photon/x-ray transmission scanning," *Phys. Med. Biol.*, **19**, 511–522.

Cho, Z. H., J. K. Chan, E. L. Hall, R. P. Kruger, and D. G. McCaughey (1975). "A comparative study of 3-D image reconstruction algorithms with reference to number of projections and noise filtering," *IEEE Trans. Nucl. Sci.*, **NS-22**(1), 344–358.

Cho, Z. H., L. Eriksson, and J. Chan (1977). "A circular ring transverse axial positron camera," in *Reconstruction Tomography in Diagnostic Radiology and Nuclear Medicine*, M. M. Ter-Pogossian et al., Eds., University Park Press, Baltimore, MD, pp. 393–421.

Christiansen, W. N. (1973). "A new southern hemisphere synthesis radio telescope," *Proc. IEEE*, **61**, 1266–1270.

Christiansen, W. N. and J. A. Warburton (1953a). "The distribution of radio brightness over the solar disk at a wavelength of 21 centimeters. I. A new highly directional aerial system," *Aust. J. Phys.*, **6**, 190–202.

Christiansen, W. N. and J. A. Warburton (1953b). "The distribution of radio brightness over the solar disk at a wavelength of 21 centimeters. II. The quiet sun—One-dimensional observations," *Aust. J. Phys.*, **6**, 262–271.

Christiansen, W. N. and J. A. Warburton (1955). "The distribution of radio brightness over the solar disk at a wavelength of 21 centimeters. III. The quiet sun—Two-dimensional observations," *Aust. J. Phys.*, **8**, 472–486.

Chu, G. and K. C. Tam (1977). "Three-dimensional imaging in the positron camera using Fourier techniques," *Phys. Med. Biol.*, **22**, 245–265.

Clark, T. A. and W. C. Erickson (1973). "Long wavelength VLBI," *Proc. IEEE*, **61**, 1230–1233.

Clayton, R. W. and G. A. McMechan (1981). "Inversion of refraction data by wavefield continuation," *Geophysics*, **46**, 860–868.

Coen, S. (1981a). "The inverse problem of the shear modulus and density profiles of a layered earth," preprint, Dept. of Materials Science and Mineral Engineering, University of California, Berkeley, CA.

Coen, S. (1981b). "Inverse scattering method for determining the density and compressibility profiles of a layered medium. Part III: Point source," preprint, Dept. of Materials Science and Mineral Engineering, University of California, Berkeley, CA.

Cohen, M. H. (1973). "Introduction to very-long-baseline interferometry," *Proc. IEEE*, **61**, 1192–1197.

Cohen, M. and G. T. Toussaint (1977). "On the detection of structures in noisy pictures," *Pattern Recognition*, **9**, 95–98.

Colsher, J. G. (1977). "Iterative three-dimensional image reconstruction from tomographic projections," *Comput. Graph, Image Proc.*, **6**, 513–537.

Colsher, J. G. (1980). "Fully three-dimensional positron emission tomography," *Phys. Med. Biol.*, **25**, 103–115.

Condon, E. U. and G. H. Shortley (1970). *Theory of Atomic Spectra*, reprinted with corrections, Cambridge University, London.

Cook, B. D. (1975). "Proposed mapping of ultrasonic fields with conventional light diffraction," *J. Opt. Soc. Am.*, **65**, 682–684.

Cook, B. D. and W. I. Arnoult (1976). "Gaussian Laguerre/Hermite formulation for the nearfield of an ultrasonic transducer," *J. Acoust. Soc. Am.*, **59**, 9–11.

Cormack, A. M. (1963). "Representation of a function by its line integrals, with some radiological applications," *J. Appl. Phys.*, **34**, 2722–2727.

Cormack, A. M. (1964). "Representation of a function by its line integrals, with some radiological applications. II," *J. Appl. Phys.*, **35**, 2908–2913.

Cormack, A. M. (1973). "Reconstruction of densities from their projections, with applications in radiological physics," *Phys. Med. Biol.*, **18**, 195–207.

Cormack, A. M. (1978). "Sampling the Radon transform with beams of finite width," *Phys. Med. Biol.*, **23**, 1141–1148.

Cormack, A. M. (1980a). "Algorithms for two-dimensional reconstruction," *Phys. Med. Biol.*, **25**, 372.

Cormack, A. M. (1980b). Nobel Prize address, Dec. 8, 1979, "Early two-dimensional reconstruction and recent topics stemming from it," *Med. Phys.*, **7**, 277–282. (also in *J. Comput. Assisted Tomog.*, **4** 658–664)

Cormack, A. M. and B. J. Doyle (1977). "Algorithms for two-dimensional reconstruction," *Phys. Med. Biol.*, **22**, 994–997. [Erratum, *Phys. Med. Biol.*, **25**, 372 (1980).]

Cormack, A. M. and E. T. Quinto (1980). "A Radon transform on spheres through the origin in R^n and applications to the Darboux equation," *Trans. Am. Math. Soc.*, **260**, 575–581.

Counselman III, C. C. (1973). "Very-long-baseline interferometery techniques applied to problems of geodesy, geophysics, planetary science, astronomy, and general relativity," *Proc. IEEE*, **61**, 1225–1230.

Courant, R. and D. Hilbert (1953). *Methods of Mathematical Physics*, Vol. I, Interscience, New York.

Courant, R. and D. Hilbert (1962). *Methods of Mathematical Physics*, Vol. II, Interscience, New York.

Cramér, H. and H. Wold (1936). "Some theorems on distribution functions," *J. London Math. Soc.*, **11**, 290–294.

Crawford, C. R. and A. C. Kak (1979). "Aliasing artifacts in computerized tomography," *Appl. Opt.*, **18**, 3704–3711.

Crooks, L. E. (1980). "Selective irradiation line scan techniques for NMR imaging," *IEEE Trans. Nucl. Sci.*, **NS-27**(3), 1239–1244.

Crooks, L. E., J. Hoenninger, M. Arakawa, L. Kaufman, R. McRee, J. Watts, and J. R. Singer (1980). "Tomography of hydrogen with nuclear magnetic resonance," *Radiology*, **136**, 701–706.

Crowther, R. A. (1971). "Procedures for three-dimensional reconstruction of spherical viruses by Fourier synthesis from electron micrographs," *Philos. Trans. R. Soc. London Ser. B*, **261**, 221–230.

Crowther, R. A. and L. A. Amos (1971). "Harmonic analysis of electron microscope images with rotational symmetry," *J. Mol. Biol.*, **60**, 123–130.

Crowther, R. A. and A. Klug (1974). "Three-dimensional image reconstruction on an extended field—A fast stable algorithm," *Nature (London)*, **251**, 490–492.

Crowther, R. A. and A. Klug (1975). "Structural analysis of macromolecular assemblies by image reconstruction from electron micrographs," *Annu. Rev. Biochem.*, **44**, 161–182.

Crowther, R. A., L. A. Amos, J. T. Finch, D. J. DeRosier, and A. Klug (1970). "Three-dimensional reconstructions of spherical viruses by Fourier synthesis from electron micrographs," *Nature (London)*, **226**, 421–425.

Crowther, R. A., D. J. DeRosier, and A. Klug, (1970). "The reconstruction of a three-dimensional structure from projections and its application to electron microscopy," *Proc. R. Soc. London Ser. A.*, **317**, 319–340.

Crowther, R. A., L. A. Amos, and A. Klug (1972). "Three-dimensional image reconstruction using functional expansions," in *Proceedings of the Fifth European Congress on Electron Microscopy*, The Institute of Physics, London and Bristol, pp. 593–597.

D'Addario, L. R. (1979). "Discussion comment (1)," in *Image Formation from Coherence Functions in Astronomy*, C. van Schooneveld, Ed. Vol. 76 of *Astrophysics and Space Science Library*, D. Reidel, Dordrecht, Holland, p. 17.

Damadian, R. (1971). "Tumor detection by nuclear magnetic resonance," *Science*, **171**, 1151–1153.

Damadian, R. (1972). "Apparatus and methods for detecting cancer in tissue," U.S. Patent 3,789,832 (filed 17 March, 1972)

Damadian, R. (1980). "Field focusing NMR (FONAR) and the formation of chemical images in man," *Philos. Trans. R. Soc. London Ser. B*, **289**, 489–500.

Dändliker, R. and K. Weiss (1970). "Reconstruction of the three-dimensional refractive index from scattered waves," *Optics Commun.*, **1**, 323–328.

Das, Y. and W.-M. Boerner (1978). "On radar target shape estimation using algorithms for reconstruction from projections," *IEEE Trans. Antennas Propag.*, **AP-26**, 274–279.

Davison, M. E. (1981a). "The ill-conditioned nature of the limited angle tomography problem," *SIAM J. Appl. Math* (to be published).

Davison, M. E. (1981b). "A singular value decomposition for the Radon transform in *n*-dimensional Euclidean space," Ames Laboratory Report IS-J-553, Applied Mathematical Science, Ames Laboratory, Iowa State University, Ames, IA 50011.

Davison, M. E. and F. A. Grünbaum (1979). "Convolution algorithms for arbitrary projection angles," *IEEE Trans. Nucl. Sci.*, **NS-26**(2), 2670–2673.

Davison, M. E. and F. A. Grünbaum (1981). "Tomographic reconstructions with arbitrary directions," *Comm. Pure Appl. Math.*, **34**, 77–119.

Deans, S. R. (1977). "The Radon transform: Some remarks and formulas for two dimensions," Lawrence Berkeley Laboratory report LBL-5691, Berkeley, CA.

Deans, S. R. (1978). "A unified Radon inversion formula," *J. Math. Phys.*, **19**, 2346–2349.

Deans, S. R. (1979). "Gegenbauer transforms via the Radon transform," *SIAM J. Math. Anal.*, **10**, 577–585.

Deans, S. R. (1981). "Hough transform from the Radon transform," *IEEE Trans. Pattern Analysis and Machine Intelligence*, **PAMI-3**, 185–188.

DeJager, E. M. (1969). *Applications of Distributions in Mathematical Physics*, 2nd ed., Mathematisch Centrum, Amsterdam.

DeJong, M. L. (1966). "Structure of 3C444 from observations of lunar occultations," *Astron. J.*, **71**, 373–376.

Delannoy, J., J. Lacroix, and E.-J. Blum (1973). "An 8-mm interferometer for solar radio astronomy at Bordeaux, France," *Proc. IEEE*, **61**, 1282–1284.

Denton, R. V., B. Friedlander, and A. J. Rockmore (1979). "Direct three-dimensional image reconstruction from divergent rays," *IEEE Trans. Nucl. Sci.*, **NS-26**(5), 4695–4703.

Derenzo, S. E. (1977). "Positron ring cameras for emission-computed tomography," *IEEE Trans. Nucl. Sci.*, **NS-24**(2), 881–885.

Derenzo, S. E., T. F. Budinger, J. L. Cahoon, R. H. Huesman, and H. G. Jackson. "High resolution computed tomography of positron emitters," *IEEE Trans. Nucl. Sci.*, **NS-24**(1), 544–558.

DeRosier, D. J. (1971). "The reconstruction of three-dimensional images from electron micrographs," *Contemp. Phys.*, **12**, 437–452.

DeRosier, D. J. and A. Klug (1968). "Reconstruction of three-dimensional structures from electron micrographs," *Nature (London)*, **217**, 130–134.

DeRosier, D. J. and P. B. Moore (1970). "Reconstruction of three-dimensional images from electron micrographs of structures with helical symmetry," *J. Mol. Biol.*, **52**, 355–369.

Dey-Sarkar, S. K. and C. H. Chapman (1978). "A simple method for the computation of body-wave seismograms," *Bull. Seismol. Soc. Am.*, **68**, 1577–1593.

Di Chiro, G. and R. A. Brooks (1979). "The 1979 Nobel Prize in Physiology or Medicine," *Science*, **206**, 1060–1062.

Di Chiro, G. and R. A. Brooks (1980). "The 1979 Nobel Prize in Physiology or Medicine," *J. Comput. Assisted Tomog.*, **4**, 241–245.

Dick, D. E., P. L. Carson, E. J. Bayly, T. V. Oughton, J. E. Kubichek, and F. L. Kitson (1977). "Technical evaluation of an ultrasound CT scanner," in *1977 Ultrasonic Symposium Proceedings*, October 26–28, 1977, Phoenix, Arizona (IEEE Cat. #77CH1264-1SU), IEEE, New York, pp. 176–181.

Dick, D. E., R. D. Elliott, R. L. Metz, and D. S. Rojohn (1979). "A new automated, high resolution ultrasound breast scanner," *Ultrasonic Imaging*, **1**, 368–377.

Dines, K. A. and A. C. Kak (1979). "Ultrasonic attenuation tomography of soft tissue," *Ultrasonic Imaging*, **1**, 16–33.

Dines, K. A. and R. J. Lytle (1979). "Computerized geophysical tomography," *Proc. IEEE*, **67**, 1065–1073. [Correction: *Proc. IEEE*, **67**, 1679 (1979).]

Dirac, P. A. M. (1927). "The physical interpretation of the quantum dynamics," *Proc. R. Soc. London*, **113A**, 621–641.

Dirac, P. A. M. (1930, 1958). *Quantum Mechanics*, 1st and 4th ed., Oxford University, London and New York.

Dreike, P. and D. P. Boyd (1976). "Convolution reconstruction of fan beam projections," *Comput. Graph. Image Proc.*, **5**, 459–469.

Duck, F. A. and C. R. Hill (1979). "Acoustic attenuation reconstruction from back-scattered ultrasound," in *Computer Aided Tomography and Ultrasound in Medicine*, J. Raviv, J. F. Greenleaf, and G. T. Herman, Eds., North-Holland, Amsterdam, New York, and Oxford, pp. 137–149.

Duda, R. O. and P. E. Hart (1972). "Use of the Hough transform to detect lines and curves in pictures," *Commun. Assoc. Comput. Mach.*, **15**, 11–15.

Duda, R. O., D. Nitzan, and P. Barrett (1979). "Use of range and reflectance data to find planar surface regions," *IEEE Trans. Pattern Analysis and Machine Intelligence*, **PAMI-1**, 259–271.

Dudany, S. and A. L. Luk (1978). "Locating straight-line edge segments on outdoor scenes," *Pattern Recognition*, **10**, 145–157.

Duerinckx, A. J. and A. Macovski (1978). "Polychromatic streak artifacts in computed tomography images," *J. Comput. Assisted Tomog.*, **2**, 481–487.

Duerinckx, A. J. and A. Macovski (1979a). "Nonlinear polychromatic and noise artifacts in x-ray computed tomography images," *J. Comput. Assisted Tomog.*, **3**, 519–526.

Duerinckx, A. J. and A. Macovski (1979b). "Classification of artifacts in x-ray CT images due to nonlinear shadows," *IEEE Trans. Nucl. Sci.*, **NS-26**(2), 2848–2852.

Duerinckx, A. J. and A. Macovski (1980). "Information and artifact in computed tomography image statistics," *Med. Phys.*, **7**, 127–134.

Durrani, T. S. and C. E. Goutis (1980). "Optimization techniques for digital image reconstruction from projections," *IEEE Proc.*, **127**, Part E. 161–169.

Edelheit, L. S., G. T. Herman, and A. V. Lakshminarayanan (1977). "Reconstruction of objects from diverging x-rays," *Med. Phys.*, **4**, 226–231.

Eggermont, P. P. B. (1975). "Three-dimensional image reconstruction by means of two-dimensional Radon inversion," T.H.-Report 75-WSK-04, Dept. of Mathematics, Technological University Eindhoven, The Netherlands.

Ein-Gal, M. (1974). "The shadow transform: An approach to cross-sectional imaging," Technical Report SEL-74-050, Information Systems Laboratory, Stanford University, Stanford, CA.

Ellingson, W. A. and H. Berger (1980). "Three-dimensional radiographic imaging," in Vol. 4 of *Research Techniques in Nondestructive Testing*, R. S. Sharpe, Ed., Academic, New York, pp. 1–38.

Elliot, J. L. (1979): "Stellar occultation studies of the solar system," *Annu. Rev. Astron. Astrophys.*, **17**, 445–475.

Elliot, J. L., J. Veverka, and J. Goguen (1975). "Lunar occultation of Saturn I. The diameters of Tethys, Dione, Rhea, Titan, and Iapetus," *Icarus*, **26**, 387–407.

Erdélyi, A. (1961). "From delta functions to distributions," in *Modern Mathematics for the Engineer*, 2nd series, E. F. Beckenback, Ed. McGraw-Hill, New York.

Erdélyi, A. (1962). *Operational Calculus and Generalized Functions*, Holt, Rinehart and Winston, New York.

Erdélyi, A., W. Magnus, F. Oberhettinger, and F. G. Tricomi (1953a). *Higher Transcendental Functions*, Vol. I, McGraw-Hill, New York.

Erdélyi, A., W. Magnus, F. Oberhettinger, and F. G. Tricomi (1953b). *Higher Transcendental Functions*, Vol. II, McGraw-Hill, New York.

Erdélyi, A., W. Magnus, F. Oberhettinger, and F. G. Tricomi (1954). *Tables of Integral Transforms*, Vol. II, McGraw-Hill, New York.

Erickson, W. C. (1973). "The Clark Lake array," *Proc. IEEE*, **61**, 1276–1277.

Evens, R. G. (1980). "Nuclear magnetic resonance: Another new frontier for radiology?" *Radiology*, **136**, 795–796.

Falconer, K. J. (1979). "Determination of a function from its projections," *Math. Proc. Camb. Philos. Soc.*, **85**, 351–355.

Farrar, T. C. and A. D. Becker (1971). *Pulse and Fourier Transform NMR*, Academic, New York.

Farrell, E. J. (1978). "Processing limitations of ultrasonic image reconstruction," in *Proceedings of the 1978 IEEE Computer Society Conference on Pattern Recognition and Image Processing*, May 31–June 2 (1978), Chicago, Illinois, IEEE, New York, pp. 8–15.

Fitzgerald, J. and H. Hörster (1971). "The temperature distribution in gas-filled incandescent lamps," *Phillips Tech. Rev.*, **32**, 206–209.

Fjeldbo, G., A. Kliore, B. Seidel, D. Sweetnam, and P. Woiceshyn (1976). "The Pioneer II radio occultation measurements of the Jovian ionosphere," in *Jupiter*, T. Gehrels, Ed., University of Arizona, Tucson, pp. 238–246.

Fomalont, E. B. (1973). "Earth-rotation aperture synthesis," *Proc. IEEE*, **61**, 1211–1218.

Fomalont, E. D. (1979). "Fundamentals and deficiencies of aperture synthesis," in *Image Formation from Coherence Functions in Astronomy*, C. van Schooneveld, Ed., Vol. 76 of *Astrophysics and Space Sciences Library*, Reidel, Dordrecht, Holland, pp. 3–18.

Forgues, P. M., M. Goldberg, A. Smith, and S. S. Stuchly (1980). "Medical computed tomography using microwaves," in *IEEE 1980 Frontiers of Engineering in Health Care*, IEEE/Engineering in Medicine and Biology Society Second Annual Conference, Sept. 28–30, Washington, D.C., pp. 270–273.

Frazier, L. N. and R. A. Phinney (1980). "The theory of finite frequency body wave synthetic seismograms in inhomogeneous elastic media," *Geophys. J. R. Astron. Soc.*, **63**, 691–717.

Frei, W. (1978). "Display processing of computed tomography images," *IEEE Trans. Nucl. Sci.*, **NS-25**(2), 939–943.

Friedman, A. (1963). *Generalized Functions and Partial Differential Equations*, Prentice-Hall, Englewood Cliffs, NJ.

Friedman, B. (1969), *Lectures on Applications—Oriented Mathematics*, V. Twersky, Ed., Holden-Day, San Francisco, CA.

Friedman, M. I., J. W. Beattie, and J. S. Laughlin (1974). "Cross-sectional absorption density reconstruction for treatment planning," *Phys. Med. Biol.*, **19**, 819–830.

Friedrich, J. (1959). "Zur Auswertung seitlicher Beobachtungen and zylindrischen Bögen." *Ann. Phys. (Leipzig)*, **3**, Ser. 7, 327–333.

Fuglede, B. (1958). "An integral formula," *Math. Scand.*, **6**, 207–212.

Funk, P. (1913). "Über Flächen mit Lauter geschlossnen geodätischen Linien," *Math. Ann.*, **74**, 278–300.

Funk, P. (1916). "Über eine geometrische Anwendung der Abelschen Integralgleichung," *Math. Ann.*, **77**, 129–135.

Gabillard, R. (1951a). "Dispositif simplifié pour l'étude de l'absorption paramegnétique nucléaire," *Comptes Rendus*, **232**, 324–326.

Gabillard, R. (1951b). "Interprétation theorique de la forme des signaux de résonance nucléaire dans le cas le plus général," *Comptes Rendus*, **232**, 1477–1479.

Gabillard, R. (1951c). "Mesure du temps de relaxation T_2 en présence d'une inhomogénéite de champ magnétique supérieure á la largeur de raie," *Comptes Rendus*, **232**, 1551–1553.

Gabillard, R. (1952). "A steady state transient technique in nuclear resonance," *Phys. Rev.*, **85**, 694–695.

Gabor, D. (1948). "A new microscopic principle," *Nature (London)*, **161**, 777–778.

Gabor, D. (1949). "Microscopy by reconstructed wave fronts," *Proc. R. Soc. London Ser. A*, **197**, 454–487.

Gabor, D. (1951). "Microscopy by reconstructed wave fronts, II." *Proc. Phys. Soc. London Ser. B*, **64**, 449–469.

Garmany, J., J. A. Orcutt, and R. L. Parker (1979). "Travel time inversion: A geometrical approach," *J. Geophys. Res.*, **84**, 3615–3622.

Gel'fand, I. M. and G. E. Shilov (1964). *Generalized Functions*, Vol. 1, Academic, New York.

Gel'fand, I. M., M. I. Graev, and N. Ya. Vilenkin (1966). *Generalized Functions*, Vol. 5, Academic, New York.

Gilbert, B. K., S. K. Kenue, R. A. Robb, A. Chu, A. H. Lent, and E. E. Swartzlander (1981). "Rapid execution of fan beam image reconstruction algorithms using efficient computational techniques and special-purpose processors," *IEEE Trans. Biomed. Eng.*, **BME-28**, 98–116.

Gilbert, P. F. C. (1972a). "The reconstruction of a three-dimensional structure from projections and its application to electron microscopy II. Direct methods," *Proc. R. Soc. London Ser B*, **182**, 89–102.

Gilbert, P. F. C. (1972b). "Iterative methods for the three-dimensional reconstruction of an object from projections," *J. Theor. Biol.*, **36**, 105–117.

Gilbert, W. M. (1955). "Projections of probability distributions," *Acta Math. Acad. Sci. Budapest*, **6**, 195–198.

Glover, G. H. and R. L. Eisner (1979a). "Theoretical resolution of computed tomography systems." *J. Comput. Assisted Tomog.*, **3**, 85–91.

Glover, G. H. and R. L. Eisner (1979b). "Consistent projection sets: Fan-beam *geometry*," in *Computer Aided Tomography and Ultrasonics in Medicine*, J. Raviv, J. F. Greenleaf, and G. T. Herman, Eds., North-Holland, Amsterdam, New York, and Oxford, pp. 235–251.

Glover, G. H. and J. C. Sharp (1977). "Reconstruction of ultrasound propagation speed distributions in solf tissue: Time-of-flight tomography," *IEEE Trans. Sonics Ultrason.*, **SU-24**, 229–234.

Goitein, M. (1972). "Three-dimensional density reconstruction from a series of two-dimensional projections," *Nucl. Instrum. Methods*, **101**, 509–518.

Good, I. J. (1970). "The interpretation of x-ray shadowgraphs," *Phys. Lett.*, **31A**, 155.

Gooden, J. S. (1950). "Nuclear resonance and magnetic field changes of 1 in 10^6," *Nature (London)*, **165**, 1014–1015.

Goodman, J. W. (1968). *Introduction to Fourier Optics*, McGraw-Hill, New York.

Gordon, R. (1974). "A tutorial on ART," *IEEE Trans. Nucl. Sci.*, **NS21**(3), 78–93.

Gordon, R. (1976). "Dose reduction in computerized tomography," *Invest. Radiol.*, **11**, 508–517.

Gordon, R. (1979). "Reconstruction from projections in medicine and biology," in *Image Formation from Coherence Functions in Astronomy*, C. van Schooneveld, Ed. Vol. 76 of *Astrophysics and Space Sciences Library*, Reidel, Dordrecht, Holland, pp. 317–325.

Gordon, R. and G. T. Herman (1971). "Reconstruction of pictures from their projections." *Commun. ACM.*, **14**, 759–768.

Gordon, R. and G. T. Herman (1974). "Three-dimensional reconstruction from projections: A review of algorithms," *International Review of Cytology*, **38**, 111–151.

Gordon, R., R. Bender, and G. T. Herman (1970). "Algebraic reconstruction technique (ART) for three-dimensional microscopy and x-ray photography," *J. Theor. Biol.*, **29**, 471–481.

Gordon, R., G. T. Herman, and S. A. Johnson (1975). "Image reconstruction from projections," *Sci. Am.*, **233(4)**, October, 56–68.

Gore, J. C. and P. S. Tofts (1978). "Statistical limitations in computed tomography," *Phys. Med. Biol.*, **23**, 1176–1182.

Gradshteyn, I. S., and I. M. Ryzhik (1965). *Table of Integrals, Series, and Products*, 4th ed., Academic, New York.

Green, J. W. (1958). "On the determination of a function in the plane by its integrals over straight lines." *Proc. Am. Math. Soc.*, **9**, 758–762.

Greenleaf, J. F. and R. C. Bahn (1981). "Clinical imaging with transmissive ultrasonic computerized tomography," *IEEE Trans. Biomed. Eng.*, **BME-28**, 177–185.

Greenleaf, J. F., S. A. Johnson, S. L. Lee, G. T. Herman, and E. H. Wood (1974). "Algebraic reconstruction of spatial distributions of acoustic absorption within tissue from their two-dimensional acoustic projections," in Vol. 5 of *Acoustical Holography*, P. S. Green, Ed. Plenum, New York, pp. 591–603.

Greenleaf, J. F., S. A. Johnson, W. F. Samayoa, and F. A. Duck (1975). "Algebraic reconstruction of spatial distributions of acoustic velocities in tissue from their time-of-flight profiles," in Vol. 6 of *Acoustical Holography*, N. Booth, Ed., Plenum, New York, pp. 71–90.

Greenleaf, J. F., S. A. Johnson, and A. H. Lent (1978). "Measurement of spatial distribution of refractive index in tissues by ultrasonic computer assisted tomography," *Ultrasound in Med. & Biol.*, **3**, 327–339.

Greenleaf, J. F., S. A. Johnson, R. C. Bahn, B. Rajagopalan, and S. Kenue (1979). "Introduction to computed ultrasound tomography," in *Computer Aided Tomography and Ultrasonics in Medicine*, J. Raviv, J. F. Greenleaf, and G. T. Herman, Eds., North-Holland, Amsterdam, pp. 125–136.

Greenleaf, J. F., S. K. Kenue, B. Rajagopalan, R. C. Bahn, and S. A. Johnson (1980). "Breast imaging by ultrasonic computer-assisted tomography," in Vol. 8 of *Acoustical Imaging*, A. F. Metherell, Ed., Plenum, New York, pp. 599–614.

Groen, F. C. A., P. W. Verbeek, G. A. van Zee, and A. Oosterlinck (1976). "Some aspects concerning the computation of chromosome banding profiles." in *Proceedings of the Third International Joint Conference on Pattern Recognition*, Nov. 8–11, Coronado, California, IEEE, Piscataway, NJ, pp. 547–550.

Guelin, M. (1973). "Pulsers as probes of the interstellar medium," *Proc. IEEE*, **61**, 1298–1302.

Guenther, R. B., C. W. Kerber, E. K. Killian, K. T. Smith, and S. L. Wagner (1974). "Reconstruction of objects from radiographs and location of brain tumors," *Proc. Natl. Acad. Sci. USA*, **71**, 4884–4886.

Guillemin, V. and S. Sterberg (1977). *Geometric Asymptotics*, Mathematical Surveys, Number 14, American Mathematical Society, Providence, RI.

Gullberg, G. T. (1975). "Entropy and transverse section reconstruction," in *Information Processing in Scintigraphy*, C. Raynaud and A. Todd-Pokropek, Eds., Proceedings of the Fourth International Conference, July 15–16, Orsay, France, pp. 325–332.

Gullberg, G. T. (1976). "Entropy and transverse section reconstruction, *International J. Nucl. Med. Biol.*, **3**, 170–171.

Gullberg, G. T. (1979a). "The reconstruction of fan-beam data by filtering the back-projection," *Comput. Graph. Image Proc.*, **10**, 30–47.

Gullberg, G. T. (1979b). *The Attenuated Radon Transform: Theory and Application in Medicine and Biology*. Ph.D. thesis, University of California, Berkeley, CA (Lawrence Berkeley Laboratory report LBL-7486, Berkeley, CA).

Gullberg, G. T. (1980). "The attenuated Radon transform: Application to single-photon emission computed tomography in the presence of a variable attenuating medium," Lawrence Berkeley Laboratory report LBL-10276, Berkeley, CA.

Gullberg, G. T. and T. F. Budinger (1980). "Single-photon emission computed tomography: Compensation for constant attenuation." Lawrence Berkeley Laboratory report LBL-10713, Berkeley, CA.

Gullberg, G. T. and T. F. Budinger (1981). "The use of filtering methods to compensate for constant attenuation in single-photon emission computed tomography," *IEEE Trans. Biomed. Eng.*, **BME-28**, 142–157.

Gustafson, D. E., M. J. Berggren, M. Singh, and M. K. Dewanjec (1978). "Computed transaxial imaging using single gamma emitters," *Radiology*, **129**, 187–194.

Hachenberg, O., B. H. Grahl, and R. Wielebinski (1973). "The 100-meter radio telescope at Effelsberg," *Proc. IEEE*, **61**, 1288–1295.

Hagfors, T. and D. C. Campbell (1973). "Mapping of planetary surfaces by radar," *Proc. IEEE*, **61**, 1219–1225.

Hagfors, T., B. Nanni, and K. Stone (1968). "Aperture synthesis in radar astronomy and some applications to lunar and planetary studies," *Radio Sci.*, **3**, 491–509.

Hall, E. L. (1979). *Computer Image Processing and Recognition*, Academic, New York.

Halperin, I. (1952). *Introduction to the Theory of Distributions*, University of Toronto, Toronto.

Hamaker, C. and D. C. Solmon (1978). "The angles between the null spaces of x-rays," *J. Math. Anal. Appl.*, **62**, 1–23.

Hamaker, C. K. T. Smith, D. C. Solmon, and S. L. Wagnor, (1980). "The divergent beam x-ray transform," *Rocky Mountain J. Math.*, **10**, 253–283.

Hamming, R. W. (1977). *Digital Filters*, Prentice-Hall, Englewood Cliffs, NJ.

Hansen, E. W. (1981). "Theory of circular harmonic image reconstruction," *J. Opt. Soc. Am.*, **71**, 304–308.

Hansen, G. L. E. Crooks, P. Davis, J. DeGroot, R. Herfkens, A. R. Margulis, C. Gooding, L. Kaufman, J. Hoenninger, M. Arakawa, R. McRee, and J. Watts (1980). "In vivo imaging of the rat anatomy with nuclear magnetic resonance, *Radiology*, **136**, 695–700.

Hanson, K. M. (1977). "Detectability in the presence of computed tomographic reconstruction noise," *Proc. SPIE*, **127**, 304–312.

Hanson, K. M. (1979a). "Detectability in computed tomographic images, *Med. Phys.*, **6**, 441–451.

Hanson, K. M. (1979b). "Proton computed tomography," in *Computer Aided Tomography and Ultrasonics in Medicine*, J. Raviv, J. F. Greenleaf, and G. T. Herman, Eds. North-Holland, Amsterdam, New York, and Oxford, pp. 97–106.

Hanson, K. M. (1980). "On the optimality of the filtered backprojection algorithm," *J. Comput. Assisted Tomog.*, **4**, 361–363.

Hanson, K. M., J. N. Bradbury, T. M. Cannon, R. L. Hutson, D. B. Laubacher, R. Macek, M. A. Paciotti, and C. A. Taylor (1978). "The application of protons to computed tomography," *IEEE Trans. Nucl. Sci.*, **NS-25**(1), 657–660.

Harris, F. J. (1978). "On the use of windows for harmonic analysis with the discrete Fourier transform," *Proc. IEEE*, **66**, 51–83.

Hart, R. G. (1968). "Electron microscopy of unstained biological materials: The polytropic montage," *Science*, **159**, 1464–1467.

Havlice, J. F. and J. C. Taenzer (1979). "Medical ultrasonic imaging: An overview of principles and instrumentation," *Proc. IEEE*, **67**, 620–641.

Hawkes, P. W. (1978). "Electron image processing: A survey," *Comput. Graphics Image Proc.*, **8**, 406–446.

Hawkes, R. C., G. N. Holland, W. S. Moore, and B. S. Worthington (1980). "Nuclear magnetic resonance (NMR) tomography of the brain: A preliminary clinical assessment with demonstration of pathology," *J. Comput. Assisted Tomog.*, **4**, 577–586.

Hazard, C. (1962). "The method of lunar occultations and its application to a survey of the radio source 3C 212," *Mon. Not. R. Astron. Soc.*, **124**, 343–357.

Hazard, C., M. B. Mackey, and A. J. Shimmins (1963). "Investigation of the radio source 3C273 by the method of lunar occultations," *Nature (London)*, **197**, 1037–1039.

Heflinger, L. O., R. F. Wuerker, and R. E. Brooks (1966). "Holographic interferometry," *J. Appl. Phys.*, **37**, 642–649.

Helgason, S. (1965). "The Radon transform on Euclidean spaces, compact two-point homogeneous spaces and Grassmann manifolds," *Acta. Math.*, **113**, 153–180.

Helgason, S. (1973). "Functions on symmetric spaces," in *Harmonic Analysis on Homogeneous Spaces, Proceedings of Symposia in Pure Mathematics*, Vol. 26, American Mathematical Society, Providence.

Helgason, S. (1980). *The Radon Transform*, Birkhäuser, Boston, Basel, Stuttgart.

Helmberger, D. V. and L. J. Burdick (1979). "Synthetic seismograms," *Annu. Rev. Earth Planet. Sci.*, **7**, 417–442.

Henrich, G. (1980). "A simple computational method for reducing streak artifacts in CT images," *Computerized Tomography*, **4**, 67–71.

Herlitz, S. I. (1963). "A method for computing the emission distribution in cylindrical light sources," *Ark. Fys.*, **23**, 571–574.

Herman, G. T. (1972). "Two direct methods for reconstructing pictures from their projections: A comparative study," *Comput. Graph. Image Proc.*, **1**, 123–144.

Herman, G. T. (1979). "On modifications to the algebraic reconstruction techniques," *Comput. Biol. Med.*, **9**, 271–276.

Herman, G. T. (1980a). "On the noise in images produced by computed tomography." *Comput. Graph. Image Proc.*, **12**, 271–285.

Herman, G. T. (1980b). *Image Reconstruction from Projections: The Fundamentals of Computerized Tomography*, Academic, New York.

Herman, G. T. and A. Lent (1976a). "Iterative reconstruction algorithms," *Comput. Biol. Med.*, **6**, 273–294.

Herman, G. T. and A. Lent (1976b). "A computer implementation of a Bayesian analysis of image reconstruction," *Inf. Control*, **31**, 364–384.

Herman, G. T. and A. Lent (1976c). "Quadratic optimization for image reconstruction I," *Comput. Graph. Image Proc.*, **5**, 319–332.

Herman, G. T. and A. Naparstek (1977). "Fast image reconstruction based on a Radon inversion formula appropriate for rapidly collected data," *SIAM J. Appl. Math.*, **33**, 511–533.

Herman, G. T. and F. Natterer (1981). *Mathematical Aspects of Computerized Tomography*, Vol. 8 of *Lecture Notes in Medical Informatics*, G. T. Herman and F. Natterer, Eds., Springer-Verlag, Berlin, Heidelberg, and New York.

Herman, G. T. and S. Rowland (1971). "Resolution in ART: An experimental investigation of the resolving power of an algebraic picture reconstruction technique," *J. Theor. Biol.*, **33**, 213–223.

Herman, G. T. and S. W. Rowland (1973). "Three methods for reconstructing objects from x-rays: A comparative study." *Comp. Graph. Image Proc.*, **2**, 151–178.

Herman, G. T. and S. W. Rowland (1977). "SNARK 77: A programming system for the reconstruction of pictures and projections," *Technical Report 130*, Department of Computer Science, SUNY—Buffalo, Amherst, NY.

Herman, G. T., A. V. Lakshminarayanan, A. Naparstek, E. L. Ritman, R. A. Robb, and E. H. Wood (1976). "Rapid computerized tomography," in *Medical Data Processing*, M. Laudet, J. Anderson, and S. Begon, Eds., Taylor and Francis, London, pp. 581–598.

Herman, G. T., A. V. Lakshminarayanan, and A. Naparstek (1976). "Convolution reconstruction techniques for divergent beams," *Comput. Biol. Med.*, **6**, 259–271.

Herman, G. T., H. Hurwitz, and A. Lent (1977). "A Bayesian analysis of image reconstruction," in *Reconstruction Tomography in Diagnostic Radiology and Nuclear Medicine*, M. M. Ter.-Pogossian et al., Eds., University Park Press, Baltimore, MD, pp. 85–103.

Herman, G. T., A. V. Lakshminarayanan, and A. Naparstek (1977). "Reconstruction using divergent-ray shadowgraphs," in *Reconstruction Tomography in Diagnostic Radiology and Nuclear Medicine*, M. M. Ter-Pogossian et al., Eds., University Park Press, Baltimore, MD, pp. 105–117.

Herman, G. T., A. Lent, and P. H. Lutz (1978). "Relaxation methods for image reconstruction," *Commun. Assoc. Comput. Mach.*, **21**, 152–158. [Corrigendum, *Commun. Assoc. Comput. Mach.*, **21**, 872 (1978).]

Herman, G. T., H. Hurwitz, A. Lent, and H.-P. Lung (1979). "On the Bayesian approach to image reconstruction," *Inf. Control*, **42**, 60–71.

Herman, G. T., S. W. Rowland, and M. Yau (1979). "A comparative study of the use of linear and modified cubic spline interpolation for image reconstruction." *IEEE Trans. Nucl. Sci.*, **NS-26**(2), 2879–2894.

Hey, J. S. (1946). "Solar radiations in the 4–6 metre radio wave-length band," *Nature* (*London*), **156**, 47–48.

Hey, J. S. (1971). *The Radio Universe*, Pergamon, New York, p. 91.

Hildebrand, B. P. and D. E. Hufferd (1976). "Computerized reconstruction of ultrasonic velocity fields for mapping of residual stress," in Vol. 7 of *Acoustical Holography*, L. W. Kessler, Ed., Plenum, New York, pp. 245–262.

Hill, T. C., R. D. Lovett, and B. J. McNeil (1980). "Observations on the clinical value of emission tomography," *J. Nucl. Med.*, **21**, 613–616.

Hills, R. E., M. A. Janssen, D. D. Thornton, and W. J. Welch (1973). "The Hat Creek millimeter-wave interferometer." *Proc. IEEE*, **61**, 1278–1282.

Hinshaw, W. S. (1974a). "Spin mapping," *Phys. Lett. A*, **48**, 87–88.

Hinshaw, W. S. (1974b). "The application of time dependent filed gradients to NMR spin mapping," in *Proceedings of the Eighteenth Ampere Congress (Nottingham)*, P. S. Allen, E. R. Andrew, and C. A. Bates, Eds., North-Holland, Amsterdam, pp. 433–434.

Hinshaw, W. S. (1976). "Image formation by nuclear magnetic resonance: The sensitive point method," *J. Appl. Phys.*, **47**, 3709–3721.

Hochstadt, H. (1971). *The Functions of Mathematical Physics*, Wiley, New York.

Hochstadt, H. (1973). *Integral Equations*, Wiley-Interscience, New York.

Högbom, J. A. (1974). "Aperture synthesis with a non-regular distribution of interferometer baselines," *Astron. Astrophys. Suppl. Ser.*, **15**, 417–426.

Högbom, J. A. and W. N. Brouw (1974). "The synthesis radio telescope at Westerbork. Principles of operation, performance and data reduction," *Astron. Astrophys.*, **33**, 289–301.

Holden, J. E. and W. R. Ip (1978). "Continuous time-dependence in computed tomography," *Med. Phys.*, **5**, 485–490.

Holland, G. N., P. A. Bottomley, and W. S. Hinshaw (1977). "^{19}F magnetic resonance imaging," *J. Magn. Reson.*, **28**, 133–136.

Holland, G. N., R. C. Hawkes, and W. S. Moore (1980). "NMR tomography of the brain, coronal and sagittal sections," *J. Comput. Assisted Tomog.*, **4**, 429–433.

Holland, G. N., W. S. Moore, and R. C. Hawkes (1980). "Nuclear magnetic resonance tomography of the brain," *J. Comput. Assisted Tomog.*, **4**, 1–3.

Holley, W. R., R. P. Henke, G. E. Gauger, B. Jones, E. V. Benton, J. I. Fabrikant, and C. A. Tobias (1979). "Heavy particle computed tomography," in *Proceedings of the Sixth Conference on Computer Applications in Radiology and Computer/Aided Analysis of Radiological Images*, June 18–21 (1979), Newport Beach, California, IEEE, New York, pp. 64–70.

Hollis, D. P. (1980). "Nuclear magnetic resonance of phosphorus in the perfused heart," *IEEE Trans. Nucl. Sci.*, **NS-27**(3), 1250–1254.

Hoppe, W. (1979). "Three-dimensional electron microscopy of individual structures: Crystallography of "crystals" consisting of a single unit cell," *Chem. Scr.*, **14**, 227–243.

Hoppe, W. and R. Hegerl (1980). "Three-dimensional structure determination by electron microscopy (nonperiodic specimens)," in *Computer Processing of Electron Microscope Images*, P. W. Hawkes, Ed., Vol. 13 of *Topics in Current Physics*, Springer-Verlag, Berlin, Heidelberg, New York, pp. 127–185.

Hoppe, W. and Typke (1978). "Three-dimensional reconstruction of aperiodic objects in electron microscopy," in Vol. 7 of *Advances in Structure Research by Diffraction Method*, W. Hoppe and R. Mason, Eds., Pergamon, New York, pp. 137–190.

Hoppe, W., R. Langer, G. Knesch, and C. H. Poppe (1968). "Protein-Kristallstrukturanalyse mit elektronenstrahlen," *Naturwissenschaften*, **55**, 333–336.

Horman, M. H. (1965). "An application of wavefront reconstruction to interferometry," *Appl. Opt.*, **4**, 333–336.

Horn, B. K. P. (1978). "Density reconstruction using arbitrary ray-sampling schemes," *Proc. IEEE*, **66**, 551–562.

Horn, B. K. P. (1979). "Fan-beam reconstruction methods," *Proc. IEEE*, **67**, 1616–1623.

Hough, P. V. C. (1962). "Method and means for recognizing complex patterns," U. S. Patent 3 069 654.

Hoult, D. I. (1979). "Rotating frame zeugmatography." *J. Magn. Reson.*, **33**, 183–197.

Hoult, D. I. (1980a). "NMR imaging. Rotating frame selective pulses," *J. Magn. Reson.*, **38**, 369–374.

Hoult, D. I. (1980b). "Rotating frame zeugmatography," *Philos. Trans. R. Soc. London Ser. B*, **289**, 543–547.

Hoult, D. I. (1980c). "Medical imaging by NMR," in *Magnetic Resonance in Biology*, J. S. Cohen, Ed., Wiley-Interscience, New York, pp. 70–109.

Hounsfield, G. N. (1972). "A method of and apparatus for examination of a body by radiation such as x- or gamma-radiation." British patent No. 1283915, The Patent Office, London (filed 1968).

Hounsfield, G. N. (1973). "Computerized transverse axial scanning (tomography): Part 1. Description of system," *Br. J. Radiol.*, **46**, 1016–1022.

Hounsfield, G. N. (1980). Nobel Prize address, Dec. 8, 1979. "Computed medical imaging." *Med. Phys.*, **7**, 283–290. (also in *J. Comput. Assisted Tomog.*, **4**, 665–674 and *Science*, **210**, 22–28)

House, W. V. (1980). "Introduction to the principles of NMR." *IEEE Trans. Nucl. Sci.*, **NS-27**(3), 1220–1226.

Hsieh, R. C. and W. G. Wee (1976). "On methods of three-dimensional reconstruction from a set of radioisotope scintigrams," *IEEE Trans. Syst. Man. Cybern.*, **SMC-6**, 854–862.

Huang, H. K. and S. C. Wu (1976). "The evaluation of mass densities of the human body in vivo from CT scans," *Comput. Biol. Med.*, **6**, 337–343.

Huesman, R. H. (1977). "The effects of a finite number of projection angles and finite lateral sampling of projections on the propagation of statistical errors in transverse section reconstruction," *Phys. Med. Biol.*, **22**, 511–521.

Huesman, R. H., G. T. Gullberg, W. L. Greenberg, and T. F. Budinger (1977). *RECLBL Library Users Manual—Donner Algorithms for Reconstruction Tomography*, Technical Report PUB 214, Lawrence Berkeley Laboratory, Berkeley, California.

Hunten, D. M. and J. Veverka (1976). "Stellar and spacecraft occultations by Jupiter: A critical review of derived temperature profiles," in *Jupiter*, T. Gehrels, Ed., University of Arizona, Tucson, pp. 247–283.

Hurwitz, H. (1975). "Entropy reduction in Bayesian analysis of measurements," *Phys. Rev. A*, **12**, 698–706.

Hutchinson, J. M. S., W. A. Edelstein, and G. Johnson (1980). "A whole-body NMR imaging machine," *J. Phys. E.*, **13**, 947–955.

Iannino, A. and S. D. Shapiro (1978). "A survey of the Hough transform and its extensions for curve detection," in *Proceedings of the IEEE Computer Society Conference on Pattern Recognition and Image Processing*, May 31–June 2, Chicago, IEEE, Piscataway, NJ, pp. 32–38.

Inouye, T. (1979). "Image reconstruction with limited angle projection data," *IEEE Trans. Nucl. Sci.*, **NS-26**(2), 2666–2669.

Iwata, K. and R. Nagata (1970). "Calculation of three-dimensional refractive-index distribution from interferograms," *J. Opt. Soc. Am.*, **60**, 133–135.

Iwata, K. and R. Nagata (1975). "Calculation of refractive index distribution from interferograms using the Born and Rytov's approximation," *Jpn. J. Appl. Phys.*, **14** (Suppl. 14-1), 379–383.

Jackson, D. (1936). "Formal properties of orthogonal polynomials in two variables," *Duke Math. J.*, **2**, 423–434.

Jackson, J. D. (1975). *Classical Electrodynamics*, 2nd ed., Wiley, New York.

Jagota, R. C. and D. J. Collins (1972). "Finite fringe holographic interferometry applied to a right circular cone at angle of attack," (Trans. ASME, **94**, Ser. E) *J. Applied Mech.*, **39**, 897–903.

Jaszczak, R. J., R. E. Coleman, and C. B. Lim (1980). "SPECT: Single photon emission computed tomography," *IEEE Trans. Nucl. Sci.*, **NS-27**(3), 1137–1153.

John, F. (1934). "Bestimmung einer Funktion aus ihren Integralen über gewisse Mannigfaltigkeiten," *Math. Ann.*, **109**, 488–520.

John F. (1955). *Plane Waves and Spherical Means Applied to Partial Differential Equations*, Interscience, New York.

Johnson, S. A. and J. F. Greenleaf (1979). "New ultrasound and related imaging techniques," *IEEE Trans. Nucl. Sci.*, **NS-26**(2), 2812–2816.

Johnson, S. A., J. F. Greenleaf, B. Rajagopalan, R. C. Bahn, B. Baxter, and D. Christensen (1979). "High spatial resolution ultrasonic measurement techniques for characterization of static and moving tissues," in *Ultrasonic Tissue Characterization II*, M. Linzer, Ed. National Bureau of Standards, Spec. Publ. 525, U.S. Government Printing Office, Washington, DC, pp. 235–246.

Johnson, S. A., J. F. Greenleaf, B. Rajagopalan and M. Tanaka (1980). Algebraic and analytic inversion of acoustic data from partially or fully enclosing apertures," in Vol. 8 of *Acoustical Imaging*, (A. Metherell, Ed. Plenum, New York, pp. 577–598.

Johnson, S. A., J. F. Greenleaf, W. A. Samayoa, F. A. Duck, and J. Sjostrand (1975). "Reconstruction of three-dimensional velocity fields and other parameters by acoustic ray tracing," in *1975 Ultrasonic Symposium Proceedings,* IEEE Cat. #75CH0994-4SU, IEEE, New York, pp. 46–51.

Jones, S. M., F. L. Kitsen, P. L. Carson, and E. J. Bayly (1979). "Investigations of phase incoherent and other signal processing with a simulated array for ultrasonic CT," in *Frontiers of Engineering in Health Care*, Proceedings of the First Annual Conference IEEE/Engineering in Medicine and Biology Society, Oct. 6–7, Denver, Colorado, IEEE, New York, pp. 73–76.

Joseph, P. M. and R. A. Schulz (1980). "View sampling requirements in fan beam computed tomography," *Med. Phys.*, **7**, 692–702.

Joseph, P. M. and R. D. Spital (1978). "A method for correcting bone induced artifacts in CT scanners," *J. Comput. Assisted Tomog.*, **2**, 100–108.

Joseph, P. M., S. K. Hial, R. A. Schulz, and F. Kelcz (1980). "Clinical and experimental investigation of a smoothed CT reconstruction algorithm," *Radiology*, **134**, 507–516.

Joseph, P. M., R. D. Spital, and C. D. Stockham (1980). "The effects of sampling on CT images," *Computerized Tomography*, **4**, 189–206.

Judd, B. R. (1975). *Angular Momentum Theory for Diatomic Molecules*, Academic, New York.

Judy, P. F., R. G., Swensson, and M. Szulc (1981). "Lesion detection and signal-to-noise ratio in CT images," *Med. Phys.*, **8**, 13–23.

Junginger, H.-G. and W. van Haeringen (1972). "Calculation of three-dimensional refractive-index field using phase integrals," *Opt. Commun.*, **5**, 1–4.

Kaczmarz, M. S. (1937). "Angenäherte Auflösung von Systemen Linearer Gleichungen," *Bull. Acad. Polonaise Sci. Lett. Classe Sci. Math. Natur. Series A*, 355–357.

Kak, A. C. (1979). "Computerized tomography with x-ray, emission, and ultrasound sources," *Proc. IEEE*, **67**, 1245–1272.

Kak, A. C. and K. A. Dines (1978). "Signal processing of broadband pulsed ultrasound: Measurement of attenuation of soft biological tissues," *IEEE Trans. Biomed. Eng.*, **BME-25**, 321–344.

Kak, A. C., C. V. Jakowatz, N. A. Baily, and R. A. Keller (1977). "Computerized tomography using video recorded fluoroscopic images." *IEEE Trans. Biomed. Eng.*, **BME-24**, 157–169.

Kashyap, R. L. and M. C. Mittal (1975). "Picture reconstruction from projections," *IEEE Trans. Comput.*, **C-24**, 915–923.

Katz, M. B. (1978). *Questions of Uniqueness and Resolution in Reconstruction from Projections*, Vol. 26 of *Lecture Notes in Biomathematics*, Springer-Verlag; Berlin, Heidelberg, New York.

Katz, M. B. (1979). "Rigorous error bounds in computerized tomography," *IEEE Trans. Nucl. Sci.*, **NS-26**(2), 2691–2692.

Kaveh, M., R. K. Mueller, and R. D. Iverson (1979). "Ultrasonic tomography based on perturbation solutions of the wave equation," *Comput. Graph. Image Proc.*, **9**, 105–116.

Kay, D. B., J. W. Keyes, and W. Simon (1974). "Radionuclide tomographic image reconstruction using Fourier transform techniques," *J. Nucl. Med.*, **15**, 981–986.

Kellerman, K. I. and I. I. K. Pauliny-Toth (1973). "Extraglactic radio sources," *Proc. IEEE*, **61**, 1174–1182.

Kenue, S. K. and J. F. Greenleaf (1979a). "High speed convolving kernels for CT having triangular spectra and/or binary values," *IEEE Trans. Nucl. Sci.*, **NS-26**(2), 2693–2696.

Kenue, S. K. and J. F. Greenleaf (1979b). "Efficient convolution kernels for computerized tomography," *Ultrasonic Imaging*, **1**, 232–244.

Kerr, F. J. (1973). "Neutral hydrogen and galactic structure," *Proc. IEEE*, **61**, 1182–1192.

Keys, W. I. (1979). "Current status of single photon emission computed tomography," *IEEE Trans. Nucl. Sci.*, **NS-26**(2), 2752–2755.

Kijewski, D. K. and B. E. Bjarngard (1978). "Correction for beam hardening in computed tomography," *Med. Phys.*, **5**, 209–214.

Klepper, J. R., G. H. Brandenburger, J. L. Busse, and J. G. Miller (1977). "Phase cancellation, reflection, and refraction effects in quantitative ultrasonic attenuation tomography," in *1977 Ultrasonics Symposium Proceedings,* Oct. 26–28, Phoenix (IEEE Cat. #77CH1264-ISU) IEEE, New York, pp. 182–187.

Klepper, J. R., G. H. Brandenburger, J. W. Mimbs, B. E. Sobel, and J. G. Miller (1981). "Application of phase-insensitive detection and frequency-dependent measurements to computed ultrasonic attenuation tomography," *IEEE Trans. Biomed. Eng.*, **BME-28**, 168–201.

Kliore, A. J. and P. M. Woiceshyn (1976). "Structure of the atmosphere of Jupiter from Pioneer 10 and 11 radio occultation measurements" in *Jupiter*, T. Gehrels, Ed., University of Arizona, Tucson, pp. 216–237.

Klug, A. (1971). "III. Applications of image analysis techniques in electron microscopy. Optical diffraction and filtering and three-dimensional reconstruction from electron micrographs," *Philos. Trans. R. Soc. London Ser. B*, **261**, 173–179.

Klug, A. and R. A. Crowther (1972). "Three-dimensional image reconstruction from the viewpoint of information theory," *Nature (London)*, **238**, 435–440.

Klug, A., F. H. C. Crick, and H. W. Wyckoff (1958). "Diffraction by helical structures," *Acta. Crystallogr.*, **11**, 199–213. [Correction, see p. 560 of Zwick and Zeitler (1973).]

Knutsson, H. E., P. Edholm, G. H. Granlund, and C. U. Petersson (1980). "Ectomography—A new radiographic reconstruction method. I. Theory and error estimates," *IEEE Trans. Biomed. Eng.*, **BME-27**, 640–648.

Koeppe, R. A., R. M. Brugger, G. A. Schlapper, G. N. Larsen, and R. J. Jost (1981). "Neutron computed tomography," *J. Comput. Assisted Tomog.*, **5**, 79–88.

Koornwinder, T. H. (1982). "The representation theory of SL(2, $\mathbb{R}$). A non-infinitesimal approach," *L'Enseignement Mathématique* **28**, fasc. 1–2, 53–90.

Koral, K. F. and W. L. Rogers (1979). "Application of ART to time-coded emission tomography," *Phys. Med. Biol.*, **24**, 879–894.

Korevaar, J. (1968). *Mathematical Methods*, Vol. 1, Academic, New York.

Kosakoski, R. A. and D. J. Collins (1974). "Application of holographic interferometry to density field determination in transonic corner flow," *AIAA J.*, **12**, 767–770.

Kowalski, G. (1977a). "Reconstruction of objects from their projections. The influence of measurement errors on the reconstruction," *IEEE Trans. Nucl. Sci.*, **NS-24**(1), 850–864.

Kowalski, G. (1977b). "The influence of fixed errors of a detector array on the reconstruction of objects from their projections," *IEEE Trans. Nucl. Sci.*, **NS-24**(5), 2006–2016.

Kowalski, G. (1978). "Suppression of ring artifacts in CT fan-beam scanners," *IEEE Trans. Nucl. Sci.*, **NS-25**(5), 1111–1116.

Kowalski, G. (1979). "Multislice reconstruction from twin-cone beam scanning," *IEEE Trans. Nucl. Sci.*, **NS-26**(2), 2895–2903.

Kramer, D. M. and P. C. Lauterbur (1979). "On the problem of reconstructing images of non-scalar parameters from projections—Applications to vector fields." *IEEE Trans. Nucl. Sci.*, **NS-26**(2), 2674–2677.

Kreel, L. (1977). "Computerized transverse axial tomography with tissue density measurements," *J. Comput. Assisted Tomog.*, **1**, 1–5.

Kreel, L. (1979). *Medical Imaging*, L. Kreel, Ed. HM + M Publication, Year Book Medical Publishers, Inc., Chicago.

Krishnan, S., S. S. Prabhu, and E. V. Krishnamurthy (1973). "Probabilistic reinforcement algorithms for the reconstruction of pictures from their projections," *Int. J. Syst. Sci.*, **4**(4), 661–670.

Kuhl, D. E. and R. Q. Edwards (1963). "Image separation radioisotope scanning," *Radiology*, **80**, 653–661.

Kuhl, D. E. and R. Q. Edwards (1968a). "Digital techniques for on-site data processing," in *Fundamental Problems of Radioisotope Scanning*, A. Gottschalk and P. N. Beck, Eds., Thomas, Springfield, IL, pp. 250–266.

Kuhl, D. E. and R. Q. Edwards (1968b). "Reorganizing data from transverse section scans of the brain using digital processing," *Radiology*, **91**, 975–983.

Kuhl, D. E., R. Q. Edwards, A. R. Ricci, R. J. Yacob, T. J. Mich, and A. Alavi (1976) "The MARK IV system for radionuclide computed tomography of the brain," *Radiology*, **121**, 405–413.

Kulagin, I. D., L. M. Sorokin, and E. A. Dubrovskaya (1972). "Evaluation of some numerical methods for solving Abel's integral equation," *Opt. Spectrosc.* (*USSR*), **32**, 459–562.

Kumar, A., D. Welti, and R. R. Ernst (1975a). "Imaging of macroscopic objects by NMR Fourier zeugmatography," *Naturwissenschaften*, **62**, 34.

Kumar, A., D. Welti, and R. R. Ernst (1975b). "NMR Fourier zeugmatography," *J. Magn. Reson.*, **18**, 69–83.

Kundu, M. R. and T. E. Gergely (1980). *Radio Physics of the Sun*, M. R. Kundu and T. E. Gergely, Eds., IAU Symposium No. 86, Reidel, Dordrecht, Holland.

Kwoh, Y. S., I. S. Reed, and T. K. Truong (1977a). "A generalized $|\omega|$-filter for 3-D reconstruction," *IEEE Trans. Nucl. Sci.*, **NS-24**(5), 1990–1998.

Kwoh, Y. S., I. S. Reed, and T. K. Truong (1977b). "Back projection speed improvement for 3-D reconstruction," *IEEE Trans. Nucl. Sci.*, **NS-24**(5), 1999–2005.

Ladenburg, R. and D. Bershader (1954). "Interferometry," in Part I of *Physical Measurements in Gas Dynamics and Combustion*, R. Ladenburg, Ed., Princeton University, Princeton, NJ, pp. 47–78.

Ladenburg, R., J. Winckler, and C. C. vanVoorhis (1948). "Interferometric studies of faster than sound phenomena. Part I. The gas flow around various objects in a free homogeneous, supersonic air stream," *Phys. Rev.*, **73**, 1359–1377.

Lager, D. L. and R. J. Lytle (1977). "Determining a subsurface electromagnetic profile from high-frequency measurements by applying reconstruction-technique algorithms," *Radio Sci.*, **12**, 249–260.

Lai, C.-M. and P. C. Lauterbur (1980). "A gradient control device for complete three-dimensional nuclear magnetic resonance zeugmatographic imaging," *J. Phys. E*, **13**, 747–750.

Lai, C.-M., J. W. Shook, and P. C. Lauterbur (1979). "Microprocessor controlled reorientation of magnetic field gradients," *Chemical, Biomedical and Environmental Instrumentation*, **9**, 1–27.

Lake, J. A. (1972). "Reconstruction of three-dimensional structures from sectional helices by deconvolution of partial data," *J. Mol. Biol.*, **66**, 255–269.

Lakshminarayanan, A. V. (1975). "Reconstruction from divergent ray data," *Dept. of Computer Science Technical Report Number 92* SUNY—Buffalo, Amherst, New York.

Lauterbur, P. C. (1973). "Image formation by induced local interactions: Examples employing nuclear magnetic resonance," *Nature (London)*, **242**, 190–191.

Lauterbur, P. C. (1974). "Magnetic resonance zeugmatography," *Pure Appl. Chem.*, **40**, 149–157.

Lauterbur, P. C. (1979). "Feasibility of NMR zeumatographic imaging of the heart and lungs," in *Medical Imaging Techniques, A Comparison*, K. Preston, Jr., K. J. W. Taylor, S. A. Johnson, and W. R. Ayers, Eds., Plenum, New York, pp. 209–218.

Lauterbur, P. C. (1980a). "Progress in zeugmatographic imaging," *Philos. Trans. R. Soc. London Ser. B*, **289**, 483–487.

Lauterbur, P. C. (1980b). "NMR zeugmatographic imaging by true three-dimensional reconstruction," *J. Computer Assisted Tomog.*, **5**, 285–287.

Lauterbur, P. C., and C.-M. Lai (1980). "Zeugmatography by reconstruction from projections," *IEEE Trans. Nucl. Sci.*, **NS-27**(3), 1227–1231.

Lauterbur, P. C., D. M. Kramer, W. V. House, and C.-N. Chen (1975). "Zeugmatographic high resolution nuclear magnetic resonance spectroscopy. Images of chemical inhomogeneity within macroscopic objects," *J. Am. Chem. Soc.*, **97**, 6866–6868.

Lax, P. D. and R. S. Phillips (1970). "The Paley–Wiener theorem for the Radon transform," *Comm. Pure Appl. Math.*, **23**, 409–424.

Lax, P. D. and R. S. Phillips (1971). "Scattering theory," *Rocky Mount. J. Math.*, **1**, 173–223.

Lax, P. D. and R. S. Phillips (1979). "Translation representation for the solution of the non-Euclidean wave equation." *Comm. Pure Appl. Math.*, **32**, 617–667. [Correction: *Comm. Pure Appl. Math.*, **33**, 685 (1980).]

Leahy, J. V., K. T. Smith, and D. C. Solmon (1982). "Uniqueness, nonuniqueness, and inversion in the x-ray and Radon problems," in *Ill-Posed Problems: Theory and Practice*, M. Z. Nashed, Ed., Reidel, Dordrecht and Hingham, MA (to be published).

Ledley, R. S. (1976). "Introduction to computerized tomography," *Comput. Biol. Med.*, **6**, 239–246.

Leith, E. N. (1976). "White light holograms," *Sci. Am.*, **235**(No. 4), 80–95.

Leith, E. N. and J. Upatnieks (1962). "Reconstructed wavefronts and communication theory," *J. Opt. Soc. Am.*, **52**, 1123–1130.

Leith, E. N., and J. Upatnieks (1964). "Wavefront reconstruction with diffused illumination and three-dimensional objects," *J. Opt. Soc. Am.*, **54**, 1295–1301.

Leith, E. N. and J. Upatnieks (1965). "Photography by laser," *Sci. Am.*, **212**(No. 6), 25–35.

Lerche, I. and E. Zeitler (1976). "Projections, reconstructions and orthogonal functions," *J. Math. Anal. Appl.*, **56**, 634–649.

Levin, S. (1980). "A frequency-dip formulation of wave-theoretical migration in stratified media," in Vol. 9 of *Acoustical Imaging*, K. Y. Wang, Ed., Plenum, New York and London, pp. 681–697.

Levitan, E. (1979). "On true 3-D object reconstruction from line integrals," *Proc. IEEE*, **67**, 1679–1680.

Levitan, E., J. Degani, and J. Zak (1979). "Regularization in Fourier-synthesis 2-D image reconstruction," *IEEE Trans. Nucl. Sci.*, **NS-26**(3), 4327–4329.

Lewis, R. M. (1969). "Physical optics inverse diffraction," *IEEE Trans. Antennas Propag.*, **AP-17**, 308–314.

Lewitt, R. M. (1979). "Ultra-fast convolution approximations for computerized tomography," *IEEE Trans. Nucl. Sci.*, **NS-26**(2), 2678–2681.

Lewitt, R. M. and R. H. T. Bates (1978a). "Image reconstruction from projections. I. General theoretical considerations," *Optik*, **50**, 19–33.

Lewitt, R. M. and R. H. T. Bates (1978b). "Image reconstruction from projections. III. Projection completion methods (theory)," *Optik*, **50**, 189–204.

Lewitt, R. M. and R. H. T. Bates (1978c). "Image reconstruction from projections. IV. Projection completion methods (computational examples)," *Optik*, **50**, 269–278.

Lewitt, R. M., R. H. T. Bates, and T. M. Peters (1978). "Image reconstruction from projections." II. Modified backprojection methods." *Optik*, **50**, 85–109.

Lighthill, M. J. (1958). *Introduction to Fourier Analysis and Generalized Functions*, Cambridge University, Cambridge, England.

Lindgren, A. G. and P. A. Rattey (1981). "The inverse discrete Radon transform with applications to tomographic imaging using projection data," in Vol. 56 of *Advances in Electronics and Electron Physics*, C. Marton, Ed. Academic, New York, pp. 359–410.

Liverman, T. P. G. (1964). *Generalized Functions and Direct Operational Methods*, Vol. 1, Prentice-Hall, Englewood Cliffs, NJ.

Locher, P. R. (1980). "Computer simulation of selective excitation in NMR imaging," *Philos. Trans. R. Soc. London Ser. B*, **289**, 537–542.

Logan, B. F. and L. A. Shepp (1975). "Optimal reconstruction of a function from its projections," *Duke Math. J.*, **42**, 645–659.

Louis, A. K. (1980). "Picture reconstruction from projections in restricted range," *Math. Meth. Appl. Sci.*, **2**, 209–220.

Louis, A. K. (1981). "Ghosts in tomography—The null space of the Radon transform," *Math. Meth. Appl. Sci.*, **3**, 1–10.

Ludwig, D. (1966). "The Radon transform on Euclidean space," *Comm. Pure Appl. Math.*, **19**, 49–81.

Luke, Y. L. (1969). *The Special Functions and Their Approximations*, Vol. 1, Academic, New York.

Lytle, R. J., E. F. Laine, D. L. Lager, and D. T. Davis (1979). "Cross-borehole electromagnetic probing to locate high-constrast anomalies," *Geophysics*, **44**, 1667–1676.

McCullough, E. C. (1975). "Photon attenuation in computed tomography," *Med. Phys.*, **2**, 307–320.

McCullough, E. C. (1980). "Specifying and evaluating the performance of computed tomography (CT) scanners," *Med. Phys.*, **7**, 291–296.

McCullough, E. C., H. L. Baker, O. W. Houser, and D. F. Reese (1974). "An evaluation of the quantitative and radiation features of a scanning x-ray transverse axial tomograph: The EMI scanner," *Radiology*, **111**, 709–715.

McDavid, W. D., R. G. Waggener, W. H. Payne, and M. J. Dennis (1975). "Spectral effects on three-dimensional reconstruction from x-rays," *Med. Phys.*, **2**, 321–324.

McKinnon, G. C. and R. H. T. Bates (1980). "A limitation on ultrasonic transmission tomography," *Ultrasonic Imaging*, **2**, 48–54.

McKinnon, G. C. and R. H. T. Bates (1981). "Toward imaging the beating heart usefully with a conventional CT scanner," *IEEE Trans. Biomed. Eng.*, **BME-28**, 123–127.

McMechan, G. A. and R. Ottolini (1980). "Direct observation of a $p - \tau$ curve in a slant stacked wave field," *Bull. Seismol. Soc. Am.*, **70**, 775–789.

McMechan, G. A. and M. J. Yedlin (1981). "Analysis of dispersive waves by wave field transformation," *Geophysics*, **46**, 869–874.

Mach, L. (1892). "Ueber einen Interferenzrefraktor," *Z. Instrumentenkde.*, **12**, 89–93.

Mader, Ph. (1927). "Über die Dartstellung von Punktfunktionen im n-dimensionalen euklidischen Raum durch Ebenenintegrale," *Math. Z.*, **26**, 646–652.

Maecker, H. (1953). "Elektronendichte und Temperatur in der Säule des Hochstromkohlebogens," *Z. Phys.* **136**, 119–136.

Maecker, H. and T. Peters (1954). "Das Elektronenkontinuum in der Säule des Hochstromkohlebogens und in anderen Bögen," *Z. Phys.*, **139**, 448–463.

Maginness, M. G. (1979). "Methods and terminology for diagnostic ultrasound imaging systems," *Proc. IEEE*, **67**, 641–653.

Maldonado, C. D. (1965). "Note on orthogonal polynomials which are 'invariant in form' to rotation of axes," *J. Math. Phys.*, **6**, 1935–1938.

Maldonado, C. D. and H. N. Olsen (1966). "New method for obtaining emission coefficients from emitted spectral intensities. Part II—Asymmetrical sources," *J. Opt. Soc. Am.*, **56**, 1305–1313.

Maldonado, C. D., A. P. Caron, and H. N. Olsen (1965). "New method for obtaining emission coefficients from emitted spectral intensities. Part I—Circularly symmetric light sources," *J. Opt. Soc. Am.*, **55**, 1247–1254.

Mallard, J., J. M. S. Hutchison, W. A. Edelstein, C. R. Ling, M. A. Foster, and G. Johnson (1980). "In vivo NMR imaging in medicine: The Aberdeen approach, both physical and biological," *Philos. Trans. R. Soc. London Ser. B*, **289**, 519–533.

Maloney, F. P. and S. T. Gottesman (1979). "Lunar occultation observations of the Crab Nebula," *Astrophys. J.*, **234**, 485–492.

Manchester, R. N. (1973). "The properties of pulsars," *Proc. IEEE*, **61**, 1205–1211.

Mansfield, P. and A. A. Maudsley (1976). "Planar spin imaging by NMR," *J. Phys. C.*, **9**, L409–L412.

Mansfield, P. and P. G. Morris (1982). *NMR Imaging in Biomedicine, Advances in Magnetic Resonance*, Supp. 2, Academic, New York.

Mansfield, P., P. G. Morris, R. J. Ordidge, I. L. Pykett, V. Bangert, and R. E. Coupland (1980). "Human whole body imaging and detection of breast tumors by NMR," *Philos. Trans. R. Soc. London Ser. B*, **289**, 503–510.

Marchand, J.-P. (1962). *Distributions, An Outline*, North-Holland, Amsterdam.

Marr, R. B. (1974). "On the reconstruction of a function on a circular domain from a sampling of its line integrals," *J. Math. Anal. Appl.*, **45**, 357–374.

Marr, R. B. (1982). "An overview of image reconstruction," in *Ill-Posed Problems: Theory and Practice*, M. Z. Nashed, Ed., Reidel, Dordrecht and Hingham, MA (*to be published*).

Mason, I. M. (1981). "Algebraic reconstruction of a two-dimensional velocity in homogeneity in the High Hazles seam Thoresby colliery," *Geophysics*, **46**, 298–308.

Matulka, R. D. and D. J. Collins (1971). "Determination of three-dimensional density fields from holographic interferograms," *J. Appl. Phys.*, **42**, 1109–1119.

Maudsley, A. A. (1980). "Multiple-line-scanning spin density imaging," *J. Magn. Reson.*, **41**, 112–126.

Mensa, D., G. Heidbreder, and G. Wade (1980). "Aperture synthesis by object rotation in coherent imaging," *IEEE Trans. Nucl. Sci.*, **NS-27**(2), 989–997.

Mersereau, R. M. (1973). "Recovering multidimensional signals from their projections," *Comput. Graphics Image Proc.*, **1**, 179–195.

Mersereau, R. M. (1976). "Direct Fourier transform techniques in 3-D image reconstruction," *Comput. Biol. Med.*, **6**, 247–258.

Mersereau, R. M. and A. V. Oppenheim (1974). "Digital reconstruction of multidimensional signals from their projections," *Proc. IEEE*, **62**, 1319–1338.

Merzkirch, W. (1974). *Flow Visualization*, Academic, New York.

Mikhailov, A. M. and B. K. Vainshtein (1971). "Electron microscope determination of the three-dimensional structure of the extended tail of the T6 bacteriophage," *Kristallografiya*, **16**(3), 505–515. [*Sov. Phys. Crystallogr.*, **16**, 428–436 (1971).]

Mikusiński, J. G. (1948). "Sur la méthode de généralisation de L. Schwartz et sur la convergence faible," *Fundamenta Math.*, **35**, 235–239.

Mikusiński, J. G. (1959). *Operational Calculus*, Pergamon, New York.

Miller, K. (1978). "An optimal method for the x-ray reconstruction problem," *Am. Math. Soc. Not.*, **25**, 161–162.

Minerbo, G. N. (1979a). "Convolutional reconstruction from cone-beam projection data," *IEEE Trans. Nucl. Sci.*, **NS-26**(2), 2682–2684.

Minerbo, G. (1979b). "MENT: A maximum entropy algorithm for reconstructing a source from projection data," *Comput. Graph. Image Proc.*, **10**, 48–68.

Minerbo, G. N. and J. G. Sanderson (1977). "Reconstruction of a source from a few (2 or 3) projections." Informal Report LA-6747-MS, Los Alamos Scientific Laboratory, Los Alamos, NM.

Minkowski, H. (1904–06). "Über die Körper konstanter Breite," in Vol. 2 of *Gesammelte Abhandlungen von Hermann Minkowski*, D. Hilbert, Ed., Chelsea, New York (1967), pp. 277–279.

Montaldi, E. (1979). "On the Radon inversion formula," *Lett. Nuovo Cimento*, **26**, 593–598.

Moore, W. E. and G. P. Garmire (1975). "The x-ray structure of the Vela supernova remnant," *Astrophys. J.*, **199**, 680–690.

Moore, W. S. and G. N. Holland (1980). "Experimental considerations in implementing a whole body multiple sensitive point nuclear magnetic resonance imaging system," *Philos. Trans. R. Soc. London Ser. B*, **289**, 511–518.

Moore, W. S., G. N. Holland, L. Kreel (1980). "The NMR CAT scanner—A new look at the brain." *Computerized Tomography*, **4**, 1–7.

Morgenthaler, D. G., R. A. Brooks, and A. J. Talbert (1980). "Noise factor of a polyenergetic x-ray beam in computed tomography," *Phys. Med. Biol.*, **25**, 251–259.

Morse, P. M. and K. U. Ingard (1968). *Theoretical Acoustics*, McGraw-Hill, New York.

Muehllehner, G. (1971). "A tomographic scintillation camera," *Phys. Med. Biol.*, **16**, 87–96.

Muehllehner, G. and Z. Hashmi (1972). "Quantification of the depth effect of tomographic and section imaging devices," *Phys. Med. Biol.*, **17**, 251–260.

Muehllehner, G. and R. A. Wetzel (1971). "Section imaging by computer calculation," *J. Nucl. Med.*, **12**, 76–84.

Mueller, R. K., M. Kaveh, and G. Wade (1979). "Reconstructive tomography and applications to ultrasonics," *Proc. IEEE*, **67**, 567–587.

Mueller, R. K., M. Kaveh, and R. D. Iverson (1980). "A new approach to acoustic tomography using diffraction techniques," in Vol. 8 of *Acoustical Imaging*, A. F. Metherell, Ed., Plenum, New York, pp. 615–628.

Müller, C. and R. Richberg (1980). "Über die Radon-Transformation kreissymmetrischer Funktionen und ihre Beziehung zur Sommerfeldschen Theorie der Hankelfunktionen," *Math. Meth. Appl. Sci.*, **2**, 108–129.

Nahamoo, D., C. R. Crawford, and A. C. Kak (1981). "Design constraints and reconstruction algorithms for transverse-continuous-rotate CT scanners," *IEEE Trans. Biomed. Eng.*, **BME-28**, 79–98.

Nalcioglu, O. and Z. H. Cho (1978). "Reconstruction of 3-D objects from cone beam projections," *Proc. IEEE*, **66**, 1584–1585.

Nalcioglu, O., Z. H. Cho, and G. F. Knoll (1979). "Forward," *IEEE Trans. Nucl. Sci.*, **NS-26**(2), 2663.

Naparstek, A. (1980). "Short-scan fan-beam algorithms for CT," *IEEE Trans. Nucl. Sci.*, **NS-27**(3), 1112–1120.

Nashed, M. Z. (1982). *Ill-Posed Problems: Theory and Practice*, M. Z. Nashed, Ed., Reidel, Dordrecht and Hingham, MA (*to be published*).

Nashed, M. Z. (1976). *Generalized Inverses and Applications*, M. Z. Nashed, Ed., Academic, New York.

Nassi, M. and W. R. Brody (1981). "Regional myocardial flow estimation using computed tomography," *Med. Phys.*, **8**, 302–307.

Nassi, M., W. R. Brody, P. R. Cipriano, and A. Macovski (1981). "A method for stop-action imaging of the heart using gated computed tomography," *IEEE Trans. Biomed. Eng.*, **BME-28**, 116–122.

Natterer, F. (1977). "The finite element method for ill-posed problems," *R.A.I.R.O. Analyse Numérique*, **11**, 271–278.

Natterer, F. (1978). "Numerical inversion of the Radon transform," *Numer. Math.*, **30**, 81–91.

Natterer, F. (1979). "On the inversion of the attenuated Radon transform," *Numer. Math.*, **32**, 431–438.

Natterer, F. (1980). "A Sobolev space analysis of picture reconstruction," *SIAM J. Appl. Math.*, **39**, 402–411.

Natterer, F. (1982). "The ill-posedness of Radon's integral equation," in *Ill-Posed Problems: Theory and Practice*, M. Z. Nashed, Ed. Reidel, Dordrecht and Hingham, MA (*to be published*).

Nestor, O. H. and H. N. Olsen (1960). "Numerical methods for reducing line and surface probe data," *SIAM Review*, **2**, 200–207.

Newton, R. G. (1966). *Scattering Theory of Waves and Particles*, McGraw-Hill, New York.

Newton, R. G. (1978). "Three-dimensional solitons," *J. Math. Phys.*, **19**, 1068–1073.

Newton, R. G. (1981). "Inversion of reflection data for layered media: a review of exact methods," *Geophys. J. R. Astron. Soc.*, **65**, 191–215.

Norton, S. J. and M. Linzer (1979a). "Ultrasonic reflectivity tomography: Reconstruction from circular transducer arrays," *Ultrasonic Imaging*, **1**, 154–184.

Norton, S. J. and M. Linzer (1979b). "Ultrasonic reflectivity imaging in three dimensions: Reconstruction with spherical transducer arrays," *Ultrasonic Imaging*, **1**, 210–231.

Norton, S. J. and M. Linzer (1981). "Ultrasonic reflectivity imaging in three dimensions: Exact inverse scattering solutions for plane, cylindrical, and spherical apertures," *IEEE Trans. Biomed. Eng.*, **BME-28**, 202–220.

O'Brien, P. A. (1953). "The distribution of radiation across the solar disk at metre wavelengths," *Mon. Not. R. Astron. Soc.*, **113**, 597–612.

Oldendorf, W. H. (1961). "Isolated flying spot detection of radiodensity discontinuities displaying the internal structural pattern of a complex object," *IRE Trans. Biomed. Elec.*, **BME-8**, 68–72.

Oldendorf, W. H. (1980). *The Quest for an Image of Brain*, Raven Press, New York.

Oldendorf, W. H. (1982). "Potential applications of magnetic resonance in brain imaging," *RNM Images* [formerly *Radiology/Nuclear Medicine Magazine*], **12**(No. 3), 40–46.

Olsen, H. N., C. D. Maldonado, and G. D. Duckworth (1968). "A numerical method for obtaining internal emission coefficients from externally measured spectral intensities of asymmetrical plasmas," *J. Quant. Spectrosc. Radiat. Transfer*, **8**, 1419–1430.

Oppenheim, A. V. and R. W. Schafer (1975). *Digital Signal Processing*, Prentice-Hall, Englewood Cliffs, NJ.

Oppenheim, B. E. (1974). "More accurate algorithms for iterative 3-dimensional reconstruction." *IEEE Trans. Nucl. Sci.*, **NS-21**(3), 72–77.

Orlov, S. S. (1975a). "Theory of three-dimensional reconstruction. I. Conditions for a complete set of projections." *Kristallografiya*, **20**, 511–515. [*Sov. Phys. Crystallogr.*, **20**, 312–314 (1975).]

Orlov, S. S. (1975b). "Theory of three-dimensional reconstruction. II. The recovery operator," *Kristallografiya*, **20**, 701–709. [*Sov. Phys. Crystallogr.* **20**, 429–433 (1975).]

Papoulis, A. (1977). *Signal Analysis*, McGraw-Hill, New York.

Partain, C. L., A. E. James, Jr., J. T. Watson, R. R. Price, C. M. Coulam, F. D. Rollo (1980). "Nuclear magnetic resonance and computed tomography," *Radiology*, **136**, 767–770.

Pasachoff, J. M. and M. L. Kutner (1978). *University Astronomy*, Saunders, Philadelphia.

Pearce, W. J. (1961). "Plasma jet temperature measurement," in *Optical Spectrometric Measurements of High Temperatures*, P. J. Dickerman, Ed. pp. 125–169, University of Chicago, Chicago.

Pennington, K. S. (1968). "Advances in holography," *Sci. Am.*, **218**(No. 2), 40–48.

Perry, R. M. (1975) Reconstructing a function by circular harmonic analysis of its line integrals," in *Image Processing for 2-D and 3-D Reconstruction from Projections: Theory and Practice in Medicine and the Physical Sciences*, Aug. 4–7, Stanford University, Technical Digest, Optical Society of America, Washington, DC, pp. ThA6-1–ThA6-4.

Peters, J. V. (1980a). "A Tauberian theorem for the Radon transform," preprint, C. W. Post Center of Long Island University, Greenvale, NY.

Peters, J. V. (1980b). "The Radon transform: An application to probability theory," preprint, C. W. Post Center of Long Island University, Greenvale, NY.

Peters, J. V. (1980c). "The ham sandwich theorem and some related results," preprint, C. W. Post Center of Long Island University, Greenvale, NY.

Peters, T. M. (1974). "Spatial filtering to improve transverse tomography," *IEEE Trans. Biomed. Eng.*, **BME-21**, 214–219.

Peters, T. M. and R. M. Lewitt (1977). "Computed tomography with fan beam geometry," *J. Comput. Assisted Tomog.*, **1**, 429–436.

Peters, T. M., P. R. Smith, and R. D. Gibson (1973). "Computer aided transverse body-section radiography," *British J. Radiology*, **46**, 314–317.

Petersen, B. E., K. T. Smith, and D. C. Solmon (1979). "Sums of plane waves and the range of the Radon transform," *Math Ann.*, **243**, 153–161.

Petersson, C. U., P. Edholm, G. H. Granlund, and H. E. Kutsson (1980). Ectomography—A new radiographic reconstruction method II. Computer simulated experiments," *IEEE Trans. Biom. Eng.*, **BME-27**, 649–655.

Pettengill, G. H. (1978). "Physical Properties of the planets and satellites from radar observations," *Annu. Rev. Astron. Astrophys.*, **16**, 265–292.

Pettengill, G. H., S. H. Zisk, and T. W. Thomspon (1974). "The mapping of lunar radar scattering characteristics," *The Moon*, **10**, 3–16.

Phelps, M. E. (1977). "Emission computed tomography," *Semin. Nucl. Med.*, **7**, 337–365.

Phelps, M. E., E. J. Hoffman, S. C. Huang, and D. E. Kuhl (1978). "ECAT: A new computerized tomographic imaging system for positron-emitting radiopharmaceuticals," *J. Nucl. Med.*, **19**, 635–647.

Phelps, M. E., E. J. Hoffman, and M. M. Ter-Pogossian (1975). "Attenuation coefficients of various body tissues, fluids, and lesions at photon energies of 18 to 136 keV," *Radiology*, **117**, 573–583.

Phinney, R. A. and D. L. Anderson (1968). "On the radio occultation method for studying planetary atmospheres," *J. Geophys. Res., Space Phys.*, 73, 1819–1827.

Phinney, R. A., K. R. Chowdhury, and L. N. Frazer (1981). "Transformation and analysis of record sections." *J. Geophys. Res.*, **86**, 359–377.

Pincus, J. D. (1964). "A mathematical reconstruction of a radioactive source density from its induced radiation pattern," AMD 359, Applied Math. Dept., Brookhaven National Laboratory, Upton, NY.

Ponsonby, J. E. B., I. Morison, A. R. Birks, and J. K. Landon (1972). "Radar images of the moon at 75 and 185 cm wavelengths," *The Moon*, **5**, 286–293.

Post, M. J. D. (1980). *Radiographic Evaluation of the Spine, Current Advances with Emphasis on Computed Tomography*, M. J. D. Post, Ed., Masson, New York.

Powell, R. L., and K. A. Stetson (1965). "Interferometric analysis by wavefront reconstruction," *J. Opt. Soc. Am.*, **55**, 1593–1598.

Pratt, W. K. (1978). *Digital Image Processing*, Wiley-Interscience, New York.

Presnyakov, Yu. P. (1976). "Calculation of two-dimensional refractive index function," *Opt. Spectrosc.* (*USSR*), **40**, 69–70.

Preston, K., Jr., K. J. W. Taylor, S. A. Johnson and W. R. Ayers, Eds. (1979). *Medical Imaging Techniques, A Comparison*, Plenum, New York.

Price, L. R. (1979a). "Imaging the electrical parameters inside a patient: A new computed tomographic (CT) technique," in *Proceedings of the Sixth Conference on Computer Applications in Radiology and Computer/Aided Analysis of Radiological Images*, June 18–21, Newport Beach, California IEEE, New York, pp. 71–73.

Price, L. R. (1979b). "Electrical impedance computed tomography (ICT): A new CT imaging technique." *IEEE Trans. Nucl. Sci.*, **NS-26**(2), 2736–2739.

Quinto, E. T. (1980a). "The dependence of the generalized Radon transform on defining measures," *Trans. Am. Math. Soc.*, **257**, 331–345.

Quinto, E. T. (1980b). "Null spaces and ranges for the classical and spherical Radon transforms," (preprint) Dept. of Mathematics, Tufts University, Medford, MA.

Quinto, E. T. (1981a). "Topological restrictions on double fibrations and Radon transforms," *Proc. Am. Math. Soc.*, **81**, 570–574.

Quinto, E. T. (1981b). "The invertibility of rotation invariant Radon transforms," (preprint) Dept. of Mathematics, Tufts University, Medford, MA.

Ra, J. B. and Z. H. Cho (1981). "Generalized true three-dimensional reconstruction algorithm" *Proc. IEEE*, **69**, 668–670.

Rabiner, L. R. and B. Gold (1975). *Theory and Application of Digital Signal Processing*, Prentice-Hall, Englewood Cliffs, NJ.

Radley, R. J. (1975). "Two-wavelength holography for measuring plasma electron density," *Phys. Fluids*, **18**, 175–179.

Radon, J. (1917). "Über die Bestimmung von Funktionen durch ihre Integralwerte längs gewisser Mannigfaltigkeiten," *Berichte Sächsische Akademie der Wissenschaften. Leipzig, Math.—Phys. Kl.*, **69**, 262–267.

Rainville, E. D. (1960). *Special Functions*, Chelsea, New York.

Ramachandran, G. N. and R. V. Lakshminarayanan (1971). "Three-dimensional reconstruction from radiographs and electron micrographs: Applications of convolutions instead of Fourier transforms," *Proc. Natl., Acad. Sci. USA*, **68**, 2236–2240.

Rattey, P. A. and A. G. Lindgren (1981). "Sampling the 2-D Radon transform," *IEEE Trans. Acoust., Speech, Signal Processing*, **ASSP-29**(5), 994–1002.

Redington, R. W. and W. H. Berninger (1981). "Medical imaging systems," *Phys. Today*, **34**(8), 36–44.

Reed, I. S., Y. S. Kwoh, T. K. Truong, and E. L. Hall (1977). "X-ray reconstruction by finite field transforms," *IEEE Trans. Nucl. Sci.*, **NS-24**(1), 843–849.

Reed, I. S., T. K. Truong, C. M. Chang, and Y. S. Kwoh (1978). "3-D reconstruction for diverging x-ray beams," *IEEE Trans. Nucl. Sci.*, **NS-25**(3), 1006–1011.

Reed, I. S., W. V. Glenn, C. M. Chang, T. K. Truong, and Y. S. Kwoh (1979). "Dose reduction in x-ray computed tomography using a generalized filter," *IEEE Trans. Nucl. Sci.*, **NS-26**(2), 2904–2909.

Reed, I. S., W. V. Glenn, Y. S. Kwoh, and T. K. Truong (1980). "A bandpass filter for the enhancement of an x-ray reconstruction of the tissue in the spinal canal," *IEEE Trans. Biomed. Eng.*, **BME-27**, 736–738.

Reed, I. S., W. V. Glenn, T. K. Truong, Y. S. Kwoh, and C. M. Chang (1980). "X-ray reconstruction of the spinal cord, using bone suppression," *IEEE Trans. Biomed. Eng.*, **BME-27**, 293–298.

Reid, M. S., R. C. Clauss, D. A. Bathker, and C. T. Stelzried (1973). "Low-noise microwave receiving systems in a worldwide network of large antennas," *Proc. IEEE*, **61**, 1330–1335.

Rényi, A. (1952). "On projections of probability distributions," *Acta. Math. Acad. Sci. Budapest*, **3**, 131–141.

Richards, P. G. (1979). "Theoretical seismic wave propagation," *Rev. Geophys. Space Phys.*, **17**, 312–328.

Riddle, A. C. (1968). *High Resolution Studies of Solar Microwave Radiation*; Ph.D. thesis, Department of Electrical Engineering, Stanford University, Stanford, California.

Riederer, S. J. (1981). "Application of the noise power spectrum to positron emission CT self-absorption correction," *Med. Phys.*, **8**, 220–224.

Riederer, S. J., N. J. Pelc, and D. A. Chesler (1978). "The noise power spectrum in computed x-ray tomography," *Phys. Med. Biol.*, **23**, 446–454.

Rivlin, T. J. (1974). *The Chebyshev Polynomials*, Wiley-Interscience, New York.

Robb, R. A., E. L. Ritman, B. K. Gilbert, J. H. Kinsey, L. D. Harris, and E. H. Wood (1979). "The DSR: A high-speed three-dimensional x-ray computed tomography system for dynamic spatial reconstruction of the heart and circulation." *IEEE Trans. Nucl. Sci.*, **NS-26**(2), 2713–2717.

Robinson, E. A. (1982). "Spectral approach to geophysical inversion by Lorentz, Fourier, and Radon transforms." *Proc. IEEE*, **70**, 1039–1054.

Roger, R. S., C. H. Costain, J. D. Lacey, T. L. Landecker, and F. K. Bowers (1973). "A supersynthesis radio telescope for neutral hydrogen spectroscopy at the dominion radio astrophysical observatory," *Proc. IEEE*, **61**, 1270–1276.

Roman, P. (1965). *Advanced Quantum Theory*, Addison Wesley, Reading, MA.

Roos, B. W. (1969). *Analytic Functions and Distributions in Physics and Engineering*, Wiley, New York.

Rosenfeld, A. and A. C. Kak (1982). *Digital Picture Processing*, 2nd ed. Vols. I and II, Academic, New York.

Rowe, R. W., P. E. Undrill, and W. I. Keyes (1980). "Synthesis and enhancement of the radionuclide tomographic section," *IEEE Proc.*, **127**, Part E, 193–196.

Rowland, S. W. (1976). "The effect of noise in the projection data on the reconstruction produced by computerized tomography." *Proc. Soc. Photo-Opt. Instrum. Eng.*, **96**, 124–130.

Rowland, S. W. (1979). "Computer implementation of image reconstruction formulas," in *Image Reconstruction from Projections*, G. T. Herman, Ed., Vol. 32 of *Topics in Applied Physics*, Springer-Verlag, New York, pp. 7–79.

Rowley, P. D. (1969). "Quantitative interpretation of three-dimensional weakly refractive phase objects using holographic interferometry." *J. Opt. Soc. Am.*, **59**, 1496–1498. [Erratum, *J. Opt. Soc. Am.*, **60**, 705 (1970).]

Rutt, B. and A. Fenster (1980). "Split-filter computed tomography: A simple technique for dual energy scanning," *J. Comput. Assisted Tomog.*, **4**, 501–509.

Ryle, M. (1972). "The 5-km radio telescope at Cambridge," *Nature*, **239**, 435–438.

Ryle, M. and A. Hewish (1960). "The synthesis of large radiotelescopes," *Mon. Not. R. Astron. Soc.*, **120**, 220–230.

Sanderson, J. G. (1979). "Reconstruction of fuel pin bundles by a maximum entropy method," *IEEE Trans. Nucl. Sci.*, **NS-26**(2), 2685–2686.

Scheuer, P. A. G. (1962). "On the use of lunar occultations for investigating the angular structure of radio sources," *Aust. J. Phys.*, **15**, 333–343.

Scheuer, P. A. G. (1965). "Lunar occultation of radio sources," *Mon. Not. R. Astron. Soc.*, **129**, 199–204.

Schlindwein, M. (1978). "Iterative three-dimensional reconstruction from twin-cone beam projections," *IEEE Trans. Nucl. Sci.*, **NS-25**(5), 1135–1143.

Schomberg, H. (1978). "An improved approach to reconstructive ultrasound tomography." *J. Phys. D*, **11**, L181–L185.

Schoor, B. and D. Townsend (1981). "Filters for three-dimensional limited-angle tomography," *Phys. Med. Biol.*, **26**, 305–312.

Schultz, P. S. and J. F. Claerbout (1978). "Velocity estimation and downward continuation by wavefront synthesis," *Geophysics*, **43**, 691–714.

Schulz, R. A., E. C. Olson, and K. S. Han (1977). "A comparison of the number of rays versus the number of views in reconstruction tomography," *Proc. Soc. Photo-Opt. Instrum. Eng.*, **127**, 313–320.

Schumacher, R. T. (1970). *Introduction to Magnetic Resonance*, Benjamin, New York.

Schwartz, L. (1957, 1959). *Théorie des Distributions*, 2nd ed., Vols. 1 and 2, Hermann, Paris.

Schwartz, L. (1966). *Mathematics for the Physical Sciences*, Hermann, Paris.

Scudder, H. J. (1978). "Introduction to computer aided tomography," *Proc. IEEE*, **66**, 628–637.

Segrè, E. (1977). *Nuclei and Particles*, Benjamin, Reading, MA.

Shapiro, S. D. (1975). "Transformations for the computer detection of curves in noisy pictures," *Comput. Graph. Image Proc.*, **4**, 328–338.

Shapiro, S. D. (1978). "Feature space transforms for curve detection," *Pattern Recognition*, **10**, 129–143.

Shapiro, S. D. (1979a). "Generalization of the Hough transform for curve detection in noisy digital images," in *Proceedings of the Fourth International Joint Conference on Pattern Recognition*, Nov. 7–10, 1978, Kyoto, Japan, IEEE, Piscataway, NJ, pp. 710–714.

Shapiro, S. D. (1979b). "Use of the Hough transform for image data compression," in *Pattern Recognition 1979*, Aug. 6–8, Chicago, IEEE, New York, pp. 576–582.

Shapiro, S. D. and A. Iannino (1979). Geometric constructions for predicting Hough transform performance," *IEEE Trans. Pattern Analysis and Machine Intelligence*, **PAMI-1**, 310–317.

Shepp, L. A. (1980). "Computerized tomography and nuclear magnetic resonance," *J. Comput. Assist. Tomog.*, **4**, 94–107.

Shepp, L. A., S. K. Hilal, and R. A. Schulz (1979). "The tuning fork artifact in computerized tomography," *Comput. Graph. Image Proc.*, **10**, 246–255.

Shepp, L. A. and J. B. Kruskal (1978). "Computerized tomography: The new medical x-ray technology," *Am. Math. Monthly*, **85**, 420–439.

Shepp, L. A. and B. F. Logan (1974). "The Fourier reconstruction of a head section," *IEEE Trans. Nucl. Sci.*, **NS-21**(3), 21–43. [Erratum: See Shepp, Hilal, and Schulz (1979).]

Shepp, L. A. and J. A. Stein (1977). "Simulated reconstruction artifacts in computerized x-ray tomography," in *Reconstruction Tomography In Diagnostic Radiology and Nuclear Medicine*, M. M. Ter-Pogossian et al., Eds. University Park Press, Baltimore, MD, pp. 33–48.

Sheridan, W. T., M. R. Keller, C. M. O'Connor, R. A. Brooks, and K. M. Hanson (1980). "Evaluation of edge-induced streaking artifacts in CT scanners," *Med. Phys.*, **7**, 108–111.

Shtein, I. N. (1972). "On the application of the Radon transform in holographic interferometry," *Radiotekh. Elektron.*, **17**, 2436–2437.

Singer, J. R. (1980). "Blood flow measurements by NMR of the intact body," *IEEE Trans. Nucl. Sci.*, **NS-27**(3), 1245–1249.

Sklansky, J. (1978). "On the Hough transform technique for curve detection," *IEEE Trans. Comput.*, **C-27**, 923–926.

Slichter, C. P. (1978). *Principles of Magnetic Resonance*, Springer-Verlag, Berlin, Heidelberg, New York.

Smerd, S. F. and J. P. Wild (1957). "The effects of incomplete resolution on surface distributions derived from strip-scanning observations, with particular reference to an application in radio astronomy," *Philos. Mag.*, Series 8, **2**, 119–130.

Smith, C. B. (1979). "A dual method for maximum entropy restoration," *IEEE Trans. Pattern Anal. Machine Intelligence*, **PAMI-1**, 411–414.

Smith, D. W., T. F. Green, and R. W. Shorthill (1977). "The upper Jovian atmosphere aerosol content determined from a satellite eclipse observation," *Icarus*, **30**, 697–729.

Smith, K. T., D. C. Solmon, and S. L. Wagner (1977). "Practical and mathematical aspects of the problem of reconstructing objects from radiographs," *Bull. Am. Math. Soc.*, **83**, 1227–1270.

Smith, K. T., D. C. Solmon, S. L. Wagner, and C. Hamaker (1978). "Mathematical aspects of divergent beam radiography," *Proc. Natl. Acad. Sci. USA*, **75**, 2055–2058.

Smith, P. R., U. Aebi, R. Josephs, and M. Kessel (1976). "Studies of the T4 bacteriophage tail sheath. I. The recovery of three-dimensional information from the extended sheath," *J. Mol. Biol.*, **106**, 243–275.

Smith, P. R., T. M. Peters, and R. H. T. Bates (1973). "Image reconstruction from finite numbers of projections." *J. Phys. A: Math. Nucl. Gen.*, **6**, 361–382.

Sneddon, I. N. (1955). "Functional analysis," in Vol. II of *Encyclopedia of Physics*, S. Flügge, Ed. Springer-Verlag, Berlin, pp. 198–348.

Sneddon, I. N. (1972). *The Use of Integral Transforms*, McGraw-Hill, New York.

Snyder, D. L. and J. R. Cox, Jr. (1977). "An overview of reconstructive tomography and limitations imposed by a finite number of projections," in *Reconstructive Tomography in Diagnostic Radiology and Nuclear Medicine*, M. M. Ter-Pogossian et al., Eds., University Park Press, Baltimore, MD, pp. 3–32.

Sobolev, S. L. (1936). "Méthode nouvelle à résoudre le probléme de Cauchy pour les équations linéaires hyperboliques normales," *Mat. Sbornik*, **1**, 39–72.

Solmon, D. C. (1976). "The x-ray transform," *J. Math. Anal. Appl.*, **56**, 61–83.

Solmon, D. C. (1979). "A note on *k*-plane integral transforms," *J. Math. Anal. Appl.*, **71**, 351–358.

South, R. (1970). "An extension of existing methods of determining refractive indices from axisymmetric interferograms," *AIAA J.*, **8**, 2057–2059.

Southon, F. C. (1981). "CT scanner comparison," *Med. Phys.*, **8**, 62–75.

Stakgold, I. (1967, 1968). *Boundary Value Problems of Mathematical Physics*, Vols. I and II, MacMillan, New York.

Stark, H. (1979a). "Bounds on errors in reconstructing from undersampled images," *J. Opt. Soc. Am.*, **69**, 1042–1043.

Stark, H. (1979b). "Sampling theorems in polar coordinates." *J. Opt. Soc. Am.*, **69**, 1519–1525.

Stark, H., I. N. Paul, and C. S. Sarna (1979). "Fourier transform reconstruction in CAT by exact interpolation," in *IEEE 1979 Frontiers of Engineering in Health Care*, First Annual Conference on Engineering in Medicine and Biology Society, Oct. 6–7 (1979) Denver, Colorado, IEEE, New York, pp. 165–168.

Stark, H., J. W. Woods, I. Paul, and R. Hingorani (1980). "Image reconstruction in computer-aided tomography by direct Fourier methods," in *IEEE 1980 Frontiers of Engineering in Health Care*, IEEE/Engineering in Medicine and Biology Society Second Annual Conference, Sept. 28–30, Washington, D.C., IEEE, New York, pp. 277–280.

Stockham, C. D. (1979). "A simulation study of aliasing in computed tomography," *Radiology*, **132**, 721–726.

Stoffa, P. L., P. Buhl, J. B. Diebold, and F. Wenzel (1981). "Direct mapping of seismic data to the domain of intercept time and ray parameter: A plane wave decomposition," *Geophysics*, **46**, 255–267.

Stonestrom, J. P., R. E. Alvarez, and A. Macovski (1981). "A framework for spectral artifact corrections in x-ray CT," *IEEE Trans. Biomed. Eng.*, **BME-28**, 128–141.

Strohbehn, J. W., C. H. Yates, H. B. Curran, and E. S. Sternick (1979). "Image enhancement of conventional transverse-axial tomograms," *IEEE Trans. Biomed. Eng.*, **BME-26**, 253–262.

Stroke, G. W. and M. Halioua (1976). "Three-dimensional reconstruction in x-ray crystallography and electron microscopy by reduction to two-dimensional holographic implementation," *Trans. Am. Crystallogr. Assoc.*, **12**, 27–41.

Stroke, G. W., M. Halioua, R. Sarma, and V. Srinivasan (1977). "Imaging of atoms: Three-dimensional molecular structure reconstruction using opto-digital computing," *Proc. IEEE*, **65**, 589–591.

Stroke, G. W., M. Halioua, F. Thon, and D. Willasch (1977). "Image improvement and three-dimensional reconstruction using holographic image processing," *Proc. IEEE*, **65**, 39–62.

Stuck, B. W. (1977). "A new proposal for estimating the spatial concentration of certain types of air pollutants." *J. Opt. Soc. Am.*, **67**, 668–678.

Swartzlander, E. E. and B. K. Gilbert (1980). "Arithmetic for ultra-high-speed tomography." *IEEE Trans. Comput.*, **C-29**, 341–353.

Swarup, G. and D. S. Bagri (1973). "An aperture-synthesis interferometer at Doty for operation at 327 MHz," *Proc. IEEE*, **61**, 1285–1287.

Sweeney, D. (1972). *Interferometric Measurement of Three-dimensional Temperature Fields*, Ph.D. thesis, Department of Mechanical Engineering, University of Michigan, Ann Arbor, Michigan.

Sweeney, D. W. and C. M. Vest (1973). "Reconstruction of three-dimensional refractive index fields from multidirectional interferometric data," *Appl. Opt.*, **12**, 2649–2664.

Sweeney, D. W. and C. M. Vest (1974). "Measurement of three-dimensional temperature fields above heated surfaces by holographic interferometry," *Int. J. Heat Mass Transfer*, **17**, 1443–1454.

Sweeney, D. W., D. T. Attwood, and L. W. Coleman (1976). "Interferometric probing of laser produced plasmas," *Appl. Opt.*, **15**, 1126–1128.

Swenson, G. W. and N. C. Mathur (1968). "The interferometer in radio astronomy," *Proc. IEEE*, **56**, 2114–2130.

Swindell, W. and H. H. Barrett (1977). "Computerized tomography: Taking sectional x-rays," *Phys. Today*, **30**(12), 32–41.

Szegö, G. (1939). *Orthogonal Polynomials*, Vol. 23 of *American Mathematical Society Colloquium Publications*, American Mathematical Society, Providence, RI.

Takahashi, S. (1969). *An Atlas of Axial Transverse Tomography and its Clinical Applications*, Springer-Verlag, New York.

Talbert, A. J., R. A. Brooks, and D. G. Morgenthaler (1980). "Optimum energies for dual-energy computed tomography," *Phys. Med. Biol.*, **25**, 261–269.

Tam, K.-C. V. Perez-Mendez, and B. Macdonald (1979). "3-D object reconstruction in emission and transmission tomography with limited angular input," *IEEE Trans. Nucl. Sci.*, **NS-26**(2), 2797–2805.

Tanabe, K. (1971). "Projection method for solving a singular system," *Numer. Math.*, **17**, 203–214.

Tanaka, E. (1979). "Generalized correction functions for convolutional techniques in three-dimensional image reconstruction," *Phys. Med. Biol.*, **24**, 157–161.

Tanaka, E. and T. A. Iinuma (1975). "Correction functions for optimizing the reconstructed image in transverse section scan," *Phys. Med. Biol.*, **20**, 789–798.

Tanaka, E. and T. A. Iinuma (1976). "Correction functions and statistical noises in transverse section picture reconstruction," *Comput. Biol. Med.*, **6**, 295–306.

Tasto, M. (1976). "Maximum likelihood reconstruction of random objects from noisy objects," in *Proceedings of the Third International Joint Conference on Pattern Recognition*, Nov. 8–11, 1976, Coronado, California, IEEE, New York, pp. 551–555.

Tasto, M. (1977). "Reconstruction of random objects from noisy projections," *Comput. Graph. Image Proc.*, **6**, 103–122.

Taylor, A. E. (1955). *Advanced Calculus*, Ginn, Boston.

Taylor, J. H. (1967). "Two-dimensional brightness distributions of radio sources from lunar occultation observations," *Astrophys. J.*, **150**, 421–426.

Taylor, J. H. and M. L. DeJong (1968). "Models of nine radio sources from lunar occultation observations," *Astrophys. J.*, **151**, 33–42.

Temple, G. (1953). "Theories and applications of generalized functions," *J. London Math. Soc.*, **28**, 134–148.

Temple, G. (1955). "The theory of generalized functions." *Proc. R. Soc. London*, **228A**, 175–190.

Ter-Pogossian, M. M., M. E. Raichle, and B. E. Sobel (1980). "Positron-emission tomography," *Sci. Am.*, **243**(4), October, 171–181.

Tewarson, R. P. (1972). "Solution of linear equations in remote sensing and picture reconstruction," *Computing*, **10**, 221–230.

Tewarson, R. P. and P. Narain (1974a). "Solution of linear equations resulting from satellite soundings." *J. Math. Anal. Appl.*, **47**, 1–14.

Tewarson, R. P. and P. Narain (1974b). "Generalized inverses and resolution in the solution of linear equations," *Computing*, **13**, 81–88.

Thompson, T. W. (1974). "Atlas of lunar radar maps at 70-cm wavelength," *The Moon*, **10**, 51–85.

Thomson, J. H. and J. E. B. Ponsonby (1968). "Two-dimensional aperture synthesis in lunar radar astronomy," *Proc. R. Soc. London*, **A303**, 477–491.

Tittman, B. R. (1980). "Imaging in NDE," in *Acoustical Imaging*, Vol. 9, K. Y. Wang, Ed., Plenum, New York and London, pp. 315–340.

Tretiak, O. J. (1978). "Noise limitations in x-ray computed tomography," *J. Comput. Assisted Tomog.*, **2**, 477–480.

Tretiak, O. J. (1980). "Emission computed tomography—A short annotated bibliography," in *IEEE 1980 Frontiers of Engineering in Health Care*, IEEE/Engineering in Medicine and Biology Society Second Annual Conference, Sept. 28–30, Washington, D.C., IEEE, New York, pp. 265–268.

Tretiak, O. J. and C. Metz (1980). "The exponential Radon transform," *SIAM J. Appl. Math.*, **39**, 341–354.

Tretiak, O. J., M. Eden, and W. Simon (1969). "Internal structures from x-ray images," in Proceedings of the Eighth International Conference on Medical and Biological Engineering, Session 12-1, Chicago, Illinois.

Tretiak, O. J., D. Ozonoff, J. Klopping, and M. Eden (1971). "Calculation of internal structure from multiple radiograms," in *Proceedings of the Two-Dimensional Digital Processing Conference*, University of Missouri, Columbia, MO, sec. 6.2.1–6.2.3.

Trolinger, J. D. (1975). "Flow visualization holography," *Opt. Eng.*, **14**, 470–481.

Trolinger, J. D. (1976). "Holographic interferometry as a diagnostic tool in reactive flows," *Combust. Sci. Technol.*, **13**, 229–244.

Tsui, E. and T. F. Budinger (1978). "Transverse section imaging and mean clearance time," *Phys. Med. Biol.*, **23**, 644–653.

Tsui, E. T., and T. F. Budinger (1979). "A stochastic filter for transverse section reconstruction," *IEEE Trans. Nucl. Sci.*, **NS-26**(2), 2687–2690.

Uhlenbeck, G. E. (1925). "Over een stelling van Lorentz en haar uitbreiding voor meerdimensionale ruimten," *Physica, Nederlandsch Tijdschrift voor Natuurkunde*, **5**, 423–428.

Vainshtein, B. K. (1970). "Finding the structure of objects from projections," *Kristallografiya*, **15**, 894–902. [*Sov. Phys. Crystallogr.*, **15**, 781–787 (1971).]

Vainshtein, B. K. (1971). "Synthesis of projecting functions," *Dokl. Akad. Nauk SSSR*, **196**, 1072–1075. [*Sov. Phys. Dokl*, **16**(2), 66–69 (1971).]

Vainshtein, B. K. (1973). "Three-dimensional electron microscopy of biological macromolecules," *Usp. Fiz. Nauk*, **109**, 455–497. [*Sov. Phys. Usp.*, **16**, 185–206 (1973).]

Vainshtein, B. K. (1978). "Electron microscopical analysis of the three-dimensional structure of biological macromolecules," in Vol. 7 of *Advances in Optical and Electron Microscopy*, V. E. Closlett and R. Barer, Eds., Academic, New York, pp. 281–377.

Vainshtein, B. K. and A. M. Mikhailou (1972). "Some properties of the synthesis of projecting functions," *Kristallografiya*, **17**, 258–263. [*Sov. Phys. Crystallography*, **17**, 217–222 (1972).]

Vainshtein, B. K. and S. S. Orlov (1972). "Theory of the recovery of functions from their projections." *Kristallografiya*, **17**, 253–257. [*Sov. Phys.-Crystallogr.*, **17**, 213–216 (1972).]

Vainshtein, B. K., V. V. Barynin, and G. V. Gurskaya (1968). "The hexagonal crystalline structure of catalase and its molecular structure," *Dokl. Akad. Nauk SSSR*, **182**, 569–572. [*Sov. Phys. Dokl.*, **13**, 838–841 (1969).]

Van der Pol, B. and H. Bremmer (1955). *Operational Calculus Based on the Two-Sided Laplace Integral*, Cambridge University, Cambridge, England.

van Schooneveld, C. (1979). *Image Formation from Coherence Functions in Astronomy*, C. van Schooneveld, Ed., Vol. 76 of *Astrophysics and Space Sciences Library*, Reidel, Dordrecht, Holland.

Varde, K. S. (1974). "An optical investigation of the combustion of a stratified mixture in a dual chamber confinement," *Can. J. Chem. Eng.*, **52**, 426–431.

Verly, J. G. and R. N. Bracewell (1979). "Blurring in tomograms made with x-ray beams of finite width," *J. Comput. Assisted Tomog.*, **3**, 662–678.

Vest, C. M. (1973). "Application of Radon transforms to multi-directional interferometry," *J. Opt. Soc. Am.*, **63**, 486 (paper WB15).

Vest, C. M. (1974). "Formation of images from projections: Radon and Abel Transforms," *J. Opt. Soc. Am.*, **64**, 1215–1218.

Vest, C. M. (1975). "Interferometry of strongly refracting axisymetric phase objects," *Appl. Opt.*, **14**, 1601–1606.

Vest, C. M. (1979). *Holographic Interferometry*, Wiley, New York.

Vest, C. M. and P. T. Radulovic (1977). "Measurement of three-dimensional temperature fields by holographic interferometry," in *Applications of Holography and Optical Data Processing*, E. Marom, A. A. Friesem, and E. Wiener-Avnear, Eds., Pergamon, Oxford, pp. 241–249.

Vest, C. M. and D. G. Steel (1978). "Reconstruction of spherically symmetric objects from slit-imaged emission: Application to spatially resolved spectroscopy," *Optics Letters*, **3**, 54–56.

Vezzetti, D. J. and S. Ø. Aks (1979). "Reconstructions from scattering data: Analysis and improvements of the inverse Born approximation," *Ultrasonic Imaging*, **1**, 333–345.

Wade, G. (1980). "Ultrasonic imaging by reconstructive tomography," in Vol. 9 of *Acoustical Imaging*, K. Y. Wang, Ed., Plenum, New York and London, pp. 379–431.

Wade, G., S. Elliott, I. Khogeer, G. Flesher, J. Eisler, D. Mensa, N. S. Ramesh, and G. Heidbreder (1980). "Acoustic echo computer tomography," in Vol. 8 of *Acoustical Imaging*, A. F. Metherell, Ed., Plenum, New York, pp. 565–576.

Wade, G., R. K. Mueller, and M. Kaveh (1979). "A survey of techniques for ultrasonic tomography," in *Computer Aided Tomography and Ultrasonics in Medicine*, J. Raviv, J. F. Greenleaf, and G. T. Herman, Eds., North-Holland, Amsterdam, New York, and Oxford, pp. 165–214.

Waggener, R. G., and W. D. McDavid (1979). "Computed Tomography," *Adv. Biomed. Eng.*, **7**, 65–100.

Wagner, W. (1976). "Reconstruction of object layers from their x-ray projections: A simulation study," *Comput. Graph. Image Proc.*, **5**, 470–483.

Wagner, W. (1979). "Reconstructions from restricted region scan data—New means to reduce the patient dose," *IEEE Trans. Nucl. Sci.*, **NS-26**(2), 2866–2869.

Walters, T. E., W. Simon, D. A. Chesler, and J. A. Correia (1981). "Attenuation correction in gamma emission computed tomography," *J. Comput. Assisted Tomog.*, **5**, 89–94.

Wang, L. (1977). "Cross-section reconstruction with a fan-beam scanning geometry," *IEEE Trans. Comput.*, **C-26**, 264–268.

Watson, G. N. (1966). *Theory of Bessel Functions*, 2nd edition reprinted, Cambridge University, London.

Weaver, K. E. and D. J. Goodenough (1979). "Imaging factors and evaluation: Computed tomography scanning," in *The Physics of Medical Imaging: Recording System Measurements and Techniques*, (A. G. Haus, Ed. American Institute of Physics, New York, pp. 309–355.

Wee, W. G. and T.-T. Hsieh (1976). "An application of the projection transform technique in image transmission," *IEEE Trans. Syst. Man Cybern.*, **SMC-6**, 486–493.

Wee, W. G. and A. Prakash (1979). "Fourier quadratic optimization technique for image reconstruction," in *Proceedings of the 1979 IEEE Computer Society Conference on Pattern Recognition and Image Processing*, Aug. 6–8, 1979, Chicago, Illinois, IEEE, New York, pp. 1–8.

Weinstein, F. S. (1978). "Accurate fan-beam reconstruction," *Proc. IEEE*, **66**, 608–609.

Weinstein, F. S. (1980). "Formation of images using fan-beam scanning and noncircular source motion," *J. Opt. Soc. Am.*, **70**, 931–935.

Weir, A. J. (1973). *Lebesgue Integration and Measure*, Cambridge University, Cambridge, England.

Wells, P. N. T. (1975). "Absorption and dispersion of ultrasound in biological tissue," *Ultrasound in Med. and Biol.*, **1**, 369–376.

Wells, P. N. T. (1977a). "Ultrasonics in medicine and biology," *Phys. Med. Biol.*, **22**, 629–669.

Wells, P. N. T. (1977b). *Biomedical Ultrasonics*, Academic, New York.

Wernecke, S. J. and L. R. D'Addario (1977). "Maximum entropy image reconstruction," *IEEE Trans. Comput.*, **C-26**, 351–364.

Wilkins, J. E., Jr. (1948). "Neumann series of Bessel functions," *Transactions Am. Math. Soc.*, **64**, 359–385.

Willis, J. R. (1971). "Interfacial stresses induced by arbitrary loading of dissimilar elastic half-spaces joined over a circular region." *J. Inst. Maths Applics*, **7**, 179–197.

Willis, J. R. (1972). "The penny-shaped crack on an interface," *Quart. J. Mech. Appl. Math.*, **25**, 367–385.

Willis, J. R. (1973). "Self-similar problems in elastodynamics," *Philos. Trans. Soc. London Ser. A* **274**, 435–491.

Witcofski, R. L., N. Karstaedt, and C. L. Partain (1982). *NMR Imaging*, The Bowman Gray School of Medicine of Wake Forest University, Winston-Salem, NC.

Wolf, E. (1969). "Three-dimensional structure determination of semi-transparent objects from holographic data," *Opt. Commun.*, **1**, 153–156.

Wolf, E. (1970). "Determination of the amplitude and phase of scattered fields by holography," *J. Opt. Soc. Am.*, **60**, 18–20.

Wolfe, S., L. E. Crooks, P. Brown, R. Howard, and R. B. Painter (1980). "Tests for DNA and chromosomal damage induced by nuclear magnetic resonance imaging," *Radiology*, **136**, 707–710.

Wood, S. W. and M. Morf (1981). "A fast implementation of a minimum variance estimator for computerized tomography image reconstruction," *IEEE Trans. Biomed. Eng.*, **BME-28**, 56–68.

Wood, E. H., J. H. Kinsey, R. A. Robb, B. K. Gilbert, L. D. Harris, and E. L. Ritman (1979). "Applications of high temporal resolution computerized tomography to physiology and medicine," in *Image Reconstruction from Projections* G. T. Herman, Ed., Vol. 32 of *Topics in Applied Physics*, Springer-Verlag, New York, pp. 247–279.

Yilmaz, Ö. and J. F. Claerbout (1980). "Prestack partial migration," *Geophysics*, **45**, 1753–1779.

Young, J. D. (1976). "Radar imaging from ramp response signatures," *IEEE Trans. Antennas Propag.*, **AP-24**, 276–282.

Zabele, G. S. and J. Koplowitz (1979). "On improving line detection in noisy images," in *Pattern Recognition 1979*, Aug. 6–8, 1979, Chicago, IEEE, New York, pp. 146–149.

Zadeh, L. A. and C. A. Desoer (1963). *Linear System Theory, The State Space Approach*, McGraw-Hill, New York.

Zehnder, L. (1891). Ein neuer Interferenzrefraktor, *Z. Instrumentenk.*, **11**, 275–285.

Zeitler, E. (1974). "The reconstruction of objects from their projections," *Optik*, **39**, 396–415.

Zemanian, A. H. (1965). *Distribution Theory and Transform Analysis*, McGraw-Hill New York.

Zernike, F. (1934). "Beugungstheorie des Schneidenverfahrens und seiner verbesserten form, der Phasenkontrastmethode." *Physica*, **1**, 689–704.

Zien, T.-F., W. C. Ragsdale, and W. C. Spring (1975). "Quantitative determination of three-dimensional density field by holographic interferometry," *AIAA J.*, **13**, 841–842.

Zisk, S. H., G. H. Pettengill, and G. W. Catuna (1974). "High-resolution radar maps of the lunar surface at 3.8-cm wavelength," *The Moon*, **10**, 17–50.

Zwick, M. and E. Zeitler (1973). "Image reconstruction from projections," *Optik*, **38**, 550–565.

Index